Lecture Notes in Bioinformatics 11095

Subseries of Lecture Notes in Computer Science

More information about this series at http://www.springer.com/series/5381

Milan Češka · David Šafránek (Eds.)

Computational Methods in Systems Biology

16th International Conference, CMSB 2018
Brno, Czech Republic, September 12–14, 2018
Proceedings

 Springer

Editors
Milan Češka
Brno University of Technology
Brno
Czech Republic

David Šafránek
Masaryk University
Brno
Czech Republic

ISSN 0302-9743   ISSN 1611-3349 (electronic)
Lecture Notes in Bioinformatics
ISBN 978-3-319-99428-4   ISBN 978-3-319-99429-1 (eBook)
https://doi.org/10.1007/978-3-319-99429-1

Library of Congress Control Number: 2018951890

LNCS Sublibrary: SL8 – Bioinformatics

This Springer imprint is published by the registered company Springer Nature Switzerland AG
The registered company address is: Gewerbestrasse 11, 6330 Cham, Switzerland

Preface

This volume contains the papers presented at CMSB 2018, the 16th Conference on Computational Methods in Systems Biology, held during September 12–14, 2018, at the Faculty of Informatics, Masaryk University, Brno, Czech Republic.

The CMSB annual conference series, initiated in 2003, provides a unique discussion forum for computer scientists, biologists, mathematicians, engineers, and physicists interested in a system-level understanding of biological processes. Topics covered by the CMSB proceedings include: formalisms for modeling biological processes; models and their biological applications; frameworks for model verification, validation, analysis, and simulation of biological systems; high-performance computational systems biology and parallel implementations; model inference from experimental data; model integration from biological databases; multi-scale modeling and analysis methods; computational approaches for synthetic biology; case studies in systems and synthetic biology.

There were 73 submissions in total for the five conference tracks. In particular, 46 submissions were submitted to the proceedings tracks and 27 submissions to the presentation tracks. The submissions were as follows: 37 regular paper submissions, five tool paper submissions, four original work poster submissions, 14 poster submissions, and 13 presentation-only submissions describing recently published work. Each regular submission and tool paper submission was reviewed by at least three Program Committee members. Each original work poster submission was reviewed by at least two Program Committee members. For the proceedings, the committee decided to accept 15 regular papers, four tool papers, and three original posters. Moreover, the committee selected three presentation-only submissions. In addition, 16 poster presentations were selected from poster submissions and rejected presentation-only submissions. Finally, out of the selected posters, the committee accepted nine posters to be presented in the form of flash talks together with the three original work poster submissions accepted for the proceedings.

In view of the broad scope of the CMSB conference series, we selected the following five high-profile invited speakers: Ilka M. Axmann (Heinrich Heine University Düsseldorf, Germany), Mustafa Khammash (ETH Zurich, Switzerland), Chris J. Myers (University of Utah, USA), Andrew Phillips (Microsoft Research, UK), and Andrew Turberfield (University of Oxford, UK). Their invited talks stimulated fruitful discussions among the conference attendees and were the highlights of the CMSB 2018 program.

Further details on CMSB 2018 are available on the following website: https://cmsb2018.fi.muni.cz/

Finally, as the program co-chairs, we are extremely grateful to the members of the Program Committee and the external reviewers for their peer reviews and the valuable feedback they provided to the authors. Our special thanks go to François Fages, Jérôme Feret, Ezio Bartocci, and all the members of the CMSB Steering Committee, for their advice on organizing and running the conference. We acknowledge the support of the EasyChair conference system during the reviewing process and the production of these proceedings. We also thank Springer for publishing the CMSB proceedings in its *Lecture Notes in Computer Science* series. Our gratitude also goes to the tool track chair, Samuel Pastva, and all the members of the Tool Evaluation Committee, for the careful checking of the submitted tools. Additionally, we would like to thank the administrative staff of the Faculty of Informatics, Masaryk University, for helping us with the financial management of the conference. Moreover, we are pleased to acknowledge the financial support kindly obtained from the National Center for Systems Biology of Czech Republic (C4SYS), Microsoft Research, and the Faculty of Informatics, Masaryk University, where this year's event was hosted. Finally, we would like to thank all the participants of the conference. It was the quality of their presentations and their contribution to the discussions that made the meeting a scientific success.

September 2018   Milan Češka
                                                        David Šafránek

Organization

Steering Committee

Ezio Bartocci (Guest)	Vienna University of Technology, Austria
Finn Drabløs	NTNU, Norway
François Fages	Inria/Université Paris-Saclay, France
Jérôme Feret (Guest)	Inria, France
David Harel	Weizmann Institute of Science, Israel
Monika Heiner	Brandenburg Technical University, Germany
Heinz Koeppl (Guest)	TU Darmstadt, Germany
Pietro Liò (Guest)	University of Cambridge, UK
Tommaso Mazza	IRCCS Casa Sollievo della Sofferenza, Mendel, Italy
Satoru Miyano	The University of Tokyo, Japan
Nicola Paoletti (Guest)	Stony Brook University, USA
Gordon Plotkin	The University of Edinburgh, UK
Corrado Priami	CoSBi/Microsoft Research, University of Trento, Italy
Carolyn Talcott	SRI International, USA
Adelinde Uhrmacher	University of Rostock, Germany

Program Committee Co-chairs

Milan Češka	Brno University of Technology, Czech Republic
David Šafránek	Masaryk University, Czech Republic

Tools Track Chair

Samuel Pastva	Masaryk University, Czech Republic

Local Organization Chair

David Šafránek	Masaryk University, Czech Republic

Program Committee

Alessandro Abate	University of Oxford, UK
Ezio Bartocci	Vienna University of Technology, Austria
Nikola Beneš	Masaryk University, Czech Republic
Luca Bortolussi	University of Trieste, Italy
Luca Cardelli	Microsoft Research, UK
Claudine Chaouiya	Insituto Gulbenkian de Ciência, Portugal
Eugenio Cinquemani	Inria, France
Milan Češka	Brno University of Technology, Czech Republic

Thao Dang	CNRS-VERIMAG, France
Hidde De Jong	Inria, France
François Fages	Inria/Université Paris-Saclay, France
Jérôme Feret	Inria, France
Christoph Flamm	University of Vienna, Austria
Tomáš Gedeon	Montana State University, USA
Radu Grosu	Stony Brook University, USA
Monika Heiner	Brandenburg Technical University, Germany
Jane Hillston	The University of Edinburgh, UK
Heinz Koeppl	TU Darmstadt, Germany
Jean Krivine	IRIF, France
Oded Maler	CNRS-VERIMAG, France
Tommaso Mazza	IRCCS Casa Sollievo della Sofferenza, Mendel, Italy
Satoru Miyano	The University of Tokyo, Japan
Andrzej Mizera	Luxembourg Institute of Health and Luxembourg Centre for Systems Biomedicine, Luxembourg
Pedro T. Monteiro	Universidade de Lisboa, Portugal
Laura Nenzi	Vienna University of Technology, Austria
Nicola Paoletti	Stony Brook University, USA
Loïc Paulevé	CNRS/LRI, France
Ion Petre	Åbo Akademi University, Finland
Tatjana Petrov	University of Konstanz, Germany
Carla Piazza	University of Udine, Italy
Ovidiu Radulescu	University of Montpellier 2, France
Olivier Roux	IRCCyN, France
Guido Sanguinetti	The University of Edinburgh, UK
Thomas Sauter	University of Luxembourg, Luxembourg
Heike Siebert	Freie Universität Berlin, Germany
Abhyudai Singh	University of Delaware, USA
David Šafránek	Masaryk University, Czech Republic
Carolyn Talcott	SRI International, USA
Chris Thachuk	California Institute of Technology, USA
P. S. Thiagarajan	Harvard University, USA
Adelinde Uhrmacher	University of Rostock, Germany
Verena Wolf	Saarland University, Germany
Boyan Yordanov	Microsoft Research, UK
Paolo Zuliani	Newcastle University, UK

Tool Evaluation Committee

Giulio Caravagna	The University of Edinburgh, UK
Matej Hajnal	Masaryk University, Czech Republic
Juraj Kolčák	LSV, Inria and ENS Paris-Saclay, Université Paris-Saclay, France
Luca Laurenti	University of Oxford, UK
Jiří Matyáš	Brno University of Technology, Czech Republic

Samuel Pastva   Masaryk University, Czech Republic
Fedor Shmarov   Newcastle University, UK
Max Whitby   University of Oxford, UK

Additional Reviewers

Backenköhler, Michael   Luisa Vissat, Ludovica
Chodak, Jacek   Magnin, Morgan
Dague, Philippe   Molyneux, Gareth
Demko, Martin   Palaniappan, Sucheendra K.
Dreossi, Tommaso   Pang, Jun
Gilbert, David   Patanè, Andrea
Gyori, Benjamin   Pires Pacheco, Maria
Hajnal, Matej   Schnoerr, David
Hasani, Ramin M.   Selvaggio, Gianluca
Helms, Tobias   Shmarov, Fedor
Hemery, Mathieu   Soliman, Sylvain
Kyriakopoulos, Charalampos   Streck, Adam
Laurenti, Luca   Troják, Matej
Lechner, Mathias   Wijesuriya, Viraj Brian

Invited Talks

What Time Is it?

Biological Oscillators to Robustly Anticipate Changes

Nicolas M. Schmelling and Ilka M. Axmann

Institute for Synthetic Microbiology, Cluster of Excellence on Plant Sciences
(CEPLAS), Heinrich Heine University Düsseldorf, Düsseldorf, Germany
ilka.axmann@hhu.de

Abstract. Even without looking at a watch, we have an inner feeling for time. How do we measure time? Our body, in particular each of our cells, has an inner clock enabling all our rhythmic biological activities like sleeping. Surprisingly, prokaryotic cyanobacteria *Synechococcus elongatus* PCC 7942, which can divide faster than ones a day, also use an inner timing system to foresee the accompanying daily changes of light and temperature and regulate their physiology and behavior in 24 hour cycles [4]. Their underlying biochemical oscillator fulfills all criteria of a circadian clock though it is made of solely three proteins, KaiC, KaiB, and KaiA [11, 12]. At its center is KaiC in its hexameric form, which runs through a complete phosphorylation and dephosphorylation cycle every 24 hours, even under fluctuating and continuous conditions. Furthermore, the clock can be entrained by light, temperature, and nutrients. Astonishingly, reconstituted from the purified protein components this cyanobacterial protein clock can tick autonomously in a test tube for weeks [6]. This apparent simplicity has proven to be an ideal system for answering questions about the functionality of circadian clocks. Over the last decade various parts of this circadian system were identified and described in detail through computational modeling: The ordered phosphorylation of KaiC and temperature compensation of the clock [1, 10], the stimulating interaction with the other core factors and effects on gene expression [2, 5, 13], as well as the influence of varying ATP/ADP ratios [9], which on the one hand entraining the clock, on the other can cause misalignments due to rapid changes at certain times during the period. In addition, the three-protein clock is embedded in a transcription translation feedback loop, similar to eukaryotic clock systems [4]. A two-loop transcriptional feedback mechanism could be identified in which only one phosphorylation form of KaiC suppresses *kaiBC* expression while two other forms activate its own expression [3]. Further, mathematical models have identified different strategies for period robustness against internal and external noise as well as uncoupling from the cell cycle that are used by cyanobacteria [7, 8]. Overall the insights gained by studying the circadian clock of cyanobacteria have substantial impact on the field of circadian computing, which can be used for the design of synthetic switches, oscillators, and clocks and the construction of new algorithms in parallel computing in the future.

References

1. Brettschneider, C., Rose, R.J., Hertel, S., Axmann, I.M., Heck, A.J.R., Kollmann, M.: A sequestration feedback determines dynamics and temperature entrainment of the KaiABC circadian clock. Mol. Syst. Biol. **6**(1) (2010)
2. Clodong, S., Dühring, U., Kronk, L., Wilde, A., Axmann, I.M., Herzel, H., Kollmann, M.: Functioning and robustness of a bacterial circadian clock. Mol. Syst. Biol. **3**(1) (2007)
3. Hertel, S., Brettschneider, C., Axmann, I.M.: Revealing a two-loop transcriptional feedback mechanism in the cyanobacterial circadian clock. PLoS Comput. Biol. **9**(3), 1–16 (2013)
4. Ishiura, M., Kutsuna, S., Aoki, S., Iwasaki, H., Andersson, C.R., Tanabe, A., Golden, S.S., Johnson, C.H., Kondo, T.: Expression of a gene cluster kaiABC as a circadian feedback process in cyanobacteria. Sci. **281**(5382), 1519–1523 (1998)
5. Kurosawa, G., Aihara, K., Iwasa, Y.: A model for the circadian rhythm of cyanobacteria that maintains oscillation without gene expression. Biophys. J. **91**(6), 2015–2023 (2006)
6. Nakajima, M., Imai, K., Ito, H., Nishiwaki, T., Murayama, Y., Iwasaki, H., Oyama, T., Kondo, T.: Reconstitution of circadian oscillation of cyanobacterial KaiC phosphorylation in vitro. Sci. **308**(5720), 414–415 (2005)
7. Paijmans, J., Bosman, M., ten Wolde, P.R., Lubensky, D.K.: Discrete gene replication events drive coupling between the cell cycle and circadian clocks. Proc. Nat. Acad. Sci. **113**(15), 4063–4068 (2016)
8. Paijmans, J., Lubensky, D.K., ten Wolde, P.R.: Period robustness and entrainability of the kai system to changing nucleotide concentrations. Biophys. J. **113**, 157–173 (2017)
9. Rust, M.J., Golden, S.S., O'Shea, E.K.: Light-driven changes in energy metabolism directly entrain the cyanobacterial circadian oscillator. Sci. **331**(6014), 220–223 (2011)
10. Rust, M.J., Markson, J.S., Lane, W.S., Fisher, D.S., O'Shea, E.K.: Ordered phosphorylation governs oscillation of a three-protein circadian clock. Sci. **318**(5851), 809–812 (2007)
11. Snijder, J., Schuller, J.M., Wiegard, A., Lössl, P., Schmelling, N.M., Axmann, I.M., Plitzko, J.M., Förster, F., Heck, A.J.R.: Structures of the cyanobacterial circadian oscillator frozen in a fully assembled state. Sci. **355**(6330), 1181–1184 (2017)
12. Tseng, R., Goularte, N.F., Chavan, A., Luu, J., Cohen, S.E., Chang, Y.-G., Heisler, J., Li, S., Michael, A.K., Tripathi, S., Golden, S.S., LiWang, A., Partch, C.L.: Structural basis of the day-night transition in a bacterial circadian clock. Sci. **355**(6330), 1174–1180 (2017)
13. van Zon, J.S., Lubensky, D.K., Altena, P.R.H., ten Wolde, P.R.: An allosteric model of circadian KaiC phosphorylation. Proc. Nat. Acad. Sci. **104**(18), 7420–7425 (2007)

Biomolecular Control Systems

Mustafa Khammash

ETH Zurich, Control Theory and Systems Biology Laboratory Basel,
Switzerland
mustafa.khammash@bsse.ethz.ch

Abstract. Humans have been influencing the DNA of plants and animals for thousands of years through selective breeding. Yet it is only over the last three decades or so that we have gained the ability to manipulate the DNA itself and directly alter its sequences through the modern tools of genetic engineering. This has revolutionized biotechnology and ushered in the era of synthetic biology. It has also made it conceivable for the first time to engineer into living cells genetic feedback control systems that automatically monitor and steer the cell's dynamic behavior. To realize the huge promise of such systems, new theory and methodologies are needed for designing controllers that function in the special and challenging environment of the cell. We refer to the resulting technology as Cybergenetics—a modern realization of Norbert Wiener's Cybernetics vision. Here I will present our theoretical framework for the design and synthesis of cybergenetic systems and discuss the main challenges in their implementation. I will then introduce the first designer gene network that attains integral feedback in a living cell and will demonstrate its tunability and disturbance rejection properties [1]. A growth control application shows the inherent capacity of this genetic control system to deliver robustness and highlights its potential use as a universal controller for regulation of biological variables in arbitrary networks [2, 3]. I will end by exploring the potential impact of Cybergenetics in industrial biotechnology and medical therapy.

References

1. Briat, C., Zechner, C., Mustafa, K.: Design of a synthetic integral feedback circuit: dynamic analysis and DNA implementation. ACS Synth. Biol. **5**(10), 1108–1116 (2016)
2. Zechner, C., Seelig, G., Rullan, M., Mustafa, K.: Molecular circuits for dynamic noise filtering. Proc. Nat. Acad. Sci. **113**(17), 4729–4734 (2016)
3. Briat, C., Gupta, A., Khammash, M.: Antithetic integral feedback ensures robust perfect adaptation in noisy biomolecular networks. Cell Syst. **2**(1), 15–26 (2016)

A Standard-Enabled Workflow
for Synthetic Biology

Chris J. Myers

University of Utah, Salt Lake City, USA
myers@ece.utah.edu

Abstract. A synthetic biology workflow is composed of data repositories that provide information about genetic parts, sequence-level design tools to compose these parts into circuits, visualization tools to depict these designs, genetic design tools to select parts to create systems, and modeling and simulation tools to evaluate alternative design choices. Data standards enable the ready exchange of information within such a workflow, allowing repositories and tools to be connected from a diversity of sources. This talk describes one such workflow that utilizes the growing ecosystem of software tools that support the Synthetic Biology Open Language (SBOL) to describe genetic designs, and the mature ecosystem of tools that support the Systems Biology Markup Language (SBML) to model these designs [1]. In particular, this presentation will demonstrate a workflow using tools including SynBioHub, SBOLDesigner, and iBioSim. SynBioHub (http://synbiohub.org) is a database designed for storing synthetic biology designs captured using the SBOL data model, and it provides both a RESTful API for computational access and a user-friendly Web-based frontend. SBOLDesigner is a sequence editor that allows the designer to fetch parts from a SynBioHub repository and compose them to construct larger designs. Finally, iBioSim is genetic modeling, analysis, and design tool that provides a means to construct SBML models for these designs that can be simulated and analyzed using a variety of techniques [2]. Both SBOLDesigner and iBioSim also support uploading these larger system designs back to the SynBioHub repository. Finally, this talk will demonstrate how this workflow can be utilized to produce a complete record of a genetic design facilitating reproducibility and reuse.

References

1. Zhang, M., McLaughlin, J.A., Wipat, A., Myers, C.J.: SBOLDesigner 2: an intuitive tool for structural genetic design. ACS Synth. Biol. **6**(7), 1150–1160 (2017)
2. Watanabe, L., Nguyen, T., Zhang, M., Zundel, Z., Zhang, Z., Madsen, C., Roehner, N., Myers, C.J.: iBioSim 3: a tool for model-based genetic circuit design. ACS Synth. Biol. (2018)

Modelling Biomimetic Structures
and Machinery Using DNA

Andrew J. Turberfield

University of Oxford, Department of Physics, Clarendon Laboratory,
Oxford, UK
a.turberfield@physics.ox.ac.uk

Abstract. Nucleic acids, archetypal biomolecules, can be used to model and study natural biomolecular assembly processes and the operation of molecular machinery. The programmability of DNA and RNA base pairing has enabled the creation of a very wide range of synthetic nanostructures through control of the interactions between molecular components. More sophisticated design techniques allow control of the kinetics as well as the thermodynamics of these interactions, creating the potential to study and control assembly pathways and allowing the construction of both dynamic systems that process information and of biomimetic molecular machinery. Techniques of simulation and verification are important in understanding and designing these increasingly complex systems. I shall present a broad review of the rapidly developing research field of dynamic DNA nanotechnology, with particular emphasis on our use of a combination of experimental synthesis and computation to study DNA origami assembly pathways [1, 2], kinetic control of strand displacement reactions [3–7], synthetic molecular motors [8–12], and molecular machinery for the creation of sequence-controlled polymers [13, 14].

References

1. Dunn, K. E., Dannenberg, F., Ouldridge, T.E., Kwiatkowska, M., Turberfield, A.J., Bath, J.: Guiding the folding pathway of DNA origami. Nat. **525**, 82–86 (2015)
2. Dannenberg, F., Dunn, K.E., Bath, J., Kwiatkowska, M., Turberfield, A.J., Ouldridge, T.E.: Modelling DNA origami self-assembly at the domain level. J. Chem. Phys. **143**, 165102 (2015)
3. Turberfield, A.J., Mitchell, J.C., Yurke, B., Mills, A.P., Jr., Blakey, M.I., Simmel, F.C.: DNA Fuel for Free-Running Nanomachines. Phys. Rev. Lett. **90**(11), 118102 (2003)
4. Genot, A.J., Zhang, D.Y., Bath, J., Turberfield, A.J.: The remote toehold, a mechanism for flexible control of DNA hybridization kinetics. J. Am. Chem. Soc. **133**, 2177–2182 (2011)
5. Genot, A.J., Bath, J., Turberfield, A.J.: Combinatorial displacement of DNA strands: application to matrix multiplication and weighted sums. Angew. Chem. Int. Ed. **52**, 1189–1192 (2013)
6. Machinek, R.R.F., Ouldridge, T.E., Haley, N.E.C., Bath, J., Turberfield, A.J.: Programmable energy landscapes for kinetic control of DNA strand displacement. Nat. Commun. **5**, 5324 (2014)

7. Haley, N.E.C. et al.: Mismatch Repair for Enhanced Kinetic Control of DNA Displacement Reactions. Submitted

8. Green, S.J., Bath, J., Turberfield, A.J.: Coordinated chemomechanical cycles: a mechanism for autonomous molecular motion. Phys. Rev. Lett. **101**, 238101 (2008)

9. Muscat, R.A., Bath, J., Turberfield, A.J.: A programmable molecular robot. Nano Lett. **11**, 982–987 (2011)

10. Wickham, S.F.J., Bath, J., Katsuda, Y., Endo, M., Hidaka, K., Sugiyama, H., Turberfield, A.J.: A DNA-based molecular motor that can navigate a network of tracks. Nat. Nanotechnol. **7**, 169–173 (2012)

11. Ouldridge, T.E., Hoare, R.L., Louis, A.A., Doye, J.P.K., Bath, J., Turberfield, A.J.: Optimizing DNA nanotechnology through coarse-grained modeling: a two-footed DNA walker. ACS Nano **7**, 2479–2490 (2013)

12. Dannenberg, F., Kwiatkowska, M., Thachuk, C., Turberfield, A.J.: DNA walker circuits: computational potential, design, and verification. Nat. Comput. **14**, 195–211 (2015)

13. Meng, W., Muscat, R.A., McKee, M.L., Milnes, P.J., El-Sagheer, A.H., Bath, J., Davis, B.G., Brown, T., O'Reilly, R.K., Turberfield, A.J.: An autonomous molecular assembler for programmable chemical synthesis. Nat. Chem. **8**, 542–548 (2016)

14. O'Reilly, R.K., Turberfield, A.J., Wilks, T.R.: The evolution of DNA-templated synthesis as a tool for materials discovery. Acc. Chem. Res. **50**, 2496–2509 (2017)

Programming Languages for Molecular and Genetic Devices

Andrew Phillips

Microsoft Research, Biological Computation Group, Cambridge, UK
andrew.phillips@microsoft.com

Abstract. Computational nucleic acid devices show great potential for enabling a broad range of biotechnology applications, including smart probes for molecular biology research, in vitro assembly of complex compounds, high-precision in vitro disease diagnosis and, ultimately, computational theranostics inside living cells. This diversity of applications is supported by a range of implementation strategies, including nucleic acid strand displacement, localisation to substrates, and the use of enzymes with polymerase, nickase and exonuclease functionality. However, existing computational design tools are unable to account for these different strategies in a unified manner. This talk presents a programming language that allows a broad range of computational nucleic acid systems to be designed and analysed [1, 2]. We also demonstrate how similar approaches can be incorporated into a programming language for designing genetic devices that are inserted into cells to reprogram their behaviour. The language is used to characterise genetic components [4] for programming populations of cells that communicate and self-organise into spatial patterns [3]. More generally, we anticipate that languages for programming molecular and genetic devices will accelerate the development of future biotechnology applications.

References

1. Chatterjee, G., Dalchau, N., Muscat, R.A., Phillips, A., Seelig, G.: A spatially localized architecture for fast and modular DNA computing. Nat. Nanotechnol. **12**(9), 920–927(2017)
2. Chen, Y.-J., Dalchau, N., Srinivas, N., Phillips, A., Cardelli, L., Soloveichik, D., Seelig, G.: Programmable chemical controllers made from DNA. Nat. Nanotechnol. **8**(10), 755–762 (2013)
3. Grant, P.K., Dalchau, N., Brown, P.R., Federici, F., Rudge, T.J., Yordanov, B., Patange, O., Phillips, A., Haseloff, J.: Orthogonal intercellular signaling for programmed spatial behavior. Mol. Syst. Biol. **12**(1), 849–849 (2016)
4. Yordanov, B., Dalchau, N., Grant, P.K., Pedersen, M., Emmott, S., Haseloff, J., Phillips, A.: A computational method for automated characterization of genetic components. ACS Synth. Biol. **3**(8), 578–588 (2014)

Contents

Tool Papers

Poster Abstracts

Regular Papers

Modeling and Engineering Promoters with Pre-defined RNA Production Dynamics in *Escherichia Coli*

Samuel M. D. Oliveira🆔, Mohamed N. M. Bahrudeen🆔,
Sofia Startceva🆔, Vinodh Kandavalli🆔, and Andre S. Ribeiro(✉)🆔

Laboratory of Biosystem Dynamics, BioMediTech Institute,
Tampere University of Technology, P.O. Box 553, 33101 Tampere, Finland
`andre.ribeiro@tut.fi`

Abstract. Recent developments in live-cell time-lapse microscopy and signal processing methods for single-cell, single-RNA detection now allow characterizing the *in vivo* dynamics of RNA production of *Escherichia coli* promoters at the single event level. This dynamics is mostly controlled at the promoter region, which can be engineered with single nucleotide precision. Based on these developments, we propose a new strategy to engineer genes with predefined transcription dynamics (mean and standard deviation of the distribution of RNA numbers of a cell population). For this, we use stochastic modelling followed by genetic engineering, to design synthetic promoters whose rate-limiting steps kinetics allow achieving a desired RNA production kinetics. We present an example where, from a pre-defined kinetics, a stochastic model is first designed, from which a promoter is selected based on its rate-limiting steps kinetics. Next, we engineer mutant promoters and select the one that best fits the intended distribution of RNA numbers in a cell population. As the modelling strategies and databases of models, genetic constructs, and information on these constructs kinetics improve, we expect our strategy to be able to accommodate a wide variety of pre-defined RNA production kinetics.

Keywords: Model of transcription initiation · Synthetic constructs
Rate-limiting steps · Gene engineering framework

1 Introduction

Several studies have determined that, in *Escherichia coli*, the main regulatory mechanisms of gene expression dynamics act at the stage of transcription initiation [1–9]. It has recently become possible to combine time-lapse live cell microscopy with single RNA detection techniques [6, 10–13], synthetic biology techniques for gene engineering at the nucleotide level [4, 6, 9], stochastic models [14–18], and signal processing methods [19–21] to study how the dynamics of gene expression in *E. coli* is tuned by the kinetics of rate-limiting steps in transcription initiation [8, 9, 20, 22].

Using this, we propose a new strategy for, from the specification of the desired dynamics of RNA production, mean and cell-to-cell variability in RNA numbers in individual cells, and the use of detailed stochastic modeling of transcription initiation

© Springer Nature Switzerland AG 2018
M. Češka and D. Šafránek (Eds.): CMSB 2018, LNBI 11095, pp. 3–20, 2018.
https://doi.org/10.1007/978-3-319-99429-1_1

[1, 14, 22], first, predict the necessary dynamics of the rate-limiting steps in transcription initiation. Next, select an existing promoter that best fit these specifications. Afterward, fine-tune the desired dynamics by single and double point mutations of the selected promoter, so as to engineer a synthetic promoter whose RNA production dynamics best fit the original specification. This fitting is analyzed at the single RNA level, by making use of MS2-GFP probes for detection of RNA numbers at the single-RNA level in live cells [9–11, 23, 24] and objective criteria to compare the dynamics of synthetic promoters with that of the stochastic model, which, here, aside from transcription dynamics, it also accommodates for RNA degradation and cell division.

The strategy has four main steps: (i) Design a stochastic model that best fits the specifications using the modelling strategy proposed in [1, 13, 22]; (ii) Select the promoter whose *in vivo* rate-limiting steps kinetics [23, 24] best fits the model dynamics; (iii) Engineer mutant promoters of the selected one (previous step) and probe their RNA production at the single-RNA, single-cell level [13]; and (iv) Select the mutant promoter whose RNA numbers (mean and variability) best fit the model [1].

Here, we describe the strategy and the methods and present a case-study of the use of this strategy to obtain promoters with pre-defined transcription dynamics.

2   Methods

2.1   Stochastic Model of Transcription and Cell Division

The stochastic model transcription used here is based on multiple studies of transcription dynamics of individual genes [1, 9, 25]. The values set for each parameter were obtained from empirical data [9, 26–30].

The multi-step transcription process of an active promoter, P_{ON}, is modeled by reaction (1) [31] and its repression mechanism by reaction (2).

In reaction (1), the closed complex (RP_c) is formed once an RNA polymerase (RNAp) binds to a free, active promoter [32]. Subsequent rate-limiting steps follow to form the open complex (RP_o) [31, 32].

Finally, elongation starts [33], clearing the promoter. Elongation is not explicitly modeled since its time-length is much smaller than that of the rate-limiting steps in initiation [9]. Further, this process only affects noise in RNA production (mildly), not its mean rate [18]. In the end, an RNA is produced and the RNAp is released.

In the multi-step reaction (1), k_1 is the rate at which an RNAp finds and binds to promoter P, k_{-1} is the rate of reversibility of the closed complex, k_2 is the rate of open complex formation, and k_3 is the rate of promoter escape (expected to be much higher than all other rates, and thus assumed to be 'negligible' [9]):

$$P_{ON} + RNAp \underset{k_{-1}}{\overset{k_1}{\rightleftharpoons}} RP_c \xrightarrow{k_2} RP_o \xrightarrow{k_3} RNAp + P_{ON} + rna \qquad (1)$$

The reaction in (1) should not be interpreted as elementary transitions. Namely, they represent the effective rates of the rate-limiting steps in the process, which is what defines the promoter strength [9]. Next, reaction (2) models the changes in the state of the promoter, from repressed (P_{OFF}.Rep) to free for transcription, i.e. active, due to the binding/unbinding of repressor proteins (Rep) to the promoter region:

$$P_{OFF}.\text{Rep} \underset{k_{off}*[\text{Rep}]}{\overset{k_{on}}{\rightleftharpoons}} P_{ON} \tag{2}$$

In general, under full induction, the cell contains sufficient inducers to render all repressors "inactive" at all times (which can be modeled by having no Repressors in the cell). Finally, the single-step reaction (3) models RNA degradation [34]:

$$rna \xrightarrow{kd_{RNA}} \varnothing \tag{3}$$

We note that in this model, as our RNA probes (MS2-GFP coating of the target RNA, see Sect. 2.4) cause the RNA to be non-degradable for a time longer than the observation time [9–11, 23, 24], reaction (3) is not included in the model.

Finally, we assume that our gene of interest is integrated into a single-copy plasmid (not anchored to the cell membrane). Thus, we assume that the accumulation of super-coiling caused by topological constraints is negligible [35].

Given this model, we define τ_{prior} as the mean expected time for a successfully closed complex formation, which depends on the mean-time and number of attempts to initiate an open complex formation (which depends on the RNAp intracellular concentration). Meanwhile, the remaining time to produce an RNA, τ_{after}, includes the steps following commitment to open complex formation (e.g. isomerization [36]), and *prior* to transcription elongation. The mean time interval between consecutive RNA productions (Δt_{active}) of a fully active promoter is thus given by:

$$\Delta t_{active} = \tau_{prior} + \tau_{after} \tag{4}$$

Relevantly, τ_{after} does not depend on the RNAp intracellular concentration [36]. This is of significance in that, e.g., changes in this concentration will only affect τ_{prior} and thus, will only partially affect Δt_{active}. Based on this, we simplify the model, so as to be in accordance with the sensitivity of the measurements of rate-limiting steps (see below), as follows. From (1), we assume the following approximate model:

$$P_{ON} + RNAp \xrightarrow{k_1^*} RP_c \xrightarrow{k_2} RP_o \xrightarrow{k_3} RNAp + P_{ON} + rna \tag{5}$$

where: $k_1^* = \tau_{prior}^{-1}$, $k_2 = \tau_{after}^{-1}$, and k_3 = fast (i.e. 'negligible' in that it does not act as a rate-limiting step in RNA production [9].

In addition, aside from transcription, note that cell division has a major effect on RNA numbers due to 'dilution', as the RNAs are partitioned in the two daughter cells. Here, we assume a near-perfect process of partitioning [37] as the RNAs are expected to be randomly distributed in the cytoplasm. Namely, we assume that, when the number of RNAs is even, each daughter cell receives half of them. If the number is odd, one daughter cell receives (half + 0.5) and the other receives (half − 0.5).

This model assumes only one copy of the promoter in each cell at any given time. This approximation is made possible by the slow division time of the bacteria strain used here (see Sect. 2.3). Specifically, in our measurement conditions, it was established that these cells spend no more than $11 \pm 1.2\%$ of their lifetime with two copies of the target promoter (in agreement with previous measurements [9]).

2.2 Stochastic Simulations

Simulations are performed by SGNS [15], a simulator of chemical reaction systems based on the Stochastic Simulation Algorithm [38] and the Delay Stochastic Simulation Algorithm [22]. It thus allows simulating multi-delayed reactions within hierarchical, interlinked compartments that can be created, destroyed and divided at runtime. During cell division, molecules are near-evenly segregated into the daughter cells. Each model cell consists of reaction (5) along with the rate constants values (see Results section) and the initial number of each of its component molecules.

Our model (described in Sect. 2.1, reaction 5), uses the following parameter values: $k_1^* = 390^{-1} \, s^{-1}$, $k_2 = 210^{-1} \, s^{-1}$. Given these, we expect $\Delta t = 600 \, s$, and $\tau_{after}/\Delta t = 0.35$. We begin simulations with 300 cells containing no RNAs. Each cell contains 1 promoter and 1 RNAp molecule. These numbers were shown to be able to reproduce realistically the RNA production kinetics of $P_{Lac-Ara-1}$ in [9]. Also, we set a mean cell division time of 1200 s (for simplicity assumed to be constant), and we analyze the RNA numbers in the cells at the end of the simulation time (3600 s). Finally, the partitioning of RNA molecules in cell division is performed as described in Sect. 2.1.

2.3 Strain, Cell Growth, and Stress Conditions

E.coli strain used is DH5α-PRO (identical to DH5α-Z1) [39], and its genotype is: *deoR, endA1, gyrA96, hsdR17(rK⁻ mK⁺), recA1, relA1, supE44, thi⁻¹, Δ(lacZYA-argF)U169, Φ80δlacZΔM15, F⁻, λ⁻, PN25/tetR, PlacIq/lacI,* and *SpR*. Plasmids construction and transformation were done by using standard molecular cloning techniques (see Sect. 2.4). From single colonies on LB agar plates, cells were cultured in LB medium with the appropriate concentration of antibiotics and incubated overnight at 30 °C and 250 rpm. The overnight cultures were then diluted to an initial optical density (OD_{600}) of 0.05 in fresh LB medium, with a culture volume of 5 ml supplemented with the antibiotics, kanamycin for the reporter gene, and chloramphenicol for the target gene (Sigma-Aldrich, USA). Cells along with antibiotics were then incubated at 37 °C with a 250 rpm agitation until reaching an OD600 of ~ 0.3.

Next, to induce the expression of the reporter MS2-GFP proteins, 100 ng/ml of aTc (Sigma-Aldrich, USA) was added and cells were incubated at 37 °C for 30 min with 250 rpm agitation. Then, cells were incubated at 37 °C (Innova® 40 incubator, New Brunswick Scientific, USA) for 15 min with agitation, before activating the target gene. Following full induction of the target gene (1 mM IPTG and 0.1% L-arabinose, Sigma-Aldrich, USA), cells were incubated for 1 additional hour at 37 °C, *prior* to image acquisition. Partial induction of the target gene was achieved by adding to the media either only 1 mM IPTG or 0.1% L-arabinose. Finally, oxidative and acidic stresses were induced by adding, respectively, 0.6 mM of H2O2 and 150 mM of MES to the culture for 1 h along with the induction of target gene during cell exponential phase, as described in [40].

2.4 Single-RNA Detection System in a Single-Copy F-plasmid

We detect individual RNA molecules using a fluorescent MS2 tagging system that, using confocal microscopy, allow sensing integer-valued RNA numbers in individual cells, as soon as they are produced [10, 11]. From these numbers in individual cells over time, we characterize the dynamics of transcription initiation of the promoter of interest [9, 23, 24]. For this, we obtain intervals between consecutive RNA production events in individual cells (here defined as 'Δt'). Also, we obtain the distribution of RNA numbers, and from them, calculate the mean (M) and coefficient of variation (CV) of RNA numbers in individual cells [4, 26].

This technique uses an RNA coding sequence of multiple MS2 binding sites [10, 11]. Here, we engineered an RNA with 48 MS2 binding sites, with unique restriction enzymes, validated by sequencing. The construction of the single-RNA, single-protein fluorescent probe was done into two steps. First, a promoter region, a coding region for a fluorescent protein (mCherry), and the RNA with binding sites for MS2d-GFP proteins were independently synthesized *de novo* (GeneScript, USA). Second, using GenEZ™ molecular cloning, these sequences were ligated until forming the sequence of the 'target gene'. Next, they were cloned into a single-copy F-plasmid (GeneScript, USA). This probe informs on the kinetics of RNA production, at the single cell level (Fig. 1).

In our case-study, the 'target gene' is controlled by a $P_{Lac-Ara-1}$ promoter controlling the expression of a mCherry fluorescent protein, followed by an array of 48 binding sites for MS2d-GFP. Figure 1 shows the complete single-RNA detection system, composed of the 'target gene' and the 'reporter gene'. The latter is on a multi-copy plasmid carrying the $P_{L-tetO1}$ promoter controlling the expression of the fused fluorescent protein 'MS2d-GFP'. This protein rapidly binds to the MS2 binding sites of the target RNA, making it visible as a fluorescent 'spot' under fluorescence microscopy in less than 1 min, provided sufficient MS2d-GFP proteins in the cell (Fig. 3). In one of the strains engineered, the target promoter is P_{tetA}. In this case, we replaced the $P_{L-tetO1}$ promoter controlling the expression of MS2d-GFP by the P_{BAD} promoter.

By combining again *de novo* synthesis of DNA fragments with common molecular cloning and DNA assembling techniques, this new fluorescent probe can be further modified in the promoter region, RBS, and in regions between arrays of 12 MS2 binding sites, since pre-defined 'cutting points' were inserted on those regions.

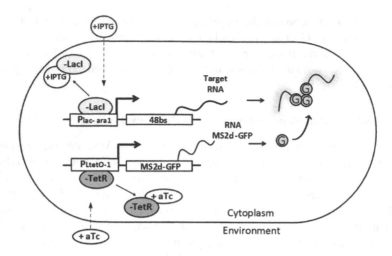

Fig. 1. Schematics of the engineered strain DH5α-PRO with the 'target gene' and its RNA tagging system, along with the intake system of one of the inducers of the 'target gene', IPTG. When in the cytoplasm, IPTG neutralizes the overexpressed LacI repressors by forming inducer-repressor complexes (LacI-IPTG). This allows the $P_{Lac-ara-1}$ promoter to express RNAs that include the array of 48 MS2d-binding sites. Meanwhile, MS2d-GFP expression is controlled by a $P_{L-tetO-1}$ promoter, which is regulated by TetR repressor, produced by its native promoter in *E. coli*'s chromosome, and the inducer anhydrotetracycline (aTc). Once an individual RNA molecule is produced, multiple tagging MS2d-GFP proteins (referred to as G) rapidly bind to it forming a visible bright spot under a confocal microscope [6, 23]. The tagging of MS2d-GFP molecules provides the RNA a long lifetime, with constant fluorescence, beyond the observation time of the measurement (see Fig. 3) [4].

2.5 Relative RNAp Quantification

To achieve different RNAp concentrations in cells, we altered their growth conditions as in [5]. For this, we used modified LB media which differed in the concentrations of some of their components. The media used are denoted as *m* x, where the composition per 100 ml are: *m* g tryptone, *m*/2 g yeast extract and 1 g NaCl (pH = 7.0). E.g. 0.25x media has 0.25 g tryptone and 0.125 g yeast extract per 100 ml.

Relative RNAp concentrations were measured using *E. coli* RL1314 cells with fluorescently-tagged β' subunits. These were grown overnight in the respective media. A pre-culture was prepared by diluting cells to an OD_{600} of 0.1 with a fresh specific medium and grown to an OD_{600} of 0.5 at 37 °C at 250 rpm. Cells were pelleted by centrifugation and re-suspended in saline. Fluorescence from the cell population was measured using a fluorescent plate-reader (Thermo Scientific Fluoroskan Ascent Microplate Fluorometer). As a control, we also measured the relative RNAp concentrations in RL1314 cells under a confocal microscope (see Sect. 2.7). Relative RNAp concentrations were estimated from the mean fluorescence of cells growing in each media. We found no differences using either method.

2.6 Mutant Target Promoters

We engineered 4 mutant target promoters from the original $P_{Lac-Ara-1}$ (referred to as 'control'). Figure 2 shows these sequences, including the control. As also shown in the Results section, single- and double-point mutations in the −35 and −10 promoter elements can affect the transcription initiation rate-limiting steps kinetics [2, 36].

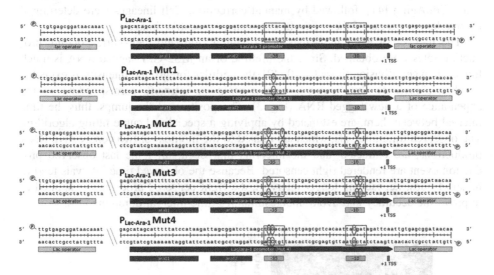

Fig. 2. Schematic representation of the target promoter's sequences: The −35 and −10 promoter elements are shown in black boxes. The transcription start site (+1 TSS) are marked in orange. Operator sites are marked as cyan and blue. In the mutants, specific nucleotide changes in the -35 and -10 regions are marked as red circles. (Color figure online)

2.7 Microscopy and Image Analysis

Cells with the target and reporter genes were grown as above. After, cells were pelleted and re-suspended in ∼ 100 μl of the remaining media. 3 ml of cells were placed on a 2% agarose gel pad of LB medium and kept in between the microscope slide and a coverslip. Cells were visualized by a Nikon Eclipse (Ti-E, Nikon) inverted microscope with a 100x Apo TIRF (1.49 NA, oil) objective. Confocal images were taken by a C2+ (Nikon) confocal laser-scanning system. MS2-GFP-RNA fluorescent spots and RNAp-GFP were visualized by a 488 nm laser (Melles-Griot) and an HQ514/30 emission filter (Nikon). Phase contrast images were taken by an external phase contrast system and DS-Fi2 CCD camera (Nikon). Phase contrast and confocal images were taken once and simultaneously by Nis-Elements software (Nikon).

For time series imaging, a peristaltic pump provided a continuous flow of fresh LB media (supplemented with inducers for the target and reporter genes and chemicals for stresses, at appropriate concentrations) to the cells, at the rate of 0.3 ml/min, through the thermal chamber (CFCS2, Bioptechs, USA). The temperature was kept as desired (at either 30, 37 or 39 °C) by a cooling/heating microfluidic system, which provides a continuous deionized water flow at a stable temperature (with no contact with cells) into the thermal chamber.

After image acquisition, cells were detected from phase contrast images as in [9, 23]. Phase contrast and fluorescence images were aligned using cross-correlation maximization and then cells were automatically segmented from phase contrast images using CellAging [41], followed by manual correction. Cell lineages were determined by overlapping areas of the segments between consecutive images.

The number of RNA molecules in individual cells and their corresponding production rates were obtained. Since the lifetime of an MS2-GFP-tagged RNA is much longer than cell division times [3, 10, 42], the cellular foreground intensity is expected to always increase (by 'jumps'), with a jump in intensity corresponding to the appearance of a new tagged RNA (Fig. 3). The position of the jumps, thus the time interval between them, are estimated by applying a specialized curve fitting algorithm. The observed time intervals, which are related to the moment of two consecutive RNA productions, are extracted, and the intervals that occur after the last observed production event are rendered right censored. Because the observed time intervals tend to be short ones, i.e. lacking longer intervals, the right censored procedure is applied to improve the accuracy and avoid underestimating time interval durations [43].

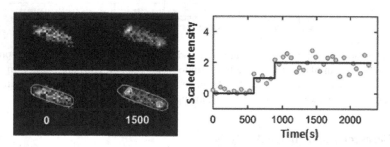

Fig. 3. (Left) Example images of an *Escherichia coli* cell expressing MS2-GFP and target RNA, taken by confocal microscopy (Top). (Middle) Segmented cells and RNA-MS2-GFP spots within. (Right) Time series of the scaled intensity of the two spots in the cell shown at the top, along with a monotone piecewise-constant fit (black line) [43].

Finally, for fluorescent RNAp studies, RNAp abundance was quantified from the total fluorescence intensity extracted from the fluorescence microscopy images.

2.8   Extracting the Duration of the Rate-Limiting Steps in Transcription

This method established in [9] and used in [4, 23, 24, 26] is based on the assumption that, increasing in the concentration of active RNAp leads to an increase in the rate of RNA production, in accordance with the model in reaction (5) and validated by recent measurements *in vivo* (see e.g. [9]).

Visibly, from reaction (5), this increase is due to the increased rate of the steps *prior* to commitment to the open complex formation, while the rate of the subsequent steps remains unaltered [36]. Note that, given this, as one further increases the amount of RNAp concentration, the model assumes that, at some point, the duration of the first step becomes negligible, when compared to the time-length of the second step. In such regime, one expects the rate of transcription to equal the inverse of the rate of the steps after commitment to the open complex formation [9, 36].

Given the above, by conducting measurements of transcription rates at different intracellular concentrations of RNAp, it is possible, by linear fitting, to infer the duration of the steps after commitment to the open complex formation [9, 36]. For this, from microscopy images, using the 'jump detection' method described above, we first obtained the mean duration of the time intervals (Δt) between consecutive RNA production events in individual cells.

Next, to estimate $\tau_{after}/\Delta t$, we plot the inverse of the RNA production rate (Δt) against the inverse of the relative RNAp concentrations, for various conditions differing in the concentration of RNAp in the cells relative to the control (these are such that cell growth rate is unaltered, as described in [9]). From this, one obtains a Lineweaver–Burk plot [44] and then fits a line to the data points to obtain the estimated rate of RNA production for "infinite" RNAp concentration (i.e. for an amount of RNAp sufficient for the steps *prior* to open complex formation to have negligible duration). This method is valid if the increase in the rate of RNA production is linear with the increase in RNAp concentration, within the range of conditions used [9] (which was shown to be true in [9, 26].

Given the model of transcription (Eq. 4), one can write the mean time interval between consecutive RNA productions (Δt) as:

$$\Delta t = \tau_{prior} + \tau_{after} \qquad (6)$$

τ_{prior} includes the time taken by multiple attempts to form a stable closed complex, whose kinetics depends on the RNAp intracellular concentrations, whereas τ_{after} does not depend on RNAp intracellular concentrations. As such, after a change in the RNAp concentration, since only τ_{prior} is affected, the new mean time interval is:

$$\Delta t^{new} = \tau_{prior}^{new} + \tau_{after} \qquad (7)$$

Where $\tau_{prior}^{new} = S^{-1} \times \tau_{prior}$ with, $S = \frac{[RNAp]^{new}}{[RNAp]}$. From this, one can write:

$$\frac{\Delta t^{new}}{\Delta t} = \frac{S^{-1} \times \tau_{prior} + \tau_{after}}{\tau_{prior} + \tau_{after}} \qquad (8)$$

Assuming a condition where cells contain an infinite concentration of RNAp, S^{-1} becomes null and Eq. 8 can be rewritten as:

$$\frac{\tau_{after}}{\tau_{prior} + \tau_{after}} = \frac{\Delta t^{new}(RNAp = \infty)}{\Delta t} \quad (9)$$

Given this, from measurements of Δt from single-cell time-lapse microscopy measurements and the corresponding RNAp concentrations in a few conditions differing in intracellular RNAp concentrations, one can extrapolate $\tau_{after}/\Delta t$. From the value of Δt, one can use Eq. (9) to obtain, τ_{after}, and subsequently, τ_{prior}.

Knowing these values, it is possible to then simulate a model (reaction (5)), which, if accounting for RNA dilution due to cell division, is expected to provide estimations of the expected mean and CV of RNA numbers in individual cells, at any moment t following the induction of the target promoter.

2.9 Assessing the Similarity Between the Rate-Limiting Steps Kinetics of the Model and the Rate-Limiting Steps Kinetics of the Promoter of Interest

To determine the best fitting promoter to the desired rate-limiting steps kinetics, we calculate as follows the Euclidean distance between the vectors (Δt, τ_{after}) of the constructs and of our preferred values (Δt_0, $\tau_{after,0}$), as determined by the model:

$$D = \sqrt{\alpha \times (\Delta t - \Delta t_0)^2 + \beta \times (\tau_{after} - \tau_{after,0})^2} \quad (10)$$

Here, for simplicity, the 'weights' α and β are set to 1, which implies that 'similar importance' is given to fitting the values of Δt_0, $\tau_{after,0}$ (as they have the same order of magnitude). Other methods of calculating this distance could be used, depending on the importance of fitting Δt_0 and $\tau_{after,0}$, by tuning the values of α and β.

2.10 Assessing the Similarity Between the RNA Numbers in Synthetic Mutant Promoters and the Desired RNA Numbers at the Single Cell Level

To best fit the mean rate of production and noise in RNA production (here assessed by the CV of RNA numbers in individual cells), we assess the "goodness of fit" of our construct dynamics to the "desired dynamics", by calculating the Euclidian distance between (M, CV) of our 'best fit' construct to the desired (M_0, CV_0) as follows:

$$D_{(M,CV)} = \sqrt{\alpha \times (M - M_0)^2 + \beta \times (CV - CV_0)^2} \quad (11)$$

Here, for simplicity, the 'weights' α and β are set to 1, which implies that 'more importance' is given to fitting the value of M_0, since, as seen in the Results section (Table 2), the values of M are 1 order of magnitude higher than the values of CV. Other methods of calculating this distance could be used, depending on the importance of fitting M_0 and CV_0, by tuning the values of α and β.

3 Results and Conclusions

To show how the framework performs and assess its performance (which depends on our library of genes available), we first created a hypothetical specification of a gene with a given RNA production dynamics that would result in a mean number of produced RNA molecules in each cell (M(RNA)) equal to 2.5 and a coefficient of variation (CV(RNA)) equal to 0.6, after 60 min of induction.

We selected these criteria with prior knowledge that they should be more suited by a $P_{Lac-Ara-1}$ promoter than, e.g. P_{BAD} or P_{tetA} [26]. The fact that the methodology chooses the former rather than the latter two (see below), is a means of assessing its effectiveness. We further note that, in this example, the construct is to be implemented on a single-copy plasmid in *E. coli* DH5α-PRO cells grown at 37 °C (Methods, Sect. 2.3), and that the intended distribution is to be reached 1 h after induction of the target gene, responsible for producing the RNA target the MS2-GFP proteins.

Finally, we note that one could overstep the modeling stage, by instead testing a large number of promoters and mutant ones, until satisfying the specifications. The purpose of modeling is to assist in the selection of the 'most promising' promoter, so as to minimize time not only in the genetic constructs but perhaps more importantly, in the measurements (including image analysis, etc.) that are required to determine whether a promoter is suitable for the goals.

3.1 Design of a Stochastic Model that Best Fits the Intended Distribution of RNA Numbers in Individual Cells

We first calculate the necessary values of τ_{prior} and τ_{after} in transcription initiation that would produce such RNA numbers, in the conditions defined, in accordance with the model of transcription assumed. First, the value of Δt is obtained from the expected mean RNA numbers in the cells, taking into consideration the cells division rate in the pre-established conditions. For this, we used the following formula:

$$\Delta t = D \times \left(1 - 2^{\left(\frac{-t}{D}\right)}\right) \times (M \times \log 2)^{-1} \qquad (12)$$

where D is the mean cell division time and M is M(RNA) in a population observed at time *t* after induction of the target gene. We measured D to be ~ 20 min and we have set our measurement duration (*t*) to 60 min. From these values, using Eq. (12), we find that we require a model whose mean interval between consecutive RNA production events (Δt) equals 10.1 min.

Next, and most importantly, we tuned the values of k_1^* and k_2 such that CV(RNA) equals 0.6 (bounded by the requirement that $\Delta t = 10.1$ min).

For this, using SGNS [15], we performed simulations of 300 model cells per condition, each with, at the start, 1 promoter and no RNA. From these, we found that our goal ($\Delta t = 10.1$ min and CV = 0.6 after 1 h of simulation time and accounting for cell divisions as described in Sects. 2.1 and 2.2) can be achieved in good approximation by setting [RNAP] $* k_1^*$ and k_2 such that $\tau_{after}/\Delta t$ equals 0.35. There are several solutions that fit these criteria. Here, we set [RNAP] $* k_1^* = 390^{-1} s^{-1}$ and $k_2 = 210^{-1} s^{-1}$.

3.2   Select a Known Promoter Whose Kinetics Best Fits the Expected Rate-Limiting Steps Kinetics

Having established the model, we next searched for a best fitting promoter. This could be done using a pre-defined library of promoters whose rate-limiting steps kinetics has been previously dissected [9, 26].

Here, as an example, we dissected *de novo* the rate-limiting steps kinetics of RNA production of 3 promoters ($P_{Lac-Ara-1}$, P_{tetA}, and P_{BAD}) from microscopy measurements (Methods). We also measured the duration of τ_{after} for each of these promoters, using τ plots (Sect. 2.8). Results are shown in Table 1.

For this, we first inserted each of the promoters of interest in the single-copy plasmid carrying the RNA coding for 48 MS2-GFP binding sites as described in Sect. 2.4. Aside from inserting this plasmid, we also inserted the multi-copy plasmid coding for MS2-GFP (Sect. 2.4). In the case of P_{tetA}, the expression of MS2-GFP in the multi-copy plasmid was controlled by a P_{BAD} promoter (Methods).

We then performed microscopy measurements in order to measure Δt and $\tau_{after}/\Delta t$ (Methods), from which we obtain also τ_{after}, for each of these 3 promoters (Methods, Sects. 2.5 and 2.8). These values are also shown in Table 1.

To determine the best fitting promoter (Sect. 2.9), applying Eq. (10), we calculated the Euclidean distance between the vectors (Δt, τ_{after}) of the constructs and the pre-established values obtained by the stochastic model. Results are shown in Table 1.

Table 1. The promoter name, the mean (Δt) and τ_{after} as measured by microscopy, and the Euclidean distance between each promoter and model RNA production kinetics.

Promoter	Mean Δt (min)	τ_{after} (min)	Euclidean distance to model
Model	10.00	3.5	–
$P_{Lac-Ara-1}$	9.82	2.4	1.11
P_{tetA}	19.23	18.1	17.3
P_{BAD}	12.03	2.9	2.12

From Table 1, the best fitting promoter, of those tested, is $P_{Lac-Ara-1}$, given that it is the one whose Euclidean distance to the model transcription kinetics is minimal.

3.3   Design and Engineering Mutant Promoter Sequences with a Fluorescent RNA-Sensor and Comparison of the Construct and the Model Dynamics

Having selected the promoter ($P_{Lac-Ara-1}$) which best fits the desired values of Δt and τ_{after}, next, we fine tune our construct by engineering mutant promoters (from $P_{Lac-Ara-1}$) with differing kinetics. The mutant promoters differ in sequence from the original $P_{Lac-Ara-1}$ in the -35 and -10 regions (see Sect. 2.6) as these were shown to control, to some extent, the kinetics of the rate-limiting steps in transcription initiation [9].

Our aim is to find a mutant promoter that can 'outperform' $P_{Lac-Ara-1}$ regarding the resulting single-cell RNA numbers (Mean and CV) when compared to these numbers obtained by the stochastic model. For this, we induced each of these mutant promoters as in the case of the original $P_{Lac-Ara-1}$ construct (Methods) and measured by microscopy the mean and CV of RNA numbers in individual cells, 1 h after induction, at 37 °C (Methods). We also obtained these numbers from simulations of model cells. Next, we estimated uncertainties of these features (Mean and CV) using a non-parametric bootstrap method [45]. Results are shown in Table 2. To assess which construct best fits the model numbers, using Eq. (11), we calculated the Euclidean distance between the vectors (M(RNA), CV(RNA)) of the constructs and the model $(M_0(RNA), CV_0(RNA))$ (Sect. 2.10). These distances are also shown in Table 2.

Table 2. Shown are the promoter name, the mean number of RNAs in each cell (M) and the coefficient of variance (CV) of RNA numbers in individual cells 60 min. after induction of the target gene $P_{Lac-Ara-1}$ promoter, referred to as 'LA', and its four mutations (in order, here referred to as 'Mu1', 'Mu2',' Mu3', and' Mu4'), with cells grown in the same induction scheme and environment conditions (Full induction: 0.1% Arabinose, 1 mM IPTG; 1x LB media, 37 °C) (see Sects. 2.3 and 2.5). Also shown are M and CV from the simulations of the model, along with the Euclidean distance between each promoter's resulting RNA numbers in individual cells and the model RNA numbers. Error bars represent the standard error, calculated as the standard deviation of the bootstrapped distributions (200 cells each) from 1000 random resamples with replacement.

Condition	M(RNA)	CV (RNA)	Euclidian distance to model	KS test (mean P value)
LA (control)	3.59 ± 0.16	0.63 ± 0.04	1.35	<0.01
Mut1	2.02 ± 0.09	0.65 ± 0.04	0.22	0.05
Mut2	1.54 ± 0.06	0.56 ± 0.02	0.71	<0.01
Mut3	1.61 ± 0.11	0.97 ± 0.02	0.71	<0.01
Mut4	1.23 ± 0.03	0.39 ± 0.02	1.04	<0.01
Model	2.24 ± 0.1	0.65 ± 0.04	–	–

From Table 2, we find $P_{Lac-Ara-1}$ Mut1 to be the construct that best fits the model-based pre-established single-cell RNA numbers.

Finally, we assessed 'how well' the best fitting construct fits the original criteria set by the stochastic model. For this, we obtained the distribution of RNA numbers in individual model cells and in measurements of RNA numbers in individual cells, 1 h after induction. Next, we performed Kolmogorov-Smirnov tests of statistical significance comparing the distributions of RNA numbers in model and measurements. In general, the higher the P value, the more 'similar' are the two distributions compared, as the more likely it is that the two sets of data are drawn from the same distribution. The results, for all constructs, are shown in Table 2. From these, we find that the RNA numbers distribution resulting from 'Mut1' has the highest probability of being drawn from the same distribution as the one drawn from the model data, confirming that this construct is the one that best fits the model criteria.

Further, its p-value equals 0.05, which in general allows assuming that the two sets of data are drawn from the same distribution and, thus, cannot be distinguished in a statistical sense. We thus conclude that the construct fits the necessary criteria.

3.4 Achieving the Desired Dynamics by Changing Promoter Induction Scheme and Environment Conditions

In general, genes are already equipped with several regulatory mechanisms (such as the repression mechanism modeled above) (see, e.g., [46]). Further, cells interact with the environment, in that they have global regulatory mechanisms of gene expression, such as σ factors, and responsiveness to changes in the chemical composition of the media. As a result of these systems, cells express genes with a dynamics that is environment-dependent. One can make use of this, to further enhance the goodness of fit between the observed expression dynamics and the desired dynamics, using the same methodology as in the section above.

As an example, we measured the kinetics of transcription of the $P_{Lac-Ara-1}$ promoter ('LA') when subject to different induction schemes, when under oxidative and acidic stresses, at different temperatures, and at different media conditions. Results of these measurements are shown in Table 3.

Table 3. Shown are the promoter, mean number of RNAs in each cell (M) and coefficient of variance (CV) of RNA numbers in individual cells 60 min. after induction of the target gene $P_{Lac-Ara-1}$ promoter, referred to as 'LA', when cells are grown in various environment conditions (see Sects. 2.3 and 2.5). Also shown are M and CV from the simulations of the model, along with the Euclidean distance between each promoter's resulting RNA numbers in individual cells and the model RNA numbers. Error bars represent the standard error, calculated as the standard deviation of the bootstrapped distributions (200 cells each) from 1000 random resamples with replacement.

Condition	M(RNA)	CV (RNA)	Euclidian distance to model	KS test (mean P value)
LA 37 °C 0.1% Ara, 1x LB	1.83 ± 0.07	0.55 ± 0.04	0.42	<0.01
LA 37 °C 1 mM IPTG, 1x LB	1.85 ± 0.08	0.62 ± 0.04	0.39	<0.01
LA 37 °C Oxidative, Full Ind., 1x LB	1.39 ± 0.05	0.51 ± 0.03	0.86	<0.01
LA 37 °C Acidic, Full Ind., 1x LB	1.62 ± 0.07	0.62 ± 0.04	0.62	<0.01
LA 30 °C, Full Ind., 1x LB	2.38 ± 0.1	0.61 ± 0.03	0.15	0.24
LA 39 °C, Full Ind., 1x LB	2.72 ± 0.11	0.52 ± 0.02	0.50	<0.01
LA 37 °C, Full Ind., 0.5x LB	2.66 ± 0.13	0.61 ± 0.04	0.42	<0.01
LA 37 °C, Full Ind., 2x LB	3.16 ± 0.18	0.82 ± 0.13	0.94	<0.01
Model	2.24 ± 0.1	0.65 ± 0.04	–	–

Again, from the empirical data and model dynamics, we calculated the Euclidean distance between the kinetics at each measurement condition and the model dynamics. We also performed KS tests to assess the goodness of fit of each construct dynamics to the original criteria set by the model. Results are also shown in Table 3.

Similarly to when testing different promoter sequences, we find that one condition ('LA 30 °C, Full Ind., 1x LB') fits the model in a statistical sense (P value equals 0.24).

4 Discussion

We proposed a new strategy for designing genes with a predefined dynamics of RNA production (mean and CV of the single-cell RNA numbers in a population). In short, based on the desired RNA numbers in individual cells, we first produce a stochastic model that informs on the necessary promoter initiation kinetics (rate-limiting steps kinetics). From that information, we search for a promoter that best fits this initiation kinetics from τ-plot measurements [9] and/or from a database already containing such information. Once reducing the state-space of our search to a 'best-possible' promoter, next we fine-tune its RNA production kinetics. Namely, mutant promoters are engineered from this best-possible promoter and their RNA production kinetics is analyzed at the single-cell, single-RNA level (namely, we obtain mean and variability in RNA numbers at the single cell level).

In addition to this, we showed that the RNA production dynamics can be further (or alternatively) tuned by changes in the promoter induction scheme (i.e. inducer concentrations) and environment conditions (e.g. temperature and media richness).

Combining these multiple strategies (mutations, multiple environmental conditions, and various induction schemes) will provide a significantly large array of various transcription dynamics, which will greatly enhance the chances of finding a 'best fitting' dynamics to that of the desired model.

At the moment, the main limitation of this strategy is the limited number of studies on the *in vivo* underlying kinetics of the rate-limiting steps of promoters in *E. coli* using live single-RNA sensitivity [6, 7, 11, 23, 24, 26]. As the library of the genetic constructs whose dynamics have been analyzed with this level of detail increases, we expect the impact of our strategy to increase as well, as the chances to match a predefined kinetics of RNA production increase.

Other means to further improve this strategy include the development of more detailed models of transcription that incorporate phenomena such as promoter-proximal pausing, σ factor competition for RNA polymerase core enzymes, cell-cycle stage, tuning cell division rates [47] based on media richness, etc.

Overall, as the databases of models and the library of genetic constructs are enlarged, and as the techniques in synthetic biology, fluorescent microscopy, and stochastic modeling are enhanced, we will be able to accommodate an increasingly wide variety of predefined RNA production dynamics with enhanced precision.

We expect that the engineering of genes with tailored RNA production dynamics, while currently a rare practice (except in a few state-of-the-art basic scientific projects making use of Synthetic Biology), will become a key strategy in Biomedicine and

Biotechnology for regulating cellular behavior for medicinal purposes, bioreactors output enhancement, etc. Also, we expect it to become a key step in synthetic engineering of genetic circuits with tailored dynamics [48, 49].

Acknowledgments. Work supported by Academy of Finland (295027 and 305342 to ASR), Jane and Aatos Erkko Foundation (610536 to ASR), Finnish Academy of Science and Letters (SO), TUT President Ph.D. grants (MB, SS). The funders had no role in study design, data collection and analysis, decision to publish, or preparation of the manuscript.

References

1. Jones, D.L., Brewster, R.C., Phillips, R.: Promoter architecture dictates cell-to-cell variability in gene expression. Science **346**, 1533–1537 (2014)
2. Shih, M.-C., Gussin, G.: Mutations affecting two different steps in transcription initiation at the phage λ PRM promoter. Proc. Natl. Acad. Sci. USA **80**, 496–500 (1983)
3. Golding, I., Cox, E.C.: RNA dynamics in live *Escherichia coli* cells. Proc. Natl. Acad. Sci. USA **101**(31), 11310–11315 (2004)
4. Goncalves, N.S.M., Startceva, S., Palma, C.S.D., Bahrudeen, M.N.M., Oliveira, S.M.D., Ribeiro, A.S.: Temperature-dependence of the single-cell kinetics of transcription activation in *Escherichia coli*. Phys. Biol. **15**(2), 026007 (2017)
5. Liang, S., et al.: Activities of constitutive promoters in *Escherichia coli*. J. Mol. Biol. **292**, 19–37 (1999)
6. Ribeiro, A.S.: Kinetics of gene expression in live bacteria: from models to measurements, and back again. Can. J. Chem. **91**(7), 487–494 (2013)
7. Mäkelä, J., Lloyd-Price, J., Yli-Harja, O., Ribeiro, A.S.: Stochastic sequence-level model of coupled transcription and translation in prokaryotes. BMC Bioinform. **12**, 121 (2011)
8. Häkkinen, A., Tran, H., Yli-Harja, O., Ribeiro, A.S.: Effects of rate-limiting steps in transcription initiation on genetic filter motifs. PLoS ONE **8**(8), e70439 (2013)
9. Lloyd-Price, J., et al.: Dissecting the stochastic transcription initiation process in live *Escherichia coli*. DNA Res. **23**(3), 203–214 (2016)
10. Peabody, D.S.: The RNA binding site of bacteriophage MS2 coat protein. EMBO J. **12**, 595–600 (1993)
11. Golding, I., Paulsson, J., Zawilski, S.M., Cox, E.C.: Real-time kinetics of gene activity in individual bacteria. Cell **123**, 1025–1036 (2005)
12. Peabody, D.S.: Role of the coat protein-RNA interaction in the life cycle of bacteriophage MS2. Mol. Gen. Genet. **254**, 358–364 (1997)
13. Fusco, D., et al.: Single mRNA molecules demonstrate probabilistic movement in living mammalian cells. Curr. Biol. **13**(2), 161–167 (2003)
14. Ribeiro, A.S., Zhu, R., Kauffman, S.A.: A general modeling strategy for gene regulatory networks with stochastic dynamics. J. Comput. Biol. **13**(9), 1630–1639 (2006)
15. Lloyd-Price, J., Gupta, A., Ribeiro, A.S.: SGNS2: a compartmentalized stochastic chemical kinetics simulator for dynamic cell populations. Bioinformatics **28**, 3004–3005 (2012)
16. Ribeiro, A.S., Häkkinen, A., Mannerström, H., Lloyd-Price, J., Yli-Harja, O.: Effects of the promoter open complex formation on gene expression dynamics. Phys. Rev. E **81**(1), 011912 (2010)
17. Rajala, T., Häkkinen, A., Healy, S., Yli-Harja, O., Ribeiro, A.S.: Effects of transcriptional pausing on gene expression dynamics. PLoS Comput. Biol. **6**(3), e1000704 (2010)

18. Bahrudeen, M.N.M., Startceva, S., Ribeiro, A.S.: Effects of extrinsic noise are promoter kinetics dependent. In: The 9th International Conference on Bioinformatics and Biomedical Technology on Proceedings, ICBBT 2017, Lisbon, Portugal, pp. 44–47 (2017)
19. Häkkinen, A., Ribeiro, A.S.: Estimation of GFP-tagged RNA numbers from temporal fluorescence intensity data. Bioinformatics **31**(1), 69–75 (2015)
20. Häkkinen, A., Ribeiro, A.S.: Identifying rate-limiting steps in transcription from RNA production times in live cells. Bioinformatics **32**(9), 1346–1352 (2016)
21. Häkkinen, A., Tran, H., Ingalls, B., Ribeiro, A.S.: Effects of multimerization on the temporal variability of protein complex abundance. BMC Syst. Biol. **7**(Suppl. 1), S3 (2013)
22. Ribeiro, A.S.: Stochastic and delayed stochastic models of gene expression and regulation. Math. Biosci. **223**(1), 1–11 (2010)
23. Oliveira, S.M.D., Häkkinen, A., Lloyd-Price, J., Tran, H., Kandavalli, V., Ribeiro, A.S.: Temperature-dependent model of multi-step transcription initiation in *Escherichia coli* based on live single-cell measurements. PLoS Comput. Biol. **12**, e1005174 (2016)
24. Mäkelä, J., Kandavalli, V., Ribeiro, A.S.: Rate-limiting steps in transcription dictate sensitivity to variability in cellular components. Sci. Rep. **7**, 10588 (2017)
25. Taniguchi, Y., Choi, P.J., Li, G.-W., et al.: Quantifying *E. coli* proteome and transcriptome with single-molecule sensitivity in single cells. Science **329**, 533–538 (2010)
26. Kandavalli, V.K., Tran, H., Ribeiro, A.S.: Effects of σ factor competition are promoter initiation kinetics dependent. Biochim. Biophys. Acta (BBA)-Gene Regul. Mech. **1859**, 1281–1288 (2016)
27. Mitarai, N., Sneppen, K., Pedersen, S.: Ribosome collisions and translation efficiency: optimization by codon usage and mRNA destabilization. J. Mol. Biol. **382**, 236–245 (2008)
28. Bremer, H., Dennis, P.P.: Modulation of Chemical composition and other parameters of the cell by growth rate. In: Neidhardt, F.C. (ed.) *Escherichia coli* and Salmonella, 2nd edn, pp. 1553–1569. ASM Press, Washington, DC (1996)
29. Kennel, D., Riezman, H.: Transcription and translation initiation frequencies of the *Escherichia coli* lac operon. J. Mol. Biol. **114**(1), 1–21 (1977)
30. Cormack, B.P., Valdivia, R.H., Falkow, S.: FACS-optimized mutants of the green fluorescent protein (GFP). Gene **173**(1), 33–38 (1996)
31. Saecker, R.M., Record, M.T., Dehaseth, P.L.: Mechanism of bacterial transcription initiation: RNA polymerase - promoter binding, isomerization to initiation-competent open complexes, and initiation of RNA synthesis. J. Mol. Biol. **412**, 754–771 (2011)
32. Chamberlin, M.: The selectivity of transcription. Annu. Rev. Biochem. **43**, 721–775 (1974)
33. deHaseth, P.L., Zupancic, M.L., Record, M.T.: RNA polymerase promoter interactions: the comings and goings of RNA polymerase. J. Bacteriol. **180**, 3019–3025 (1998)
34. Bernstein, J.A., Khodursky, A.B., Pei-Hsun, L., Lin-Chao, S., Cohen, S.N.: Global analysis of mRNA decay and abundance in *Escherichia coli* at single-gene resolution using two-color fluorescent DNA microarrays. Proc. Natl. Acad. Sci. USA **99**, 9697–9702 (2002)
35. Chong, S., Chen, C., Ge, H., Xie, X.S.: Mechanism of transcriptional bursting in bacteria. Cell **158**, 314–326 (2014)
36. McClure, W.R.: Mechanism and control of transcription initiation in prokaryotes. Annu. Rev. Biochem. **54**, 171–204 (1985)
37. Abhishekh, G., Lloyd-Price, J., Ribeiro, A.S.: *In silico* analysis of division times of Escherichia coli populations as a function of the partitioning scheme of non-functional proteins. Silico Biol. **12**, 9–21 (2014)
38. Gillespie, D.T.: Exact stochastic simulation of coupled chemical reactions. J. Phys. Chem. **81**(25), 2340–2361 (1977)

39. Lutz, R., Bujard, H.: Independent and tight regulation of transcriptional units in *Escherichia coli* via the LacR/O, the TetR/O and AraC/I1-I 2 regulatory elements. Nucleic Acids Res. **25**, 1203–1210 (1997)

40. Muthukrishnan, A.-B., Martikainen, A., Neeli-Venkata, R., Ribeiro, A.S.: n Vivo transcription kinetics of a synthetic gene uninvolved in stress-response pathways in stressed *Escherichia coli* Cells. PLoS ONE **9**, e109005 (2014)

41. Häkkinen, A., Muthukrishnan, A.B., Mora, A., Fonseca, J.M., Ribeiro, A.S.: Cell aging: a tool to study segregation and partitioning in division in cell lineages of Escherichia coli. Bioinformatics **29**(13), 1708–1709 (2013)

42. Tran, H., Oliveira, S.M.D., Goncalves, N.S.M., Ribeiro, A.S.: Kinetics of the cellular intake of a gene expression inducer at high concentrations. Mol. BioSyst. **11**, 2579–2587 (2015)

43. Häkkinen, A., Ribeiro, A.S.: Characterizing rate-limiting steps in transcription from RNA production times in live cells. Bioinformatics **32**(9), 1346–1352 (2016)

44. Lineweaver, H., Burk, D.: The determination of enzyme dissociation constants. J. Am. Chem. Soc. **56**, 658–666 (1934)

45. Carpenter, J., Bithell, J.: Bootstrap confidence intervals: when, which, what? A practical guide for medical statisticians. Stat. Med. **19**(9), 1141–1164 (2000)

46. Schleif, R.: Regulation of the L-arabinose operon of *Escherichia coli*. Trends Gen. **16**(12), 559–565 (2000)

47. Lloyd-Price, J., Tran, H., Ribeiro, A.S.: Dynamics of small genetic circuits subject to stochastic partitioning in cell division. J. Theor. Biol. **356**, 11–19 (2014)

48. Mannerström, H., Yli-Harja, O., Ribeiro, A.S.: Inference of kinetic parameters of delayed stochastic models of gene expression using a Markov chain approximation. EURASIP J. Bioinform. Syst. Biol. **2011**(1), 572876 (2011)

49. Oliveira, S.M.D., et al.: Single-cell kinetics of the repressilator when implemented in a single-copy plasmid. Mol. BioSyst. **11**, 1939–1945 (2015)

Deep Abstractions of Chemical Reaction Networks

Luca Bortolussi[1]([envelope]) and Luca Palmieri[1,2]

[1] DMG, University of Trieste, Trieste, Italy
lbortolussi@units.it, contact@lpalmieri.com
[2] SISSA, Trieste, Italy

Abstract. Multi-scale modeling of biological systems, for instance of tissues composed of millions of cells, are extremely demanding to simulate, even resorting to High Performance Computing (HPC) facilities, particularly when each cell is described by a detailed model of some intra-cellular pathways and cells are coupled and interacting at the tissue level. Model abstraction can play a crucial role in this setting, by providing simpler models of intra-cellular dynamics that are much faster to simulate so to scale better the analysis at the tissue level. Abstractions themselves can be very challenging to build ab-initio. A more viable strategy is to learn them from single cell simulation data.

In this paper, we explore this direction, constructing abstract models of chemical reaction networks in terms of Discrete Time Markov Chains on a continuous space, learning transition kernels using deep neural networks. This allows us to obtain accurate simulations, greatly reducing the computational burden.

Keywords: Deep learning · Chemical reaction networks Model abstraction · Stochastic simulation

1 Introduction

Computational modeling is a central ingredient in the quest for understanding and predicting the dynamics of complex biological systems [8]. A wide range of interesting biological processes can be modeled as a complex network of biochemical reactions. These reactions take place inside cells, which are themselves part of networks of intercellular interaction. This is the case, for instance, of tumoral tissues [11], a typical example of multi-scale system, for which *in silico modelling* also plays a central role.

At the intra-cellular level, there are fairly established techniques to model and simulate biochemical reaction networks, taking into account the intrinsic variability and noise due to the small number of molecules involved and a certain degree of randomness in their distribution [8]. The most prominent approach to analyze such models is stochastic simulation, starting from the well known Gillespie algorithm [1], and moving to more efficient but approximate methods like tau-leaping [6] and hybrid simulation [9].

© Springer Nature Switzerland AG 2018
M. Češka and D. Šafránek (Eds.): CMSB 2018, LNBI 11095, pp. 21–38, 2018.
https://doi.org/10.1007/978-3-319-99429-1_2

When dealing with a multi-scale system, it would be desirable to explicitly model the detailed intracellular mechanics. Unfortunately, simulations times become quickly unfeasible when working with large number of cells (e.g. 10^5 or 10^6), even using state of the art HPC infrastructures and techniques with approximate simulation algorithms.

Models of phenomena happening at the tissue level at this scale of complexity require a considerable reduction of single cell simulation times, simplifying cellular models. This calls model abstraction into play.

While manual crafting of abstract models is always possible, a more scalable approach is to learn abstractions from single-cell simulation data. The basic idea is to start from a suitable number of simulated system trajectories, using available simulation algorithms, and then learn a simpler probabilistic model from such data. Such model should allow us to generate approximate trajectories in a significantly faster way than the original detailed model, while retaining a reasonable accuracy. We can also limit the number of intra-cellular variables taken into account by the abstract model to simplify it further.

Related Work. This idea was employed by Liu et al. in [13] to approximate an ODE dynamic and was further refined by Palaniappan et al. in [25] to deal with stochastic dynamics. In this work, the authors select a subset of relevant variables and discretize them using information theoretic tools, and then build an approximate model based on a dynamic Bayesian network (DBN). Palaniappan et al. were able to perform accurate simulations using their abstract DBN, reducing simulation times by an order of magnitude compared to the original model. An alternative approach is that of [24], in which authors learn a simplified model of the bacteria chemotaxis mechanisms in bacteria exploiting Gaussian Processes (GP). In particular, they consider a fast equilibrating internal process, and use GP to model the choice of movement strategy as a function of the environmental state.

Contributions. In this paper, we focus on the abstraction procedure, starting from [25]. Their approach is mostly driven by information theoretic considerations, which is extremely effective in isolating the core components needed to perform an accurate approximation of the original process. Moreover, models given in [13, 25] are discrete in time and space: the domain of each tracked chemical species is partitioned in a certain number of subintervals and, at each time step, the system may change its state by jumping to another node of the discrete state space. The jumping probabilities, properly factored according to the DBN topology, have to be stored in memory. This becomes troublesome when a high number of chemical species is involved or when a high level of resolution on the state of certain variables is required. In this case, the size of the discrete state space is doomed to explode.

Instead of working with a DBN, our insight is to resort to a different probabilistic model: a discrete-time Markov chain on a *continuous state space*, bypassing the difficulties arising from the state space discretization. A Markov chain

$\{\eta_k\}_{k\in\mathbb{N}}$ is completely determined once its transition kernel has been specified, i.e. a function which maps the previous state s_n of the chain to the probability distribution on the state space describing the possible outcomes for η_{n+1} conditioned on $\eta_n = s_n$. Transition kernels are the continuous equivalent of a transition matrix for a Markov chain with a finite number of states.

Our proposal is to model transition kernels as *probability mixtures*, which are weighted combinations of a certain number of elementary probability distributions components (normal, log-normal, etc.). Sampling from a mixture is fast, while the model itself is extremely flexible and can be used to approximate fairly complicated probability distributions. Our components are chosen from the exponential family and are thus completely determined by a fixed number of parameters. Everything then boils down to an optimization problem: we need to properly tune the number of components and their parameters in order to produce a good transition kernel which maximizes the likelihood of our data.

Real world chemical reaction networks involve a high number of different species, which means that our components are high dimensional probability distributions: the parameter space we want to explore is expected to be complicated. Furthermore, parameters will depend on the previous state η_n visited by the discrete abstraction, hence they have to be modeled as (continuous) functions of this state.

We tackled this supervised learning problem using Neural Networks (NN). In particular, we exploit Mixture Density Networks, [2], and use different NN body architectures . We provide a software implementation of the described approach as a reusable Python library, and show on some case studies the effectiveness of our method, capable of reducing the computational complexity of several orders of magnitude.

Paper Structure. The paper is organized as follows: In Sect. 2, we introduce the relevant background notions, while in Sect. 3 we discuss the abstraction procedure. Section 4 is devoted to present the implementation, and Sect. 5 to experimental evaluation. Conclusions are drawn in Sect. 6.

2   Background

Chemical Reaction Networks. Chemical Reaction Networks (CRNs) are the standard formalism to describe dynamical models of biological systems. Under the well-stirred assumption, they can be interpreted stochastically as a Continuous Time Markov Chain (CTMC) [1,8,26] on a discrete state space S. We denote by $\mathcal{X} = \{X_1, \ldots, X_n\}$ the chemical species involved in our CRN and by $\eta_t = (\eta_{t,1}, \ldots, \eta_{t,n}) \in S = \mathbb{N}^n$ the state vector at time t: $\eta_{t,i}$ is the number of X_i molecules in the system at time t.

The dynamics is encoded by a set $\mathcal{R} = \{R_1, \ldots, R_m\}$ of reactions, each being a tuple (f_{R_i}, ν_i). f_{R_i} is the *propensity function*, and gives the rate at which reaction R_i fires, while $\nu_i \in S$ is the update vector: the firing of reaction R_i changes the system state from η_t to $\eta_t + \nu_i$.

$\mathbb{P}(\eta_t = s \,|\, \eta_{t_0} = s_0) =: \mathbb{P}_{s_0}(\eta_t = s)$ is the probability of finding our system in state s at time t given that it was in state s_0 at time t_0. It satisfies a system of ODEs known as *Chemical Master Equation* (CME):

$$\partial_t \mathbb{P}_{s_0}(\eta_t = s) = \sum_{i=1}^{m} [\mathbb{P}_{s_0}(\eta_t = s - \nu_i) f_{R_i}(s - \nu_i) - \mathbb{P}_{s_0}(\eta_t = s) f_{R_i}(s)] \qquad (1)$$

Solving the CME numerically is very challenging (see [26] for further details), and typically CRNs are simulated. The most commonly used simulation algorithm is Gillespie's SSA [1].

Example: The SIR Model. The SIR epidemiological model describes the spread of an infectious disease that grants immunity to those who recover from the acute phase. Despite not being properly a molecular system, it can be modeled as a CRN. The SIR model describes a population of N individuals divided in three mutually exclusive groups: **S**usceptible, **I**nfected and **R**ecovered/**R**emoved. The system state vector is given by $\eta_t = (S_t, I_t, R_t)$, each component standing for the total number of individuals in the corresponding population group. The interactions considered by the SIR model are the following:

$$R_1 : \ S + I \longrightarrow 2I \qquad (infection) \qquad (2)$$
$$R_2 : \ I \longrightarrow R \qquad (recovery/death) \qquad (3)$$

The propensity functions are of mass action type [8]: $f_{R_1}(S_t, I_t, R_t) = k_1 \frac{I_t S_t}{N}$ and $f_{R_2}(S_t, I_t, R_t) = k_2 I_t$, with corresponding update vectors $\nu_1 = (-1, +1, 0)$ and $\nu_2 = (0, -1, +1)$. The ratio k_1/k_2 is called basic reproduction number and its value is strongly connected with the overall behavior of the system: minor outbreak, serious outbreak, pandemic (see [10]). The SIR model dynamic is well understood and we shall use it as testing ground for our abstraction protocol.

Neural Networks. Machine learning can be loosely defined as a collection of algorithms and techniques employed to *automatically* tune (*learn*) an analytical model from a (possibly huge) set of data [5,23]. Learned models are used to perform a wide range of different tasks: image recognition, time series analysis, machine translation, speech recognition, etc.

Artificial neural networks (NN) date back to the 50s and have recently conquered the spotlight thanks to an impressive set of achievements [23]. The basic (*feed-forward*) NN architecture can be represented as a weighted directed graph where each vertex stands for a node (or *neuron*). It can be divided in three blocks: an input layer, a set of hidden layers[1] and an output layer. Data are fed to the

---

[1] A node is said to be hidden if it does not belong to the input or the output layer. A layer is, roughly, a collection of nodes at the same depth level with respect to the input layer. A layer composed of hidden nodes is called hidden layer.

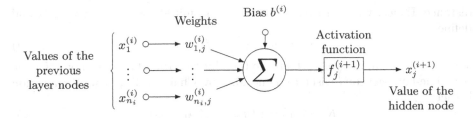

Fig. 1. A visual summary of what happens at hidden and output nodes.

input layer and flow through the network undergoing a certain number of (non-linear) transformations encoded by the hidden layers. The transformed data are then returned to the user through the output layer. This procedure is called *forward propagation*. The value of an output node is thus the result of the stacked application of linear transformations and a (non-linear) function. Despite their mathematical simplicity, NNs can approximate any measurable function from the input to the output nodes with arbitrary precision [23] (Fig. 1).

In a supervised machine learning problem we are given a training dataset: each input data is associated with the desired output. Training a NN is a (stochastic) optimization problem: we try to minimize an error function, which is computed using the values of the output nodes, with respect to the NN weights and biases.

Mixture Density Networks. A typical regression approach would try to learn the value of η_t using η_{t-1} as input minimizing a sum-of-squares error function. This corresponds, implicitly, to the task of learning the mean of a multivariate normal variable via maximum likelihood estimation (see Sect. 5.2 of [5]). This scheme, however, does not capture information about the distribution of η_t given η_{t-1}, and is going to perform particularly poorly if the system exhibits a certain degree of multimodality.

To circumvent such problem, we are going to learn the parameters needed to specify a mixture distribution: these are going to be our output nodes, as anticipated in Sect. 1. The error function associated with the output layer is naturally the negative log-likelihood of the mixture distribution. This type of NNs are called Mixture Density Networks (MDN). All the abstract models in this work are based on MDNs. A detailed explanation of MDNs can be found in [2], where they made their first appearance. MDNs are also well explained in Sect. 5.6 of [5].

3   MDN-Based Abstraction Procedure

Model Abstraction. Let $\{\eta_t\}_{t \geq 0}$ be a CTMC describing a CRN system with state space $S = \mathbb{N}^m$. To construct our abstraction, we assume to be interested only in the behavior of the model in a grid of time points at a fixed temporal distance. Hence, we fix a time step Δt and an initial time instant $t_0 \in \mathbb{R}$ and define

$$\tilde{\eta}_i := \eta_{t_0 + i\Delta t} \qquad \forall i \in \mathbb{N}. \tag{4}$$

The stochastic process $\{\tilde{\eta}_i\}_i$, thanks to the Markov property enjoyed by CTMCs, is a time-homogeneous Discrete Time Markov Chain (DTMC) with transition kernel

$$K_d(s \mid s_0) = \mathbb{P}(\eta_{\Delta t} = s \mid \eta_0 = s_0) \tag{5}$$

for all $s, s_0 \in S$.

Our model abstraction procedure introduces two approximations:

1. The state space $S = \mathbb{N}^m$ is embedded into the continuous space $\tilde{X} = \mathbb{R}^m_{\geq 0}$. The abstract model takes values in $\tilde{X}$.
2. The kernel K_d is approximated by a new kernel $K(x \mid x_0)$ taking values in the continuous space $\tilde{X}$.

In constructing the approximate kernel $K(x \mid x_0)$, rather than trying to preserve the full behavior of the process, we restrict our attention to a time-bounded *reward function* r from S^M to an arbitrary space T (i.e. $\mathbb{R}$, $\mathbb{N}$, $\mathbb{B}$, or $\mathbb{R}^k$). Here M is an upper bound on the duration of discrete time trajectories we consider to evaluate the reward; we indicate time-bounded trajectories by $\tilde{\eta}_{[0,M]}$. Such a function r can be a projection, thus monitoring the number of molecules belonging to a certain subset of chemical species at a certain time step, or it can take Boolean values in $\mathbb{B} = \{0, 1\}$, representing the truth of a linear temporal property, for example checking if the system has entered into a dangerous region. Note that $r(\tilde{\eta}_{[0,M]})$ is a probability distribution on T.

The second ingredient we need is a way to measure the error introduced by the abstract model, i.e. how much the abstract distribution differs from $r(\tilde{\eta}_{[0,M]})$. This can be accomplished by fixing a distance among distributions. In our experiments, we rely on the L1 norm, typically used to measure the goodness of fit:

$$d(X, Y) := \int_{\mathbb{R}^k} |p_X(z) - p_Y(z)| \, dz \tag{6}$$

This metric will be practically evaluated statistically, resulting in the so called histogram distance [7]. We now give a formal definition of model abstraction.

Definition 1. *Let $\eta = \{\eta_i\}_{i=0}^M$ be a discrete time stochastic process over an arbitrary state space S, with $M \in \mathbb{N}_+$ a time horizon, and let $r : S^M \to T$ be the associated reward function. An abstraction of (η, r) is a triple $(\bar{S}, p, \bar{r}, \bar{\eta} = \{\bar{\eta}_i\}_{i=0}^M)$ where:*

- *$\bar{S}$ is the* abstract state space;
- *$p : S \to \bar{S}$ is the* abstraction function;
- *$\bar{r} : \bar{S}^M \to T$ is the* abstract reward;
- *$\bar{\eta} = \{\bar{\eta}_i\}_{i=0}^M$ is the* abstract discrete time stochastic process over $\bar{S}$.

Let $\varepsilon > 0$. $\bar{\eta}$ is said to be ε-close to η with respect to d if, for almost any $s_0 \in S$,

$$d\big(r(\eta_{[0,M]}), \bar{r}(\bar{\eta}_{[0,M]})\big) < \varepsilon \qquad conditioned \ on \ \eta_0 = s_0, \ \bar{\eta}_0 = p(s_0) \tag{7}$$

It is common enough to choose a projection over a subset of chemical species as abstraction function p - as in [25], possibly identified by information theoretic criteria to be those most influencing the reward of interest. Alternatively, we could follow [13] and use a projection over a certain number of sub-regions of the original state space in order to get a finite abstract state space $\bar{S}$. Inequality 7 is typically experimentally verified simulating a sufficiently high number of trajectories from both the original system η and the abstraction $\bar{\eta}$ starting from a common initial setting. There is no way to ensure, with this experimental procedure, that inequality 7 holds for almost every s_0 in S. What can be done, instead, is to choose a high number of different initial settings which were not in the training set and check if the condition holds for them - the classical training/validation procedure which is used in Machine Learning to estimate the generalization error of a model.

Dataset Generation. We build our model abstraction reframing the situation as a supervised learning problem. Choose N random starting states $\{s_0^{(j)}\}_{j=1}^N$ from (a bounded subset of) $\tilde{X}$. For each $s_0^{(j)}$ run a simulation from t_0 to $t_1 := t_0 + \Delta t$. Denote by $\eta_{t_1}^{(j)}$ the system state at time t_1 for each one of these simulations. Define: $x^{(j)} := s_0^{(j)}$ and $y^{(j)} := \eta_{t_1}^{(j)}$ for all $j \in \{1, \ldots, N\}$.

We have thus built $\mathcal{D} := \{(x^{(j)}, y^{(j)})\}_{j=1}^N$, where each $y^{(j)}$ is a sample from the probability distribution $\mathbb{P}(\eta_{\Delta t} \mid \eta_0 = x^{(j)})$. We can as well simulate trajectories from t_0 to $t_h := t_0 + h\Delta t$, $h \in \mathbb{N}_+$. It is then sufficient to extract the system state at time instants $\{t_0, t_0 + \Delta t, \ldots, t_0 + h\Delta t\}$ in order to consider consecutive datapoints $(\eta_{t_0+i\Delta t}, \eta_{t_0+(i+1)\Delta t})$, $i \in \{0, \ldots, h-1\}$, as an (x, y) pair, like we described above.

Model Training. Fix a parametrized family of mixture distributions $\mathcal{M}$. Let g_θ be a MDN with $g_\theta(x) \in \mathcal{M}$ for each feature vector x, where θ are the network weights. g_θ is trained on the dataset $\mathcal{D}$, the simulation data, to learn the desired approximation K of K_d:

$$K_d(s \mid s_0) = \mathbb{P}(\eta_{\Delta t} = s \mid \eta_0 = s_0) \approx \mathbb{P}(g_\theta(s_0) \in B_s) := K(B_s \mid s_0) \quad (8)$$

where $B_s := \left\{ x \in \tilde{X} \mid \|x - s\|_\infty < \frac{1}{2} \right\}$ is the ball with respect to the infinity norm of radius $1/2$, centered in s. Note that the use of the infinity norm ball B_s, centered in s with radius $\frac{1}{2}$ is needed to properly compare a discrete distribution (given by the original kernel K_d) with a continuous distribution (given by the approximating kernel K). The idea is to partition the space into hypercubes centered in each point s of the original state space, each hypercube B_s representing the probability of s under the original model.

Training g_θ has a cost: neural networks are computationally intensive models. Nonetheless, once the network has been tuned, its evaluation is extremely fast, considering that all modern deep learning frameworks expose highly parallelized algorithms running on GPUs.

We remark that the MDN is trained to minimize the error in approximating the kernel for a given Δt. In this sense, we try to find an abstract model as accurate as possible to describe stepwise the dynamics. However, in our framework we evaluate the performance of the approximation using a more general and flexible criterion, based on a reward function and on a longer time span than Δt. The idea is that a good local approximation of the kernel should reflect in a good global approximation on longer time scales. An alternative we will explore in the future is to craft a loss function for the MDN based on the reward itself, to see if this could improve the approximation.

Abstract Model Simulation. In order to simulate the abstract model, we just need to sample up to time horizon $M > 0$ from the approximate kernel K, starting from the initial state s_0 and initial time t_0. While sampling, we are not restricted to the discrete state space S, but rather simulate trajectories on the continuous state space $\tilde{X}$. Each timestep of our simulations has thus a fixed computational cost, and requires us to evaluate g_θ at the current state x and draw a sample from the resulting distribution. This means that, choosing Δt equal to the timescale of interest, we can simulate arbitrary long trajectories at the needed level of time resolution without wasting computational resources. This algorithm can be easily employed in a multi-scale setting: we just need to train g_θ once, while a high number of agents can be simulated in parallel levering the computational power of one or more GPUs.

4   Implementation

Our implementation is in Python, and builds on several available tools and libraries, developed in the research communities of deep learning and computational systems biology. Integrating tools of different communities in a common pipeline has not been straightforward and a considerable amount of work is required to design a robust experimental setup. This was the main reason behind the development of *StochNet*, which hides the difficulties of such integration to the user, which can then focus on the design and tuning of models. The library, at the present stage of development, offers the following functionalities through a high-level API:

- Concurrent simulation of CRN models starting from different initial states using SSA/τ-leaping;
- A wide collection of implemented random variables to be used on their own or in mixtures to approximate complex probability distributions. Mixtures can be build using random variables from different families, as long as their samples have the same dimensionality;
- Seamless integration with Keras and Tensorflow to train and deploy Mixture Density Networks - there is no low-level scripting required to define, train and run a complete model;
- Simulation of trajectories from a Mixture Density Networks model using a GPU-based concurrent sampling strategy;

- A ready-to-run fault-tolerant pipeline to manage long-running numerical experiments (abstraction building). The pipeline can be easily customized and extended with custom computational tasks.

Most of these functionalities are built on top of existing Python packages: Tensorflow [18] and Keras [20] for neural networks, Gillespy [22] as an interface to StochKit 2.0 [14] for CRN simulations and Luigi [15] as work-flow manager.

StochNet is currently hosted on GitHub, where we are going to provide a detailed package documentation: https://github.com/LukeMathWalker/StochNet.

5 Experimental Results

In this section, we validate our approach on two case studies: the simple SIR model, and a more computationally intensive genetic network. Our focus is in the accuracy of the abstract model and on its computational efficiency.

Experimental Setting. To perform all the simulations described in the next sections we used a desktop personal computer, equipped with an AMD Ryzen 1700 CPU (3 GHz - 8 cores), an Nvidia GeForce GTX 1080Ti GPU (11 GB - 3'584 CUDA cores) and 32 GB of RAM (DDR4).

Data Preparation. Neural network convergence is enhanced if each component of the training dataset has zero mean and unit variance (cfr. Sect. 4.3 in [3] or Chap. 12 in [23]). We have thus followed this established procedure for all our datasets. No other forms of data cleaning or preprocessing have been performed.

5.1 SIR Model

We start by presenting experimental results on the SIR model, first discussing the MDN architecture we used to build the approximate kernel, and then presenting the experimental results.

Abstract Model. The abstract model $(\bar{S}, p, \bar{r}, \bar{\eta} = \{\bar{\eta}_i\}_{i=0}^M)$ is built as follows:

- The abstract state space $\bar{S}$ is $\mathbb{R}^3$, i.e. the continuous approximation of the state space $S = \mathbb{N}^3$;
- The abstract function p maps each point of S into its embedding into $\bar{S}$, i.e. p is essentially the identity function;
- The abstract reward $\bar{r}$ is just a projection function on one of the species;
- The abstract discrete time stochastic process $\bar{\eta} = \{\bar{\eta}_i\}_{i=0}^M$ will be a DTMC with a MDN-based kernel, learned from simulation data.

MDN Architecture. Considering the simple dynamic of the SIR model we opted for a fairly straight-forward architecture: our final setup uses a single hidden layer composed of 150 units with a ReLU[2] activation function; the output layer is designed to learn the parameters of a mixture of 2 multivariate normal random variables with a diagonal covariance matrix. We used the Adam algorithm to optimize our loss function, with standard settings (stepsize and decay rates) as suggested in [16].

In order to avoid overfitting we introduced two forms of regularization:

- *Max-norm regularization* [17];
- *Early stopping* (cfr. Sect. 7.8 of [23]).

As far as max-norm regularization is concerned, we constrained the Euclidean norm of the weights of the net to be smaller than 3.

For early stopping, in each epoch we used 30'000 training datapoints, processed in batches of 32 datapoints each. Our early stopping patience was set to 6 epochs and we evaluated the validation error on 5'000 held-out validation datapoints.[3]

The number of nodes in the hidden layer, the number of normal random variables in the mixture, the max-norm for weights and the usage of other regularization techniques (such as Dropout [17] or Gaussian noise injection) have been compared using the loss on the validation dataset as a measure of the generalization error. At this stage, we set hyperparameters and network structure by manual tuning, though we plan to implement automated methods (e.g. grid search or random search) in a future release.

It is worth to point out that even though the system dynamic is essentially unimodal in the chosen time step Δt, we experienced a significant worsening in the MDN performances when using a single multivariate normal random variable as output scheme, with several training attempts failed due to a divergent behavior in the loss minimization procedure. This never happened when using a mixture of two normal random variables: the possibility to distribute "errors" on two different Gaussians seems to stabilize the learning phase, leading to more consistent results. This is consistent with current practice in deep learning, which prefers a high model-capacity [23] coupled with a strong regularization. This strategy has also the advantage to require to the modeler only a limited knowledge of the output distribution to approximate, by picking a large number of mixture components.

Results. The execution time of whole pipeline takes slightly more than 3 min, of which roughly the 16% is used to train the MDN - see Table 1.

---

[2] Rectified Linear Unit: $f(x) = \max\{0, x\}$, cfr. [12].

[3] We generated training and validation sets with the same number of datapoints. A t each epoch, we subsample from these sets without replacement (until the whole dataset has been consumed) to have a finer control on overfitting using early stopping regularization.

Table 1. Execution time required to complete each step of the abstraction pipeline for the SIR model and the gene network model.

Task	Time: SIR	Time: GeneNet
GenerateDataset (training)	35 s	2,328 s (~39 min)
GenerateDataset (validation)	35 s	2,319 s (~39 min)
FormatDataset (training)	0.015 s	0.07 s
FormatDataset (validation)	0.015 s	0.07 s
GenerateHistogramData (training)	45 s	26,650 s (~7 h & 24 min)
GenerateHistogramData (validation)	36 s	27,299 s (~7 h & 30 min)
TrainNN	31 s	1,337 s (~22 min)

Our MDN has been trained to predict the system state after 0.5 units of simulation time. We computed the mean histogram distance on 1-step and 5-steps predictions between the MDN and the SSA algorithm with respect to the projection on the S population. A sample of the histograms after 5 steps can be seen in Fig. 2. The mean histogram (over 25 different initial conditions) after 1 step is 0.3 and after 5 steps is 0.33.

We can see from the figures that the MDN model captures the system dynamic quite accurately - the mode is almost always perfectly aligned with the SSA mode, while the reconstructed variance is sometimes slightly defective or excessive. Even though 5-steps performances are slightly worse than their respective single-step analogues, we do not observe any significant error-propagation phenomenon in place, considering the level of resolution we are using for our histograms (200 bins on $[0, 200]$). Nonetheless we cannot affirm that the MDN distribution is consistently indistinguishable from the SSA distribution: the upper-bound on the mean of the self-distance using 200 bins and 10'000 samples (cfr. [7]) is ~0.16 with a variance upper-bound of ~0.015; the lowest mean histogram distance achieved by our MDN on the training histogram dataset, with single-step simulations, is 0.24.

We remark that we tested the abstraction starting from random initial states, picking points also from regions with a low density of training points, in order to test the generalization capabilities of our method.

5.2   Gene Regulatory Network

Introduction. Stochasticity plays a prominent role in CRN dynamics whenever some key chemical species in the network have low molecular counts: adding or removing a single copy might result in significant fluctuations. A typical example of this behavior can be observed in a simple self-regulated gene network [19]: a single gene G is transcribed to produce copies of a mRNA signal molecule M, which are in turn translated into copies of a protein P; P acts as a repressor with respect to G - it binds to a DNA-silencer region, inhibiting gene transcription.

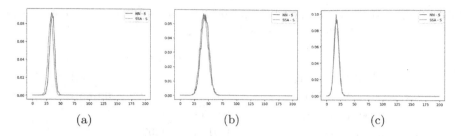

Fig. 2. Comparison of SSA and MDN histograms after 5 simulation steps for the SIR model over 3 different initial settings sampled from the validation dataset. $K = 200$ equally sized bins are being used. The mean histogram distance, over 25 different initial settings, is 0.33.

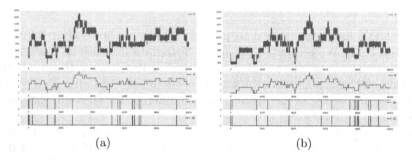

Fig. 3. Two sample trajectories for the gene regulatory network.

In other words, the gene activity is regulated through a negative-feedback loop, a common pattern in biological systems. The relevant chemical equations of the system are the following:

$$R_1 : \quad G^{ON} \xrightarrow{k_{prodM}} G^{ON} + M \qquad\qquad (transcription) \qquad\qquad (9)$$

$$R_2 : \quad M \xrightarrow{k_{prodP}} M + P \qquad\qquad (translation) \qquad\qquad (10)$$

$$R_3 : \quad G^{ON} + P \xrightarrow{k_{act}} G^{OFF} \qquad\qquad (protein\ binding) \qquad\qquad (11)$$

$$R_4 : \quad G^{OFF} \xrightarrow{k_{deact}} G^{ON} + P \qquad\qquad (protein\ unbinding) \qquad\qquad (12)$$

$$R_5 : \quad M \xrightarrow{k_{degM}} \varnothing \qquad\qquad (mRNA\ degradation) \qquad\qquad (13)$$

$$R_6 : \quad P \xrightarrow{k_{degP}} \varnothing qquad \qquad\qquad (protein\ degradation) \qquad\qquad (14)$$

All propensity functions are assumed to be of mass-action type, with the name of their respective rate specified on the reaction arrow. The system dynamic varies significantly with respect to the choice of reaction rates. The parameters values for our simulations are reported in Table 2 while Fig. 3 shows some of the simulated system trajectories: the systems exhibits several well-separated *stable configurations* which are roughly determined by the number of available mRNA

Table 2. Reaction rates used in our simulations of the gene regulation network.

k_{prodM}	k_{prodP}	k_{act}	k_{deact}	k_{degM}	k_{degP}
350	300	1	166	0.001	1.5

molecules. It is worth mentioning that, on a smaller scale, each stable point is actually noisy - we can in fact observe a high number of small amplitude oscillations. These characteristics can be easily detected looking at the probability density function of $\mathbb{P}_{s_0}(\eta_t)$, for $t > t_0$: we have 5 or 6 distinguishable modes, with a certain amount of Gaussian noise affecting each one of them (see Fig. 5).

Approximating this system dynamics with a normal regression NN is an impossible task, considering that the system dynamic is definitely not unimodal: this makes of this model a perfect testing ground for our MDN-based abstraction. In particular, we consider an abstract model similar to the one defined for SIR: we embed the discrete state space into a continuous one, and learn an MDN kernel on this space, considering unidimensional projections as our reward functions.

MDN Architecture. To model the dynamics of the gene regulation network we devised a more sophisticated architecture than for the SIR case, see Fig. 4. We are no longer dealing with a shallow neural network: our architecture uses two hidden layers. The learning capabilities of deep neural networks are significantly higher, even though depth increases the chance of overfitting and introduces the vanishing/exploding gradient problem, an issue affecting the minimization of the loss function (cfr. [4]).

ReLU activation functions avoid the problem of exploding gradients (ReLU derivative is either 0 or 1) but they still do not solve the issue of vanishing gradients. A variety of different strategies have been devised (changing weight initialization, gradient clipping, etc.): we chose to follow in the footsteps of [21] - the first to introduce the ResNet architecture. Their proposal is strikingly simple: instead of stacking hidden layers directly on top of each other we use *skip-connections* to regularize the network behavior. In other words, instead of learning a map of the form $NN(x) = f_1(x)$ we try to learn $NN(x) = f_1(x) + x$. If $H(x)$ is the *true* function we are trying to fit, then f_1 is actually trying to approximate $H(x) - x$ which is called *residual function*. Despite of its simplicity, this adjustment allowed [21] authors to train effectively neural networks with 152 hidden layers. We do not need such extreme configurations, and it's difficult to go deeper with fully connected hidden layers, but deeper architectures can help improve performance of our approximations.

In terms of regularization, we constrained the Euclidean norm of layer weights to be below 3 (max-norm regularization), we used early stopping with 6 epochs of patience and we injected Gaussian noise (0 mean, 0.01 variance) between the input layer and the first hidden layer. Noise addition is another common regularization technique which tries to force the network to learn a more robust representation, i.e. a mapping robust enough to be insensible to small perturbation

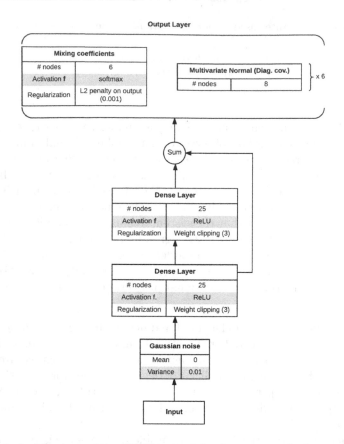

Fig. 4. Architecture of the MDN trained for the gene regulation network.

of the inputs. Each epoch used 30′000 training datapoints, which were processed in batches of 32 datapoints each. We evaluated the validation error at the end of each epoch using 5′000 held-out validation datapoints.[4] All model hyper-parameters have been tuned using the loss on the validation dataset as a measure of the generalization error.

Results. Executing the whole pipeline requires ∼16 hours and 30 min: the training of the MDN is responsible for a mere 2% of the overall computational cost - see Table 1. Our MDN has been trained to predict the system state after 400.0 units of simulation time. We computed the mean histogram distance on 1-step and 5-steps predictions between the MDN and the SSA algorithm with respect to the projection on the protein P, just like we did for the SIR model. A sample of the resulting histograms for the 5 step case can be seen in Fig. 5. The mean

---

[4] As in the SIR case, training set and validation set have the same number of datapoints, and we subsample without replacement at each epoch to have a finer control on overfitting using early stopping regularization.

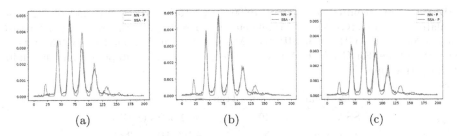

(a)   (b)   (c)

Fig. 5. Comparison of SSA and MDN histograms of species P after 5 steps of simulation time over 3 different initial settings sampled from the validation dataset. $K = 200$ equally sized bins are being used. The mean histogram distance, over 25 different initial settings, is 0.28.

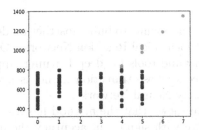

Fig. 6. Plot of initial conditions for the training of the GRN model (blue) and its test (orange). X axis, mRNA. Y axis, Protein. (Color figure online)

histogram distance after 1 step is 0.34, while it stabilizes to 0.28/0.29 after 5, 10 and 25 steps. Although the distributions are not indistinguishable (mean histogram self distance is ~ 0.16), the true and the approximate distributions are quite similar, and the main qualitative characteristics of the process are well captured. In particular, the MDN model is able to identify quite consistently the four modes associated with the highest probabilities, with a quite accurate reconstruction of the respective variances (prone to be overestimated, more than underestimated). The two rarest modes, instead, are usually ignored. Note that we tested the abstract model starting from random initial conditions, including some points coming from low coverage areas of the training set, as can be observed in Fig. 6.

The speed gain, though, is quite remarkable (see Table 3): simulating 10′000 trajectories for 25 different initial settings until final time 10′000 takes ~ 17 h using SSA, while the MDN model is capable of doing it in 35 s - it is roughly 1730 times faster than SSA. We also compared our MDN approach to τ-leaping. Using $\tau = 400.0$ in order to achieve the greatest possible speed-up, τ-leaping produces trajectories that are indistinguishable from SSA-generate trajectories, achieving a mean histogram distance of ~ 0.1. However, it took τ-leaping ~ 3 hours to generate 10′000 trajectories for 25 different initial settings with final time 10′000: almost 6 times faster than SSA but still 300 slower than our MDN abstraction.

Table 3. Time required to simulate 10'000 trajectories of the SIR and of the gene regulatory network (GRN) for 25 different initial settings with final time 5 for SIR and 10'000 for the GRN. For SIR: The MDN returns a datapoint every 0.5 units of simulation time. For GRN: the MDN and τ-leaping return a datapoint every 400 units of simulation time.

Algorithm	SIR - Time	GRN - Time
SSA	40 s	~60'650 s (~17 h)
τ-leaping	NA	~10'750 s (~3 h)
MDN (trained)	7 s	35 s

6 Conclusions

In the paper we presented a pipeline to build abstract models in order to increase simulation efficiency of Biochemical Reaction Networks. Our approach leverages recent advances in theory and tools for deep learning, approximating a CTMC as a Discrete Time process whose transition kernel is learned from simulation data using Mixture Density Neural Networks.

In the paper, we have shown that the method has a significant potential: we have been able to capture with significant accuracy the qualitative behavior of a genuinely multimodal CRN without introducing any kind of prior knowledge into our procedure. The achieved speed-up is impressive and it would enable, on similar systems, to actually perform multi-scale simulations with population of 10^6 or 10^7 cells. Future work is mostly on performing further experiments and studies in this direction, and integrating these approaches for multi-scale models.

Moreover, even though MDNs are quite an old model (they were first introduced in 1991) they have not seen wide adoption so far, possibly because they present challenges for gradient based methods (see [23] Sect. 6.2.2.4). There is not a consistent record of publications in the ML literature trying to specifically address the limitations and the best practices concerning these models. It is thus worthwhile to further research MDNs on their own, trying to work out the best way to design and train these neural networks, in particular for the applications we are concerned with.

References

1. Gillespie, D.T.: Exact stochastic simulation of coupled chemical reactions. J. Phys. Chem. **81**(25), 2340–2361 (1977). (Visited 28 Nov 11 2013)
2. Bishop, C.M.: Mixture density networks. Technical report NCRG/94/004. Neural Computing Research Group, Aston University (1994)
3. LeCun, Y.A., Bottou, L., Orr, G.B., Müller, K.-R.: Efficient BackProp. In: Montavon, G., Orr, G.B., Müller, K.-R. (eds.) Neural Networks: Tricks of the Trade. LNCS, vol. 7700, pp. 9–48. Springer, Heidelberg (2012). https://doi.org/10.1007/978-3-642-35289-8_3. http://yann.lecun.com/exdb/publis/pdf/lecun-98b.pdf

4. Hochreiter, S., et al.: Gradient flow in recurrent nets: the difficulty of learning long-term dependencies. In: Kremer, S.C., Kolen, J.F. (eds.) A Field Guide to Dynamical Recurrent Neural Networks. IEEE Press (2001)
5. Christopher, M.: Pattern Recognition and Machine Learning. Springer, New York (2006)
6. Cao, Y., Gillespie, D.T., Petzold, L.R.: Efficient step size selection for the tau-leaping simulation method. J. Chem. Phys. **124**(4), 044109 (2006). https://doi.org/10.1063/1.2159468
7. Cao, Y., Petzold, L.: Accuracy limitations and the measurement of errors in the stochastic simulation of chemically reacting systems. J. Comput. Phys. **212**(1), 6–24 (2006). https://doi.org/10.1016/j.jcp.2005.06.012
8. Wilkinson, D.J.: Stochastic Modelling for Systems Biology. Chapman & Hall, Boca Raton (2006)
9. Pahle, J.: Biochemical simulations: stochastic, approximate stochastic and hybrid approaches. Brief. Bioinform. **10**(1), 53–64 (2008). https://doi.org/10.1093/bib/bbn050
10. Greenwood, P.E., Gordillo, L.F.: Stochastic epidemic modeling. In: Chowell, G., Hyman, J.M., Bettencourt, L.M.A., Castillo-Chavez, C. (eds.) Mathematical and Statistical Estimation Approaches in Epidemiology, pp. 31–52. Springer, Dordrecht (2009). https://doi.org/10.1007/978-90-481-2313-1_2
11. Deisboeck, T.S., et al.: Multiscale cancer modeling. Annu. Rev. Biomed. Eng. **13**(1), 127–155 (2011). ISSN 1523-9829, 1545-4274. https://doi.org/10.1146/annurev-bioeng-071910-124729
12. Glorot, X., Bordes, A., Bengio, Y.: Deep sparse rectifier neural networks. In: Proceedings of the Fourteenth International Conference on Artificial Intelligence and Statistics, vol. 15, pp. 315–323 (2011). http://proceedings.mlr.press/v15/glorot11a.html
13. Liu, B., Hsu, D., Thiagarajan, P.S.: Probabilistic approximations of ODEs based bio-pathway dynamics. Theor. Comput. Sci. **412**, 2188–2206 (2011)
14. Sanft, K.R,.: StochKit2: software for discrete stochastic simulation of biochemical systems with events. Bioinformatics **27**(17), 2457–2458 (2011). https://doi.org/10.1093/bioinformatics/btr401
15. Bernhardsson, E., Freider, E.: Luigi (2012). https://github.com/spotify/luigi
16. Kingma, D.P., Ba, J.: Adam: a method for stochastic optimization (2014). https://arxiv.org/abs/1412.6980
17. Srivastava, N.: Dropout: a simple way to prevent neural networks from overfitting. J. Mach. Learn. Res. **15**, 1929–1958 (2014)
18. Abadi, M., et al.: TensorFlow: large-scale machine learning on heterogeneous systems (2015). Software: https://www.tensorflow.org/
19. Bodei, C.: On the impact of discreteness and abstractions on modelling noise in gene regulatory networks. Comput. Biol. Chem. **56**, 98–108 (2015). https://doi.org/10.1016/j.compbiolchem.2015.04.004
20. Chollet, F., et al.: Keras (2015). https://github.com/fchollet/keras
21. He, K., et al.: Deep residual learning for image recognition, December 2015. https://arxiv.org/abs/1512.03385
22. Abel, J.H., et al.: GillesPy: a Python package for stochastic model building and simulation. In: IEEE, September 2016, pp. 35–38 (2016). https://doi.org/10.1109/LLS.2017.2652448
23. Goodfellow, I., Bengio, Y., Courvilleet, A.: Deep Learning. MIT Press, Cambridge (2016). http://www.deeplearningbook.org

24. Michaelides, M., Hillston, J., Sanguinetti, G.: Statistical abstraction for multi-scale spatio-temporal systems. In: Quantitative Evaluation of Systems, QEST 2017, pp. 243–258 (2017). https://doi.org/10.1007/978-3-319-66335-7_15
25. Palaniappan, S.K., et al.: Abstracting the dynamics of biological pathways using information theory: a case study of apoptosis pathway. Bioinformatics (2017). ISSN 1367-4803, 1460-2059, https://doi.org/10.1093/bioinformatics/btx095
26. Schnoerr, D., Sanguinetti, G., Grima, R.: Approximation and inference methods for stochastic biochemical kinetics - a tutorial review. J. Phys. A: Math. Theor. **50**(9), 093001 (2017). ISSN 1751-8113, 1751-8121, https://doi.org/10.1088/1751-8121/aa54d9. Visited 20 Apr 2017

Derivation of a Biomass Proxy
for Dynamic Analysis of Whole Genome
Metabolic Models

Timothy Self[1], David Gilbert[1(✉)], and Monika Heiner[1,2]

[1] Brunel University London, Uxbridge, UK
timself101@hotmail.com, {david.gilbert,monika.heiner}@brunel.ac.uk
[2] Brandenburg Technical University, Cottbus, Germany
monika.heiner@b-tu.de

Abstract. A whole genome metabolic model (GEM) is essentially a reconstruction of a network of enzyme-enabled chemical reactions representing the metabolism of an organism, based on information present in its genome. Such models have been designed so that flux balance analysis (FBA) can be applied in order to analyse metabolism under steady state. For this purpose, a biomass function is added to these models as an overall indicator of the model's viability.

Our objective is to develop dynamic models based on these FBA models in order to observe new and complex behaviours, including transient behaviour. There is however a major challenge in that the biomass function does not operate under dynamic simulation. An appropriate biomass function would enable the estimation under dynamic simulation of the growth of both wild-type and genetically modified bacteria under different, possibly dynamically changing growth conditions.

Using data analytics techniques, we have developed a dynamic biomass function which acts as a faithful proxy for the FBA equivalent for a reduced GEM for *E. coli*. This involved consolidating data for reaction rates and metabolite concentrations generated under dynamic simulation with gold standard target data for biomass obtained by steady state analysis using FBA. It also led to a number of interesting insights regarding biomass fluxes for pairs of conditions. These findings were reproduced in our dynamic proxy function.

1 Introduction

A large amount of publicly available information, regarding whole genome metabolic reaction networks in e.g. *Escherichia coli* (*E. coli*), has been encoded as constraint-based flux-balance analysis (FBA) models. This forms a very useful resource, especially when combined with genome information, as in the BiGG collection [13]. Our overall aim is to build on this knowledge to make whole genome metabolic models (GEMs) available for dynamic simulation in order to be able to observe new and complex behaviours including, for example, under dynamically changing growth conditions. In previous work we have reported our

© The Author(s) 2018
M. Češka and D. Šafránek (Eds.): CMSB 2018, LNBI 11095, pp. 39–58, 2018.
https://doi.org/10.1007/978-3-319-99429-1_3

methodology to convert FBA models into dynamic models [7], as the first steps that we have already made in this direction.

Constraint-based FBA models are designed to analyse metabolism activity under steady state. For this purpose a biomass function is added, implemented as an abstract reaction over metabolites and serving as an overall indicator of the model's viability. However this artificial function is very complex and highly tuned in that it comprises many substrates and products, with a wide range of specific non-integer stoichiometries [22]. This tuned complexity means that we have found it impossible to directly use the existing FBA biomass function as an indicator of viability in the simulation of dynamic models.

The work reported in this paper describes a data analytics approach to derive a proxy biomass function for dynamic GEMs, relying on averaged stochastic simulation traces of both metabolite concentrations and reaction rates. This proxy has been developed to be both highly robust and accurate with respect to a wide variety of growth conditions. Such a biomass proxy will enable the estimation by dynamic simulation of the growth of both wild-type and genetically modified bacteria under different growth conditions.

Our contributions include: the development of a well-defined general method, organised as a workflow which provides guidance to derive a biomass function for any GEM. We demonstrate our method for the well-established reduced *E. coli* core model for the K12 strain [21] available in SBML format. Our workflow exploits a number of well recognised data analytics methods, including regression analysis and machine learning. The gold standard FBA data on which our work is based was generated using the *Cobra* software [25], and the dynamic simulation data was generated with the stochastic simulation algorithm Delta Leaping [23] using the *Marcie* software [11]. For this purpose, we converted the SBML model into a stochastic Petri net by help of the *Snoopy* software [10]. The robustness of our results was ensured by the use of a large number of observations generated by single and combined growth conditions. An additional unexpected result was the observation of the non-linear additive effects of certain paired growth conditions which were found in the FBA results and faithfully preserved in the predictions of our biomass proxy.

This paper is organised as follows. In the next section we review some related work, followed by a section on the data used, its generation and preparation. Next we describe the data analytics methods deployed and their application in our workflow. We then evaluate the key results, followed by conclusions and outlook. Some additional information is provided as Supplementary Materials, available at http://www-dssz.informatik.tu-cottbus.de/DSSZ/Software/Examples.

2   Related Work

A genome scale metabolic model (GEM) is essentially a reconstruction of a network of enzyme-enabled chemical reactions representing the metabolism of an organism, based on information present in its genome. It can be used to understand an organism's metabolic capabilities. The reconstruction involves a number of steps, including the functional annotation of the genome, identification of the associated reactions and determination of their stoichiometry, which is the relationship between the relative quantities of substances taking part in a reaction. It also involves determining the biomass composition, estimating energy requirements and defining model constraints [1]. A characteristic of these models is that although they describe the reactions in terms of substrates and products, they do not contain information on reaction rate constants because these cannot be determined by the current reconstruction process.

GEMs have become an invaluable tool for analysing the properties and steady state behaviour of metabolic networks, and have been especially successful for *E. coli* [9]. The most recent model iJO1366 has been accepted as the reference for *E. coli* network reconstruction. It has provided valuable insights into the metabolism of *E. coli* and been used to formulate intervention strategies for targeted modifications of the metabolism for biotechnological applications. Bacterial GEMs can comprise about 5000 reactions and metabolites, and encode a huge variety of growth conditions. The BiGG public domain database contains 92 GEMs, of which 52 are for *E. coli* [13].

However, it has been argued that as the size and complexity of genome scale models increases, limitations are placed on popular modelling techniques, such as constraint-based modelling and kinetic modelling [9]. A similar argument was put forward by [4] as a justification for developing a network reduction algorithm to derive smaller models by unbiased stoichiometric reduction, based on the view that the basic principles of an organism's metabolism can be studied more easily in smaller models. A number of reduced models have been proposed, including [9,21].

Flux balance analysis (FBA) is a constraint-based approach for analysing the flow of metabolites through a metabolic network by computing the reaction fluxes in the steady state. This enables the prediction of the growth rate of an organism or the production rate of biotechnologically important metabolites. An additional biomass objective function is added to compute an optimal network state and resulting flux distribution out of the set of feasible solutions. The growth rate as reflected by the steady state flux of the biomass function is constrained by the measured substrate uptake rates and by maintenance energy requirements [22].

The biomass function indirectly indicates how much certain reactions contribute to the phenotype. It does so by being represented as a pseudo (i.e. abstract and artificial) "biomass reaction" that drains substrate metabolites from the system at their relative stoichiometries to simulate biomass production The biomass reaction is based on experimental measurements of biomass components. This reaction is scaled so that the flux through it is equal to the exponential growth rate (μ) of the organism [20].

FBA has limitations as it is unable to predict metabolite concentrations because it does not use initial metabolite concentrations or kinetic parameters. The mathematical model incorporates the stoichiometric matrix and any biologically meaningful constraints over the flux ranges. Therefore it is only suitable for determining relative fluxes at steady state [21].

Dynamic simulation. The network described by the stoichiometric matrix can be equally read as a dynamic model to explore the temporal behaviour of the system by tracing how metabolite concentrations and reaction rates (fluxes) change over time [6,7]. For this purpose the model has to be enriched by initial metabolite concentrations and kinetic reaction rates (kinetic laws and corresponding parameters), both initially estimated and ultimately determined by experimental observation.

There are three main approaches for dynamic simulation: qualitative, stochastic, and deterministic approaches. The most abstract representation of a biochemical network is qualitative. However, biochemical systems are inherently governed by stochastic laws, though due to the computational resources required, continuous models are commonly used in place of stochastic models to approximate stochastic behaviour with a deterministic approach [8]. These approaches to dynamic simulation do not make any assumptions about steady state, unlike FBA and dynamic FBA [16], thus facilitating the analysis of the transient behaviour of the biological system.

The dynamic simulation of large and complex whole genome models has been a bottleneck in the past [26], which has presented considerable difficulties both for stochastic and deterministic methods [7]. However, stochastic simulation based on Delta Leaping [24], permits the efficient simulation of these very large GEMs, enabling the observation of new and complex behaviours [7].

There is also another limitation however, which is that the biomass function for constraint-based GEMs does not work correctly under the dynamic simulation of transient behaviour without quasi-steady state assumption, due to the complexity in terms of the number of variables and specificity in terms of the stoichiometries of the function. A systematic approach to the development of a proxy function that can be used to determine the amount of biomass produced is the main focus of the work presented here.

3   Data

Model. The research reported in this paper builds on the reduced *E. coli* core model for the K12 strain of Orth et al. [21] available in SBML format from http://systemsbiology.ucsd.edu/Downloads/EcoliCore. Its reactions and pathways have been chosen to represent the most well-known and widely studied metabolic pathways of *E. coli*.

The metabolic reconstruction of the model includes 54 unique metabolites in two compartments: cytosol and extracellular, and these metabolites may exist as SBML species in both, differentiated by appropriate tags. The cytosol contains 52 of these metabolites (of which 34 are uniquely cytosol species), and the extracellular compartment contains 20 metabolites, two of which are not found in the cytosol. By definition each of the 20 extracellular metabolites exists as two copies – 'boundary' and 'extracellular' – the latter type being used in the transport mechanism between extracellular and cytosol compartments, making 40 extracellular species. In total there are $52 + 40 = 92$ species in the SBML specification.

The model has 94 reactions of which 46 are reversible, which can be categorised into 49 metabolic reactions, 25 transport reactions between compartments, and 20 exchange reactions [20]. Exchange reactions are always reversible and exist for each extracellular metabolite (boundary condition), the directions of which can be changed using the flux constraints. Additionally there is a biomass function implemented as an abstract (irreversible) reaction, which comprises 16 substrates and 7 products with stoichiometries varying from 0.0709 to 59.81, see Table 4 in the Supplementary Materials.

The model can be configured to investigate the effect of different growth conditions using the 20 extracellular species. Of these, 14 are carbon source growth conditions including formate; we follow standard practice to ignore formate due to viability issues [20] leaving 13 carbon sources that we considered. Five of the remaining 6 boundary conditions correspond to the ingredients of a minimal growth medium based on M9 [17], namely CO_2, H+, H_2O, D-Glucose, ammonium, and phosphate. Finally, oxygen is also a boundary condition. Each of the carbon sources can be considered both aerobically and anaerobically, making in total $2 \cdot 13 = 26$ single growth conditions while ignoring formate.

For the purpose of simulation, we convert the SBML model into a stochastic Petri net (SPN), which is done with the Petri net editor and simulator *Snoopy* [10]. This involves four adjustments.

- As required for any discrete dynamic modelling approach, reversible reactions are modelled by two opposite transitions representing the two directions a reversible reaction can occur.
- Metabolites which have been declared as boundary conditions are associated with additional source and sink transitions (called boundary transitions), mimicking the FBA assumption of appropriate in/outflow. This transforms a place-bordered net into a transition-bordered net, if all boundary places (i.e., source/sink places) have been declared as boundary conditions.
- Reaction rates are assigned to all transitions following the mass-action pattern with uniform kinetic parameters of 1.
- The initial concentration is set to zero, except for those 12 metabolites involved in mass conservation (P-invariants), computed with *Charlie* [12], which were all set to the same initial amount, e.g. 10.

The Petri net model (ignoring the biomass function) comprises 180 transitions $(94 + 46 + 2 \cdot 20$ boundary transitions) and 92 places; it is shown in Fig. 6 in the Supplementary Materials. In addition, the biomass function is present, but never active.

Datasets. "Gold standard" target data for biomass was generated using FBA with the *Cobra* software [25]. Time-series data for reaction activity and metabolite concentrations were generated under dynamic simulation with the approximative stochastic simulation algorithm Delta Leaping using the *Marcie* software [11] and recorded for all species (places) and reactions (transitions) for 1,000 time points averaged over 10,000 runs. The average of the last 200 time points was calculated for each of the reactions and metabolites in the dynamic time series data for use in regression analysis, based on the assumption that this best represented the steady state. The rates of the forward and backward transitions of reversible reactions were combined and appropriate new variables introduced with suffix FwRe. Redundant variables were removed, e.g. boundary transitions introduced by the conversion of SBML into SPN, and also the original biomass function. These data preparation steps are summarised in Fig. 1.

This was initially done for the 26 different single growth conditions. However, the number of predictors (reactions and metabolites) in our model is large in relation to the number of observations (growth conditions), and regression analysis is more accurate for larger numbers of observations. Furthermore, analysis based on more conditions helps to reduce the likelihood of overfitting and allows more predictor variables to be included in the regression equation. Given a certain number of observations, there is an upper limit to the complexity of the model that can be derived with any acceptable degree of uncertainty [2]. Also there is a broadly linear relationship between the sample size (number of observations) and the number of predictors included in a multiple linear regression model used for prediction [14].

Therefore, additional observations were generated using pairwise combinations of 13 carbon sources, both aerobic and anaerobic, yielding $2 \cdot (13^2 - 13)/2 = 156$ pairwise conditions. This also enabled us to investigate the effect of paired combinations of carbon sources. Finally in order to further enhance the effectiveness of the regression analysis, we combined the 156 pairwise observations with the 26 single condition observations to create a combined dataset of 182 observations (growth conditions) with 300 variables (metabolites and reactions) and 1 dependant variable (Biomass proxy).

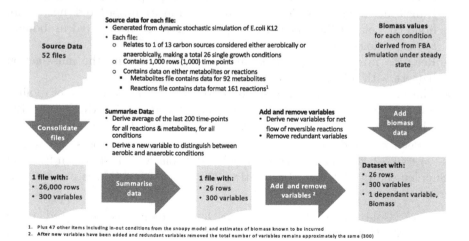

Fig. 1. Summary of data preparation steps. A dataset was created for analysis with 300 variables and 26 single growth conditions, later extended by 156 pairwise conditions.

4 Data Analytics Methods

In terms of data analytics we wish to derive a mathematical function to correctly predict w.r.t. FBA results the biomass, to be precise: the steady state flux of the biomass reaction, for different growth conditions based on the various metabolite concentrations and reaction rates in the steady state as determined by simulation of a dynamic model. In other words, the expected result is a proxy function which predicts the FBA value in the steady state. Our analysis is based on the assumption that a steady state exists for that model.

The overall workflow is illustrated in Fig. 2, and essential steps are explained below. A more detailed workflow protocol is provided in the Supplementary Materials.

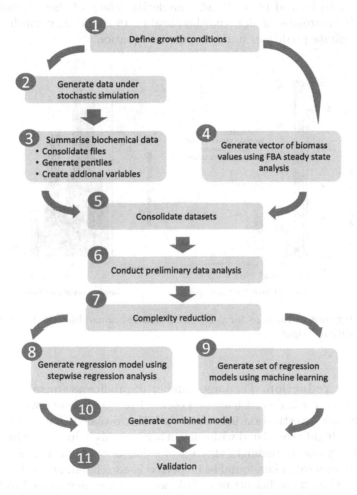

Fig. 2. Workflow of key analytical steps in the development of the proxy function. Steps 1–5 are covered in Sect. 3, steps 6–10 in Sect. 4 and step 11 in Sect. 5.

We use the term regression analysis to refer to the analysis of the relationships between a dependent variable (which in this paper is biomass) and the predictor variables (which in this paper are the metabolites and reactions).

Preliminary data analysis generated two important observations that shaped the approach for regression analysis.

(i) The biomass values of anaerobic conditions follow "zero inflated distribution" (in which a large portion of values were either zero or close to zero), whereas biomass production for aerobic conditions resembled a normal distribution, as illustrated in Fig. 3. This finding led to the creation of a dichotomous (binary) variable to distinguish between the two sets of conditions.

(ii) There are a large number of independent variables or predictors with the potential to lead to too much complexity. Many of these variables were highly correlated with each other, leading to collinearity which can cause inaccurate predictors in the regression equation.

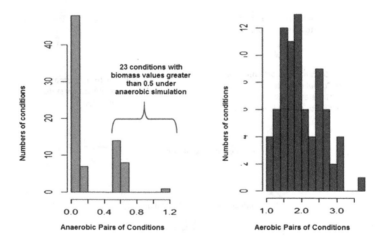

Fig. 3. Histograms of biomass for pairs of conditions; anaerobic conditions (left) and aerobic conditions (right).

Complexity reduction. The large number of predictors (metabolite concentrations and reaction fluxes) had the potential to generate overwhelming complexity. Further investigation into these variables revealed that a large number of them were highly correlated with each other and this helped to reduce some of the complexity and to highlight the risk of collinearity in the regression model. Some variables were in fact found to be perfectly correlated, given that they had Pearson correlation coefficients of 1. This was because they related to the same metabolite at different stages in transportation (or different compartments) as

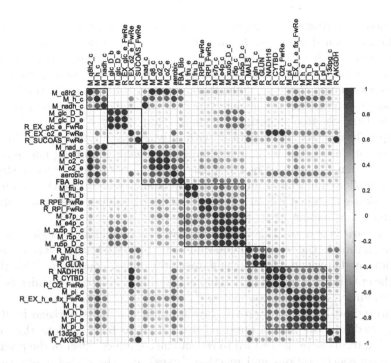

Fig. 4. Variable correlation matrix with hierarchical clustering based on complete linkage, using 37 key variables from the initial dataset of 26 single conditions.

represented by the underlying biology model. So, the concentrations of metabolite did not change irrespective of whether it was outside the *E. coli* bacterium or passing through the outer part of the *E. coli* bacterium.

Clustering techniques were used to identify groups of highly correlated variables. Figure 4 provides an illustration of hierarchical variable clustering using complete linkage for 37 key variables. Note that only a limited number of variables were used as including all 300 would not be visually effective.

We applied two approaches to regression analysis—stepwise regression and a machine learning based algorithmic approach.

Stepwise regression analysis. The decision was taken to initially develop a multiple linear regression model to predict biomass in preference to employing machine learning algorithms, due to the additional insight that statistical methods can offer in terms of inference or interpretability of the parameters (which in this case is the underlying biochemistry) as opposed to simply looking at prediction. The methodology and approach applied to the regression analysis was strongly influenced by the findings identified in the preliminary analysis described above.

An initial regression equation was derived by applying a process of stepwise regression to the small dataset of 26 single conditions. Variables were included in the stepwise regression process based on the earlier work carried-out around correlation and clustering analysis. The validity of using correlation analysis could be questioned because a linear combination of a few dependent variables that are only weakly correlated with the dependent variable may have larger correlation with the dependent variable than a linear combination of a few strongly correlated variables. However, it should be pointed out that an alternative, more formulaic approach to variable selection was applied when running the automated feature selection as outlined in the section below.

In spite of the fact that a dataset with only 26 conditions imposed limits on the scope of the regression analysis, the results were promising and served as a starting point for analysis based on the larger dataset. A process of stepwise regression followed in which different terms were successively added and removed. A regression equation with an adjusted r-squared value of 0.91 was obtained, providing a fair amount of explanatory power.

A comparison of datasets generated for paired, as opposed to single conditions identified a drop in the adjusted r-squared value from 0.92 to 0.83 when the same regression model was applied to the dataset for paired conditions. This finding led to the creation of a second dichotomous variable '*Pair*' to distinguish between paired condition and single conditions.

A procedure known as StepAIC (available in the Mass package in R) was then applied and the results obtained were used to help validate this model. Interaction terms were then added to reflect the combined effect of the predictors (metabolite concentrations and reaction fluxes) and the dichotomous variable created to distinguish between aerobic and anaerobic conditions. The inclusion of such terms in the regression model led to a significant improvement and an adjusted r-squared of 0.976 was obtained, illustrating the strong explanatory power. Furthermore, all of the coefficients and the overall model were shown to be highly statistically significant.

A machine learning based algorithmic approach to regression. One of the main challenges identified in the preliminary analysis was the need to manage the complexity created by the large number of variables (300), which is a characteristic of many modern datasets. Kursa and Rudnicki identified two main issues with large datasets. Firstly, the decrease in accuracy that can occur when too many variables are included, known as the *minimal optimal problem*. Secondly, the challenges in finding all relevant variables as opposed to just the non-redundant ones, which is known the *all-relevant problem* [15]. This is of particular importance when one wishes to understand the mechanisms related to the subject of interest, as opposed to purely building a black box predictive model. Kursa and Rudnick have developed *Boruta*, a package in R [3] for variable selection, which includes a variable selection algorithm (also called Boruta) to address the all-relevant problem. The algorithm employs a wrapper approach which is built around a random forest classifier. In a wrapper approach, the classifier (in this case a random forest classifier) is used as a black box to return output, which is used to evaluate the importance of variables. Random forest is an ensemble method used in machine learning in which classification is performed by voting on (or taking the average of) multiple unbiased weak classifiers -

decision trees. These trees are independently developed on different samples of the training dataset [15].

Diagnostics terminology. In the following we first explain the terminology of the methods that we have used, followed by their application in our approach.

Akaike's information criterion (AIC) is a diagnostic used in regression, which takes into account how well the model fits the data while adjusting for the ability of that model to fit any dataset. It seeks to strike a balance between goodness of fit and parsimony and assigns a penalty based on the number of predictors to guard against overfitting. It is defined as

$$AIC = -2 \cdot ln(L) + 2 \cdot p,$$

where L is a measure of the log likelihood and p is the number of variables in the model [18].

Bayesian information criterion (BIC) is a Bayesian extension of AIC with

$$BIC = -2 \cdot ln(L) + p \cdot ln(n),$$

where L is a measure of the log likelihood and p is the number of variables in the model as above. It is known to be a more conservative measure than AIC in the sense that it assigns a stronger penalty as more predictors are added to the model. Like AIC, the lower the value of BIC the better.

Note that AIC and BIC are used to determine the relative quality of different statistical models based on the same dataset. They cannot be used to compare models generated from different datasets.

Cross-validation is an evaluation technique, which is used to assess the accuracy of results obtained from training data on test data. In cross-validation, the number of folds 'k' is defined in advance. The data is then split equally into 'k' folds. Each fold in turn is used for testing and the remainder used for training. This procedure is repeated 'k' times so that at the end every instance has been used exactly once for testing [27]. The cross-validation residual is then derived by calculating the difference between the prediction using the 'refit' regression model and the actuals for the test dataset. Witten et al. claim that in extensive tests on numerous different datasets, with different learning techniques, 10-fold cross validation is about the right number of folds to get the best estimate of error, and there is also some theoretical evidence that backs this up [27].

The following steps were applied in our algorithmic approach to perform regression analysis in order to derive a proxy function for biomass.

(i) The Boruta package in R was used to identify 80 important independent variables from a total of 300.

(ii) Collinearity was then removed by eliminating variables with a variance inflation factor (VIF) higher than 4, using a routine developed with the 'car' package in R [5]. Collinearity refers to strong correlations between independent variables. It can result in biased coefficients in the regression equation, which means it is difficult to assess the impact of the independent variables on the dependent variable. VIF is an excellent measure of the collinearity of the ith independent variable with the other independent variables in the model, according to O'Brien [19]. He also argues against the need to apply low VIF thresholds as was the case here. In fairness a higher threshold could have been used, however as we will see below, the algorithm used to test all the combinations of linear regression models is extremely resource intensive and only a limited number could be employed.

(iii) A matrix was created to store all of the potential subsets of predictors.

(iv) Training and validation samples were created.

(v) Linear models were generated, using the 14 most important variables resulting in the creation of 16,384 (2^{14}) different models. Due to the exponential complexity of the problem we confined our analysis to a maximum of 14 variables, which took about 2 h to run.

(vi) Key diagnostics are captured for all models including: r-squared, adjusted r-squared, p-values, AIC and BIC and k-fold cross-validation mean squared error.

Combining results from stepwise regression with the machine learning based algorithmic approach. The results obtained from this algorithmic approach were inferior to the results obtained through stepwise regression. However, there were 12 predictors that appeared in the top *algorithmic models* that were also absent from the stepwise regression model, which were reviewed in order to determine whether any improvement could be made to the results of the stepwise regression analysis.

After another process of stepwise regression, two additional predicators were included and a regression model to estimate the biomass was developed leading to an improvement in the adjusted r-squared from 0.976 to 0.979:

$$
\begin{aligned}
Biomass \approx \\
&- 14.2113 \\
&+ 2.1133 \cdot M\_fru\_b + 2.1744 \cdot M\_glc\_D\_b + 4.5078 \cdot M\_o2\_b \\
&+ 13.4913 \cdot R\_GLUN \\
&+ Aerobic\ (0.7191 \cdot Pair - 0.1056 \cdot M\_h\_b \\
&\qquad + 1.8578 \cdot M\_fru\_b + 1.8466 \cdot M\_glc\_D\_b - 3.4306 \cdot M\_o2\_c \\
&\qquad + 0.8033 \cdot R\_RPI - 3.5964 \cdot R\_SUCOAS\_FwRe).
\end{aligned} \tag{1}
$$

See Table 1 for an explanation of all variables used in the function. Note that unlike the original biomass function (compare Table 4 in the Supplementary Materials), reaction rates as well as metabolite concentrations are involved.

Validation of the proxy function was undertaken. The standard diagnostics were reviewed, which included but were not limited to the following.

(i) The adjusted r-squared value of 0.979 was very high. The adjusted r-squared being the preferred measure of explanatory power as it is more

Table 1. Variables occurring in the biomass proxy, see Eq. (1). *Aerobic* represents a dichotomous variable which was added to distinguish between aerobic and anaerobic conditions; likewise for *Pair*. Prefixes: M – metabolite, R – reaction; suffixes: b – boundary condition, c – cytosol, $FwRe$ – combined rate of forward and backward direction of a reversible reaction.

Short name	Explanation
Aerobic	Dichotomous variable
Pair	Dichotomous variable
M_fru_b	Fructose
M_glc_D_b	D-Glucose
M_o2_b, *M_o2_c*	Oxygen
M_h_b	Hydrogen
R_GLUN	glutaminase
R_RPI	ribose-5-phosphate isomerase, forward reaction
R_SUCOAS_FwRe	succinyl-CoA synthetase (ADP-forming)

conservative than the r-squared value and has been adjusted for the number of predictors in the regression model.

(ii) The p-value for the F-statistic was a lot less than 0.1% (0.001), meaning that it is highly statistical significant and that there is strong evidence of a relationship between the dependent and independent variables.

(iii) All the p-values for the coefficients were statistically significant at the 0.1% (0.001) level, meaning that there is evidence that the coefficients are significant.

Finally, the reassuring results were obtained from 10-fold cross-validation. The 10 dashed lines in Fig. 5, which relate to the best fit lines for the 10 respective folds in cross-validation do not vary significantly and are parallel and close together, as would be expected in a good model. The overall mean square value, i.e. the mean squared difference between the predicted value and the actual value, is a commonly used diagnostic in cross-validation and is 0.0265 for this data.

Further analysis was undertaken to ascertain whether the regression model meets assumptions for linear regression in order to determine whether it can be used for inference in addition to prediction. Some modest violations were identified with regard to homoscedasticity and some collinearity was also identified, but it was demonstrated that this could be effectively addressed by removing two of the variables from the regression equation with only a modest drop in the adjusted r-squared value from 0.979 to 0.965.

5   Evaluation of Key Results

Preliminary data analysis generated a number of critical insights that helped to guide the approach towards the regression analysis.

(i) First, it was found that biomass production for the different anaerobic conditions followed what can be described as a 'zero inflated distribution' (in which a large portion of values were either zero or close to zero), whereas biomass production for aerobic conditions resembles a normal distribution, as illustrated in Fig. 3.

(ii) The large number of predictors (metabolite concentrations and reaction fluxes) had the potential to generate overwhelming complexity. Further investigation into these variables revealed that a large number of them where highly correlated with each other and this together with clustering analysis helped to reduce the number of dimensions and to highlight the risk of collinearity in the regression model.

(iii) The initial dataset with only 26 single conditions imposed restrictions on the scope of the regression modelling, as there were not enough conditions to incorporate all the key predictors without a risk of overfitting. This led to the generation of additional data for pairs of conditions. The benefits of obtaining this data were twofold, firstly it improved the regression model by allowing for the inclusion of more predictors, without the same risk of overfitting. Secondly, it led to some interesting insights around biomass values for pairs of conditions which will be discussed below.

Key insights from the analysis of biomass for pairs of conditions. Not only did the additional data on 156 pairs of conditions help to improve the predictive power of the regression model, but it led to some serendipitous findings. First, pairs of aerobic conditions always have biomass values that are between 1% and 7% larger than the sum of the two single conditions as illustrated in Table 2.

Secondly, it was also shown that acetaldehyde, which does not produce biomass anaerobically as a single condition, produced biomass when paired with a number of other conditions that do not produce biomass anaerobically as illustrated in Table 3.

Approach towards development of a proxy function to predict biomass using multiple linear regression. Two separate approaches were used in relation to the predictive modelling. The first was the traditional statistical approach of stepwise regression. The second was to use feature selection algorithms to select variables together with an automated process to iterate through all the different combinations of the variables. Interestingly, the stepwise regression model outperformed the model generated through the algorithmic approach. This scenario was unexpected, but analysis showed that the single most important factor in improving the predictive power of the regression model was the inclusion of interaction terms to reflect the combined effect of the predictors (metabolite concentrations and reaction fluxes) together with the dichotomous variable created to distinguish between aerobic and anaerobic conditions, that were not included in the automated algorithmic approach. The lesson here is that one should not overlook the importance of the preliminary data analysis in helping shape the approach toward predictive model building.

Table 2. Comparing the sum of aerobic single conditions with pair of conditions.

Condition 1		Condition 2		Condition 1+2	Paired value	Biomass total	Increase %
Name	Biomass	Name	Biomass				
Ethanol	0.70	Glutamine	1.16	1.86	1.97	0.104	5.6%
Ethanol	0.70	Fumarate	0.79	1.49	1.58	0.097	6.5%
Ethanol	0.70	Malate	0.79	1.49	1.58	0.097	6.5%
Fructose	1.79	Glutamine	1.16	2.95	3.05	0.096	3.2%
Glucose	1.79	Glutamine	1.16	2.95	3.05	0.096	3.2%
Ethanol	0.70	Glutamate	1.24	1.94	2.04	0.096	4.9%
Glutamine	1.16	Lactate	0.74	1.90	2.00	0.092	4.9%
Ethanol	0.70	Auccinate	0.84	1.54	1.63	0.092	5.9%
Acetaldehyde	0.61	Glutamine	1.16	1.77	1.86	0.091	5.1%
Fructose	1.79	Glutamate	1.24	3.03	3.12	0.090	3.0%
Glucose	1.79	Glutamate	1.24	3.03	3.12	0.090	3.0%
Acetaldehyde	0.61	Fumarate	0.79	1.39	1.48	0.090	6.5%
Acetaldehyde	0.61	Malate	0.79	1.39	1.48	0.090	6.5%
Acetaldehyde	0.61	Glutamate	1.24	1.85	1.94	0.090	4.8%
Acetaldehyde	0.61	Auccinate	0.84	1.45	1.53	0.087	6.0%
Glutamate	1.24	Lactate	0.74	1.98	2.07	0.085	4.3%
Ethanol	0.70	Fructose	1.79	2.49	2.57	0.083	3.3%
Ethanol	0.70	Glucose	1.79	2.49	2.57	0.083	3.3%
Fructose	1.79	fumarate	0.79	2.58	2.66	0.083	3.2%
Fructose	1.79	Malate	0.79	2.58	2.66	0.083	3.2%

Table 3. FBA values for anaerobic paired conditions.

Acetaldehyde paired with	FBA value for paired condition
Fumarate	0.145
Malate	0.145
Lactate	0.117
2-oxoglutarate	0.068
Glutamate	0.045
Glutamine	0.040
All other conditions	<0.01

Elements of the automated approach involving automated feature selection and regression model building did however help to improve the final stepwise regression model, see Eq. (1), with an adjusted r-squared of 0.98.

Interpretation of the biomass proxy. The methodology that we employed to derive the biomass proxy – incorporating regression analysis and machine learning – was by its very nature designed to derive a robust and accurate proxy along the lines of Occam's razor, without overt regard to biological interpretation.

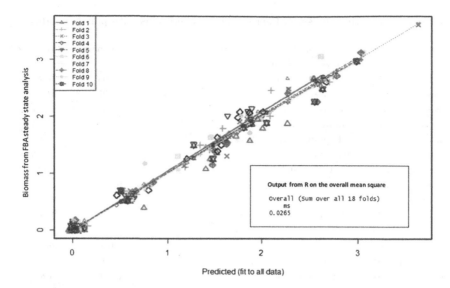

Fig. 5. Cross-validation output for the final regression model. Small symbols show predicted values; large symbols represent actuals. The 10 dashed lines relate to the best fit line for the respective folds.

Our starting point was an FBA model, lacking appropriate kinetic data. Thus, to be able to demonstrate our approach, we assumed mass-action rates with uniform kinetic parameters for all reactions. Our workflow embodies a general approach which works for any kinetic parameters; their choice, however, may influence the final outcome of the derived proxy function. Also note that the result obtained is not unique, because there are many highly correlated variables—some were even perfectly correlated, given that they had Pearson correlation coefficients of 1. The representative for an equivalence class of pairwise highly correlated variables (above an appropriate threshold) is selected according to predictive power and collinearity.

In other words, our function given in Eq. (1) is inherently not explanatory, but mimics the calculation flux value of the FBA biomass function (given in Table 4 in the Supplementary Materials). Moreover, the proxy function is not a pseudo-reaction in the way that the FBA one is, but is merely a function over a subset of the observables, both metabolite concentrations and reaction rates. It is for this reason that a mere syntactic comparison between the two is not meaningful, along the lines of comparing apples and pears, and it is the predictive power of the proxy which is of interest.

Reproducibility. Supplementary Materials can be found on our website http://www-dssz.informatik.tu-cottbus.de/DSSZ/Software/Examples, where we provide the original SBML model and its *Snoopy* version in ANDL format, which can be easily configured and simulated for the various growth conditions using the script provided. All the data analytics methods used are well recognised. Please also note, only public tools were used; thus all results presented are reproducible. Further data is also available in the form of additional tables and figures.

6 Conclusions

The research reported here describes a workflow to derive a dynamic biomass function which acts as a robust and accurate proxy for the FBA equivalent. The application of the method was illustrated for a reduced GEM for *E. coli*. Data generated by stochastic simulation for growth under a wide variety of conditions was used to develop a proxy function to predict biomass in the dynamic model, using data analytics techniques. This involved consolidating data for reactions and metabolites generated under dynamic simulation with gold standard target data for biomass generated under steady state analysis using a state of the art FBA solver.

The complexity generated by the large number of potential predictors (metabolites and reactions) was addressed through correlation and clustering analysis. In addition, the limited number of conditions in the initial dataset led to the need to generate more data using pairs of conditions. This not only improved the regression model by allowing for the inclusion of more predictors without the risk of overfitting, but led to a number of interesting insights regarding biomass for pairs of conditions. Namely, that pairs of aerobic conditions always have a biomass value that is between 1% and 7% larger than the sum of the two single conditions. In addition, it was shown that acetaldehyde, which does not produce biomass anaerobically as a single condition, produced biomass when paired with a number of other conditions that do not produce biomass anaerobically. These findings were faithfully reproduced in our dynamic proxy function.

Our workflow operates with any sets of kinetic data [7], and the biomass proxy results may be refined as more precise kinetic parameters become available.

Outlook. In further work we want to semi-automate the workflow developed and apply it to unreduced GEMs. Because of our unexpected finding that regression out-performs machine learning, we also plan to modify the algorithmic machine learning approach in such a way that we incorporate interactive terms that combine the effect of the predictors together with the dichotomous variables which distinguish between aerobic and anaerobic environments.

We also intend to investigate whether the biomass proxy function will correctly predict biomass in transient states before a steady state is reached. This would permit us to explore the effects of dynamic changes in growth conditions – for example during the process whereby a carbon source is gradually exhausted, or the availability of carbon sources in the environment fluctuates up and down, or the oxygen available is gradually used up. It would also be interesting to investigate how an active biomass function could be included in a dynamic model in order to retain its recycling properties as well as the draining of biomass components, possibly by decomposition into parts, or by employing non-mass action kinetics. This would enable us to investigate the dynamic evolution of the biomass function, i.e. to analyse more realistically at what time points the system becomes biologically non-viable under certain conditions. This would allow us to address the interpretation of the proxy function compared with the re-engineered biomass function in the context of the simulation of dynamic GEMs, i.e. whether the proxy function derived by regression analysis and machine learning can be not only predictive but also explanatory with regard to the behaviour of large-scale metabolism.

Acknowledgements. The authors would like to thank Bello Suleiman for his help in providing the FBA data, and Alessandro Pandini for his expert knowledge of R.

References

1. Baart, G.J., Martens, D.E.: Genome-scale metabolic models: reconstruction and analysis. In: Christodoulides, M. (ed.) Neisseria meningitidis, vol. 799, pp. 107–126. Springer, Heidelberg (2012). https://doi.org/10.1007/978-1-61779-346-2_7
2. Babyak, M.A.: What you see may not be what you get: a brief, nontechnical introduction to overfitting in regression-type models. Psychosom. Med. **66**(3), 411–421 (2004)
3. Chavent, M., Kuentz, V., Liquet, B., Saracco, L.: ClustOfVar: an R package for the clustering of variables. arXiv preprint arXiv:1112.0295 (2011)
4. Erdrich, P., Steuer, R., Klamt, S.: An algorithm for the reduction of genome-scale metabolic network models to meaningful core models. BMC Syst. Biol. **9**(1), 48 (2015)
5. Fox, J., et al.: The car package. R Foundation for Statistical Computing (2007)
6. Gilbert, D., et al.: Computational methodologies for modelling, analysis and simulation of signalling networks. Brief. Bioinform. **7**(4), 339–353 (2006)
7. Gilbert, D., Heiner, M., Jayaweera, Y., Rohr, C.: Towards dynamic genome-scale models. Briefings in Bioinformatics (2017)
8. Gilbert, D., Heiner, M., Lehrack, S.: A unifying framework for modelling and analysing biochemical pathways using Petri nets. In: Calder, M., Gilmore, S. (eds.) CMSB 2007. LNCS, vol. 4695, pp. 200–216. Springer, Heidelberg (2007). https://doi.org/10.1007/978-3-540-75140-3_14
9. Hädicke, O., Klamt, S.: Ecolicore2: a reference network model of the central metabolism of Escherichia coli and relationships to its genome-scale parent model. Sci. Rep. **7**, 39647 (2017)
10. Heiner, M., Herajy, M., Liu, F., Rohr, C., Schwarick, M.: Snoopy – a unifying Petri net tool. In: Haddad, S., Pomello, L. (eds.) PETRI NETS 2012. LNCS, vol. 7347, pp. 398–407. Springer, Heidelberg (2012). https://doi.org/10.1007/978-3-642-31131-4_22
11. Heiner, M., Rohr, C., Schwarick, M.: MARCIE – model checking and reachability analysis done efficiently. In: Colom, J.-M., Desel, J. (eds.) PETRI NETS 2013. LNCS, vol. 7927, pp. 389–399. Springer, Heidelberg (2013). https://doi.org/10.1007/978-3-642-38697-8_21
12. Heiner, M., Schwarick, M., Wegener, J.-T.: Charlie – an extensible Petri net analysis tool. In: Devillers, R., Valmari, A. (eds.) PETRI NETS 2015. LNCS, vol. 9115, pp. 200–211. Springer, Cham (2015). https://doi.org/10.1007/978-3-319-19488-2_10

13. King, Z., et al.: BiGG models: a platform for integrating, standardizing and sharing genome-scale models. Nucleic Acids Res. **44**(D1), D515–D522 (2016)

14. Knofczynski, G., Hadavas, P., Hoffman, L.: Effects of implementing projects in an elementary statistics class. J. Math. Sci. Math. Educ. **2**(2) (2007)

15. Kursa, M.B., Rudnicki, W.R.: Feature selection with the Boruta package. J. Stat. Softw. **36**(11), 1–13 (2010)

16. Mahadevan, R., Edwards, J.S., Doyle III, F.J.: Dynamic flux balance analysis of diauxic growth in Escherichia coli. Biophys. J. **83**(3), 1331–1340 (2002)

17. Mamiatis, T., Fritsch, E., Sambrook, J., Engel, J.: Molecular Cloning-A Laboratory Manual. New York: Cold Spring Harbor Laboratory, 1982, 545 S (1985)

18. Montgomery, D.C., Peck, E.A., Vining, G.G.: Introduction to Linear Regression Analysis, vol. 821. Wiley, Hoboken (2012)

19. O'Brien, R.M.: A caution regarding rules of thumb for variance inflation factors. Q. Quant. **41**(5), 673–690 (2007)

20. Orth, J.: Systems biology analysis of Escherichia coli for discovery and metabolic engineering. Ph.D. thesis, University Of California, San Diego (2012)

21. Orth, J., Fleming, R., Palsson, B.: Reconstruction and use of microbial metabolic networks: the core Escherichia coli metabolic model as an educational guide. EcoSal Plus **4**(1) (2010)

22. Palsson, B.: Systems Biology: Constraint-Based Reconstruction and Analysis. Cambridge University Press, Cambridge (2015)

23. Rohr, C.: Simulative analysis of coloured extended stochastic Petri nets. Ph.D. thesis, BTU Cottbus, Department of CS (2017)

24. Rohr, C.: Discrete-time leap method for stochastic simulation. Fundamenta Informaticae **160**(1–2), 181–198 (2018). https://doi.org/10.3233/FI-2018-1680

25. Schellenberger, J., et al.: Quantitative prediction of cellular metabolism with constraint-based models: the COBRA Toolbox v2. 0. Nat. Protoc. **6**(9), 1290–1307 (2011). https://doi.org/10.1038/nprot.2011.308

26. Smallbone, K., Mendes, P.: Large-scale metabolic models: from reconstruction to differential equations. Ind. Biotechnol. **9**(4), 179–184 (2013)

27. Witten, I.H., Frank, E., Hall, M.A., Pal, C.J.: Data Mining: Practical Machine Learning Tools and Techniques. Morgan Kaufmann, Burlington (2016)

Computing Diverse Boolean Networks from Phosphoproteomic Time Series Data

Misbah Razzaq[1], Roland Kaminski[2], Javier Romero[2], Torsten Schaub[2], Jeremie Bourdon[3], and Carito Guziolowski[1(✉)]

[1] Laboratoire des Sciences du Numerique de Nantes,
Ecole Centrale de Nantes, Nantes, France
`{misbah.razzaq,carito.guziolowski}@ls2n.fr`
[2] University of Potsdam, Potsdam, Germany
`{kaminski,javier,torsten}@cs.uni-potsdam.de`
[3] Laboratoire des Sciences du Numerique de Nantes, Universite de Nantes,
Nantes, France
`jeremie.bourdon@ls2n.fr`

Abstract. Logical modeling has been widely used to understand and expand the knowledge about protein interactions among different pathways. Realizing this, the *caspo-ts* system has been proposed recently to learn logical models from time series data. It uses Answer Set Programming to enumerate Boolean Networks (BNs) given prior knowledge networks and phosphoproteomic time series data. In the resulting sequence of solutions, similar BNs are typically clustered together. This can be problematic for large scale problems where we cannot explore the whole solution space in reasonable time. Our approach extends the *caspo-ts* system to cope with the important use case of finding diverse solutions of a problem with a large number of solutions. We first present the algorithm for finding diverse solutions and then we demonstrate the results of the proposed approach on two different benchmark scenarios in systems biology: (1) an artificial dataset to model *TCR signaling* and (2) the *HPN-DREAM* challenge dataset to model breast cancer cell lines.

Keywords: Diverse solution enumeration · Answer set programming
Boolean Networks · Model checking · Time series data

1 Introduction

Network analysis methods have been widely used for studying phosphoproteomic data, yielding important insights into protein interactions, functions, and evolution. Several formalisms including differential equations, Boolean logic and fuzzy logic exist for modeling signaling networks [4,26,28]. Models elucidated using differential equations require explicit specifications of kinetic parameters of the system and work well for smaller systems. Despite being highly predictive, mathematical modeling becomes computationally intensive as networks become

© Springer Nature Switzerland AG 2018
M. Češka and D. Šafránek (Eds.): CMSB 2018, LNBI 11095, pp. 59–74, 2018.
https://doi.org/10.1007/978-3-319-99429-1_4

larger. Stochastic modeling is suitable for problems of random nature but also fails to scale well with large scale systems of proteins [16, 28].

On the other hand, Boolean network (BN) modeling [14] has demonstrated to be a powerful framework for studying signaling networks [1] and for predicting novel behavior under perturbations. Phosphoproteomic data shows alteration in protein levels under different perturbation. Several methods have been proposed for learning BNs from such data. Most of the methods restrict their focus on one time point only [9, 17, 23, 27], which prevents them from capturing interesting dynamic characteristics such as loops [16]. Realizing this, methods have been proposed to model time series data [5, 20, 24]. Given noisy experimental data, most existing methods based on integer linear programming [17] and answer set programming (ASP) [19, 27] infer a family of BNs, which equally well represent the underlying signaling behavior in different pathways.

In this study, we focus on the ASP-based *caspo-ts* system which learns a family of BNs from time series data and a prior knowledge network (PKN). *caspo-ts* uses an over-approximation to learn candidate BNs, which leads to some false positive (FP) BNs. These BNs are not guaranteed to reproduce all traces of the time series data. To resolve this issue, it uses exact model checking to filter out FP BNs. The *caspo-ts* method uses the *clingo* ASP solver [7], which is able to exhaustively enumerate all solutions. The *clingo* solver by default uses an enumeration scheme, in which, once a solution is found, it backtracks to the first point from where the next solution can be found. This typically leads to the situation where successive solutions only change in a small part. As a result, *caspo-ts* may enter a solution space where FP BNs are clustered together. Given the size of the PKN and the small number of perturbations in the experimental data, the solution space of the *caspo-ts* can be very large containing billions of BNs making it difficult to enumerate true positive (TP) BNs in reasonable time if it gets stuck in a cluster of FP BNs.

To overcome this, we extend *caspo-ts* with a new enumeration scheme for breaking up clusters of similar solutions. In [6], various methods were presented for computing diverse solutions in ASP. However, these methods are not applicable to *caspo-ts*, since this system enumerates optimal (subset minimal) solutions, in order to produce simpler and more relevant solutions. Instead, we extend the approach of [21] for computing optimal diverse solutions[1] in ASP. The novelty of this extension is that we use heuristics for both the computation of optimal (subset minimal) solutions, and the diversification. By sampling the large solution space of BNs, we can retrieve a more complete set of mechanisms explaining the experimental data and better approximate biological reality.

Regarding model refinement of BNs dynamics using solvers, the works of [2, 22] propose ways to discover BNs or prune them according to experimental data related to fix points or attractors; which represent key biological functions. The objective is to find mechanisms explaining these biological functions. Their

---

[1] In the following, a diverse optimal solution is a solution which is minimal w.r.t. an objective function, there is no solution which is a subset of it, and it is different from previously enumerated solutions.

results are exhaustive, notably focusing on multiple mechanisms. Compared to [2,22], we propose a method that handles large scale networks and time-series phosphoproteomic data. The advantage of this is that such data is derived from standard experimental protocols. Moreover, the networks we handle are inferred from publicly available databases. Our method also allows us to provide optimal data, with respect to noisy or incomplete datasets. In this sense our method adapts more to current high-throughput experimental technologies as well as to massive signaling knowledge sources.

In the following, we refer to the modified *caspo-ts* as *caspo-ts^D*. We apply both systems to two datasets: (1) an artificial dataset for network signaling model, and (2) the HPN-DREAM challenge dataset. Our results show substantial improvements of *caspo-ts^D* in solution quality by discovering more signaling behaviors than *caspo-ts*. Moreover, *caspo-ts^D* is able to find solutions in cases where *caspo-ts* is unable to find any. Our method is applicable to gene or protein expression time series datasets measured upon different perturbations. Moreover, the proposed method is not specific to our biological application. It computes diverse subset minimal solutions in ASP, and therefore can be applied to any problem modeled in ASP.

The remainder of the paper is structured as follows. In Sect. 2, we describe the datasets and the algorithms. In Sect. 3, we study the performance of the modified enumeration scheme on the artificial and real datasets. In Sect. 4, we give concluding remarks and describe future work.

2   Materials and Methods

In this subsection, we describe the datasets, the *caspo-ts* system, and the new algorithm to enumerate diverse BNs implemented in *caspo-ts^D*.

2.1   Phosphoproteomic Time Series Dataset

Here, we give a brief description of the phosphoproteomic datasets used for testing the performance of the extended *caspo-ts^D* system. Phosphoproteomic data show changes in protein levels under sets of perturbations. Here, proteins are referred to by three names: (1) stimuli, (2) inhibitors, and (3) readouts. Stimuli serve as interaction points for the experimentalist. Inhibitors are blocked over all time points of the perturbation. Readout proteins are measured under sets of perturbations at different time points. Perturbations are a combination of stimuli and inhibitors. Fig. 1 depicts an example of phosphoproteomic time series data, where the values between zero and one of three proteins are shown in different colors. In this figure, we see the time series of one readout protein (blue) under a perturbation of one stimulus (green) and one inhibitor (red). Stimuli have value 1 and inhibitors have value 0 across all time points of an experimental perturbation. Readouts take continuous values in [0;1] after normalization. In some phosphoproteomic datasets an inhibitor can also act as a readout protein, which means that there are perturbations where it will be measured.

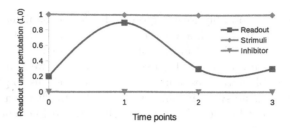

Fig. 1. Phosphoproteomic time series data. (Color figure online)

Artificial Dataset. The artificial dataset for TCR signaling was generated by [19] by simulating the PKN using logic based ODEs. This dataset consists of 4 readouts, 3 stimuli and 2 inhibitors. The readout proteins were measured at 16 time points under 10 perturbations. The PKN was derived from the TCR signaling model of [15] and consists of 16 nodes and 25 edges.

HPN-DREAM Dataset. The HPN-DREAM dataset consists of phosphoproteomic data of four breast cancer cell lines (BT20, UACC812, MCF7, BT549). This dataset was downloaded from the web portal of the HPN-DREAM challenge [11,12]. It includes temporal changes in phosphorylated proteins at seven different time points ($t_1 = 0\,\text{min}$, $t_2 = 5\,\text{min}$, $t_3 = 15\,\text{min}$, $t_4 = 30\,\text{min}$, $t_5 = 60\,\text{min}$, $t_6 = 120\,\text{min}$, and $t_7 = 240\,\text{min}$) under sets of perturbations. Maximum value based normalization was applied to the data to bring values into the range $[0; 1]$ and noisy and incomplete time series data was removed. After this, we have approximately 23 phosphorylated readout proteins per cell line. The number of perturbations varies from one cell line to another. The main goal of the HPN-DREAM challenge is to learn context specific signaling networks efficiently and effectively to predict dynamics in breast cancer.

The PKN was generated by mapping the experimentally measured phosphorylated proteins (HPN-DREAM dataset) to their equivalents from literature-curated databases and connecting them together within one network. The PKN was built using the ReactomeFIViz (Cytoscape app), which accesses the interactions existing in the Reactome and other databases [29]. The PKN consists of 64 nodes (7 stimuli, 3 inhibitors, and 23 readouts) and 178 edges.

2.2 Caspo-ts

The *caspo-ts* system is based on a combination of ASP and model checking. The ASP part of the *caspo-ts* system is used to solve the combinatorial optimization problem of finding BNs compatible with a PKN and time series data. All learned BNs are optimized using an objective function, minimizing the distance between the original and the time series data determined by the BN learned with *caspo-ts*. The ASP solver guarantees finding all optimal solutions w.r.t. an objective function. The model checking part of the system detects TP BNs by checking the reachability of time series traces given compatible BNs generated by the ASP part of the system. TP BNs are guaranteed to reproduce all the (binarized) traces under all perturbations by verifying reachability in the BN state graph. Since checking this reachability is a PSPACE-hard problem, the second step can be very time consuming for large BNs. To resolve this issue, the ASP part over-approximates solutions [19]. This over-approximation removes a large set of BNs that have no reachable traces, reducing the number of calls to the model checker.

The BNs learned by *caspo-ts* are represented by Boolean formulas in Disjunctive Normal From (DNF), i.e., as a disjunction of conjunctive clauses [2]. The BNs inferred by *caspo-ts* use the smallest DNF formulas possible, in the sense that no conjunctive clause can be removed from a DNF formula without changing the Boolean function it represents. We refer to these BNs as subset minimal BNs.

In the following, we give a brief description of ASP and the solving algorithms used by *caspo-ts*. But first let us have a look at an example Boolean formula.

Example 1. To use Boolean formulas we discretize the phosphoproteomic data: values greater or equal to .5 are set to 1, and to 0 otherwise. Let protein A have the Boolean formula $(B) \vee (\neg B \wedge C)$ containing the two conjunctive clauses (B) and $(\neg B \wedge C)$, where B and C represent proteins. This formula can be used to update the value of protein A. If the update is applied, A is set to 1 if either the value of B is 1, or the value of B is 0 and the value of C is 1. Otherwise, the value of A is set to 0.

Answer Set Programming. A *logic program* consists of *rules* of the from

$$h \leftarrow b_1 \wedge \cdots \wedge b_m \wedge \neg b_{m+1} \wedge \cdots \wedge \neg b_n$$

where h is an atom, $0 \leq m \leq n$, and each b_i is an atom. Such a logic program induces a set of stable models determined by the stable model semantics; see [8] for details. Each stable model is a subset of the atoms occurring in the logic program. Atoms appearing in this set are said to be true, and false otherwise. A rule is satisfied if its body (the part after the $\leftarrow$) is not satisfied, or its head atom h is true. A rule body is satisfied if all the atoms b_1 to b_m are true and all the atoms b_{m+1} to b_n are false. A *stable model* satisfies all rules of a logic program and also satisfies a minimality criterion. We do not go into the full details here, but this criterion requires that each atom in a stable model is proved by some rule. For this, a true atom has to appear in at least one rule head with a satisfied body. In the following, we simply refer to the stable models of a logic program as its solutions.

We use two extensions [25] to logic programs, which are frequently used in practice and ease modeling problems with ASP. A *choice rule* has form

$$\{h_1, \ldots, h_o\} \leftarrow b_1 \wedge \cdots \wedge b_m \wedge \neg b_{m+1} \wedge \cdots \wedge \neg b_n$$

---

[2] A clause can be seen as a reaction, where the proteins represented positively are available, and the proteins represented negatively are absent. A Boolean formula in DNF encompasses all possible reactions to update the value of a protein.

where $1 \leq o$ and each h_i is an atom. Unlike with the normal rule above, a choice rule can be used to prove any subset of the atoms h_1 to h_o whenever its body is satisfied. A *constraint* has form

$$\leftarrow b_1 \wedge \cdots \wedge b_m \wedge \neg b_{m+1} \wedge \cdots \wedge \neg b_n$$

and it removes all solution candidates that satisfy its body, without proving any atoms.

Example 2. Using a single choice rule, the solutions of the program

$$\{a, b, c\} \leftarrow \tag{1}$$

are all the subsets of the set $\{a, b, c\}$. To build up our running example, we further add the following constraints:

$$\leftarrow b \wedge \neg a \wedge \neg c \tag{2}$$
$$\leftarrow \neg b \wedge c \tag{3}$$
$$\leftarrow \neg b \wedge \neg c \tag{4}$$

The first constraint discards all solutions where b is true, a is false, and c is false. The second those where b is false and c is true, and the third those where both b and c are false. Hence, for the above program, we obtain the solutions $\{a, b\}$, $\{b, c\}$, and $\{a, b, c\}$.

ASP Solving. Next, we describe how the ASP solver *clingo* used by *caspo-ts* discovers solutions (BNs) using the conflict driven clause learning algorithm [7], shown in Algorithm 1.

Input: program P
1 Initialize assignment;
2 **while** *assignment is partial* **do**
3   Decide;
4   Propagate;
5   **if** *propagation let to a conflict* **then**
6   Analyze;
7   **if** *conflict can be resolved* **then**
8   Backjump;
9   **else**
10   **return** *unsatisfiable*;

11 **return** solution given by assignment;

Algorithm 1: Conflict-driven clause learning.

The algorithm works by extending a Boolean assignment over the atoms occurring in the given logic program P until a solution is found. The assignment is initialized in line 1. Then it is extended by the decision heuristic and propagation in the loop in lines 2–10. The call to Decide() in line 3 at the beginning of the loop uses a heuristic to select an atom, makes it either true or false, and adds it to the assignment. The consequences of this decision are then propagated in the following line extending the assignment accordingly. Then it is checked if propagation leads to a conflict. If this is the case, then the conflict is analyzed in line 6 and the assignment adjusted in line 8 accordingly. Note that a call to Backjump() takes back one or more decisions together with their consequences, and then adds an additional consequence to the assignment. This property ensures that the algorithm always terminates. It can also happen that a conflict cannot be recovered from. In this case, the problem is found unsatisfiable and the algorithm returns in line 10. Once the assignment is complete, the corresponding solution (set of true atoms) is returned in line 11.

Example 3. We can now apply this algorithm to our running example (c.f. Example 2). Starting with an empty assignment, we set a to false as the first decision. There are no immediate consequences and, hence, no conflict can arise. Then we decide to make b false. The consequences of this decision are that c is false via rule (3), and c is true via rule (4). Hence, we get a conflict, which is resolved and followed by a backjump. Since the conflict was caused by deciding a truth value for b, but is independent of the decision for a, the algorithm takes back all decisions and adds b as a consequence (we now know it must be true in all solutions). We can then decide to make a false again, which sets c to true via rule (2). This decision does not cause a conflict and the assignment is no longer partial. Hence, the algorithm terminates with solution $\{b, c\}$.

2.3 Caspo-ts^D

Here, we describe the algorithm used for enumerating diverse subset minimal solutions. In *caspo-ts*, the algorithm is implemented in the Python programming language using *clingo*'s multi-shot solving API [13]. The API allows us to customize the solving process, in particular, it allows us to customize the decision heuristic of the solving component, which is the key feature to find subset minimal answer sets. Note that we implemented the algorithm using the multi-shot solving API of *clingo* version 5. For that, we have upgraded the solver *clingo* of *caspo-ts* from 4.5.4 to 5.

Algorithm 2 is used to enumerate subset minimal answer sets. The idea is to configure the decision heuristic in lines 8 and 10 (see function Decide() in Algorithm 1 line 3) so that it first makes all atoms subject to subset minimization false before deciding truth values for other atoms. This modification ensures that the solution obtained from Algorithm 1 by calling Solve() is a subset minimal solution (see [3] for more details). Such a solution is output in line 12 of the algorithm. Furthermore, the algorithm calls Solve() in line 5 multiple times to find *all* subset minimal solutions. To not enumerate solutions twice, a constraint preventing to find the same solution again is added to the logic program P in the following line. This constraint is violated whenever a superset of the atoms

Input: program P and atoms T to subset minimize

1 Prepare(P);
2 **foreach** $x \in T$ **do**
3 $\quad$ SetSign(x, *false*, 1)

4 **while** *satisfiable* **do**
5 $\quad$ $S \leftarrow$ Solve();
6 $\quad$ AddConstraint($\leftarrow a_0, \ldots, a_n$ *for* $\{a_0, \ldots, a_n\} = T \cap S$);
7 $\quad$ **foreach** $x \in T \cap S$ **do**
8 $\quad\quad$ SetSign(x, *false*, 2);
9 $\quad$ **foreach** $x \in T \setminus S$ **do**
10 $\quad\quad$ SetSign(x, *false*, 1);
11 $\quad$ **if** S *is a true positive* **then**
12 $\quad\quad$ Output(S);

Algorithm 2: Diverse subset minimal solution enumeration.

in the previously found solution is true. This process is repeated in the loop in lines 4–12 until the program is no longer satisfiable and, hence, all solutions have been enumerated.

So far we only discussed how to enumerate subset minimal solutions. Now we explain how to extend the method in order to compute diverse solutions. The key idea to make the next solution different from the previous one is to assign atoms appearing in the last solution to false before assigning any other atoms. To modify the heuristic, we use function SetSign(a, t, l), which instructs the decision heuristic to assign atom a to truth value t on level l. The decision heuristic assigns free atoms with the highest level to the designated truth values before assigning atoms on lower levels. By default all atoms have level 0 and the decision heuristic is free to make them either *true* or *false*. The loop in lines 7–8 instructs the decision heuristic to assign atoms that appeared in the last solution to false on level 2. Any other atoms subject to subset minimization are assigned to false on level 1 in lines 9–10. And since all other atoms by default have level 0, they are assigned last. We see in the experiments in the next section that this strategy breaks up clusters of similar solutions in the solution sequence.

Example 4. We continue with Example 2. Let us assume that a, b, and c are the atoms subject to subset minimization. Note that in Example 3 all decisions assigned atoms to false, so the first solution $\{b, c\}$ obtained is in fact a subset minimal solution. Let us further assume that Algorithm 2 produced this solution in the first iteration (line 5). First, the constraint $\leftarrow b \wedge c$ preventing any superset of $\{b, c\}$ as solution is added in line 6. Then, the decision heuristic is configured to set atoms b and c to false on level 2 (line 8) and atom a to false on level 1 (line 10). In the next iteration, the only possible decision is to set c to false because b is already irrevocably assigned. The consequence of this decision is to set a to true via rule (2). Hence, we obtain solution $\{a, b\}$, which is a subset minimal solution. This is followed by adding the constraint $\leftarrow a \wedge b$ in line 6, which in turn makes the program unsatisfiable and causes the loop in lines 4–12 to terminate. We correctly obtain the subset minimal solutions $\{b, c\}$ and $\{a, b\}$. Solution $\{a, b, c\}$ is not subset minimal and not enumerated.

3   Results

In this section, we discuss the results of applying Algorithm 2 on two different datasets. We start with the artificial benchmark, where the solution space is small enough to compute all solutions. This allows us to study this benchmark in more detail. We can analyze how well a limited number of solutions enumerated with *caspo-ts* and *caspo-ts^D* represents the solution space. Then we move to the real dataset, where we cannot enumerate all solutions and can only consider a limited number of solutions because the solution space is too large. Nevertheless, we can show improved results with *caspo-ts^D* over *caspo-ts* by being able to enumerate more TP BNs and also more diverse BNs.

3.1   Artificial Dataset

Here, we use the TCR signaling dataset [15] to demonstrate the working of Algorithm 2 by describing two factors: (1) frequency of the clauses, and (2) true positive rate of BNs. The purpose of studying the first factor is to observe how many clauses we discover while learning a limited number of BNs. In this case, we expect to discover more clauses with *caspo-ts^D* than with *caspo-ts*. The second factor (the true positive rate) is used to study how TP solutions are distributed in the solution space. With *caspo-ts^D*, we expect TP solutions to be distributed much more evenly.

Figure 2 depicts the frequency of clauses in the solutions of the artificial benchmark. Clauses that occurred in at least one solution are depicted on the x axis. Each tick stands for one clause and the label has format $n \leftarrow c$ where c is a clause and n is a node name. Furthermore, clauses associated with the same node are grouped together by shading the background alternatingly in light gray and white. The frequencies of the clauses are depicted on the y axis. The red line depicts the frequency considering all 68338 solutions, while the green and blue lines depict the frequencies of the first 100 solutions computed by *caspo-ts* and *caspo-ts^D*, respectively. In total, there are 49 clauses appearing in the family of BNs. While *caspo-ts^D* (blue line) discovered 48 clauses, *caspo-ts* (green line) learned only 29 clauses by enumerating the same number of BNs (100). We also observe that the blue line is often much closer to the red line (with an average distance of 0.06) than the green line is (with an average distance of 0.20). This shows that the diverse enumeration scheme is able to produce solutions that are less similar to each other and better represent the solution space. The underlying ASP solver of *caspo-ts* by default uses an enumeration scheme that backtracks to the first point from which the next solution can be found. This approach typically leads to the situation where successively enumerated solutions only change in a small part. We can observe this behavior in Fig. 2, where some clauses are overrepresented.

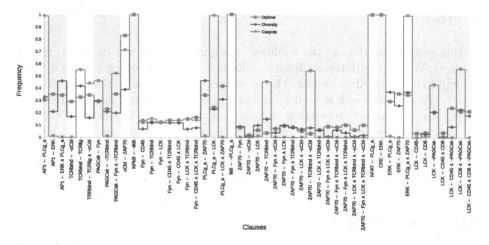

Fig. 2. Frequency of clauses per node in all 68338 BNs (red line), and in the first 100 BNs enumerated by *caspo-ts* (green line) and *caspo-ts*^{D} (blue line). (Color figure online)

Figure 3 depicts the true positive rate of blocks of successive solutions. Each tick on the x axis stands for a block of 1000 solutions. The y axis depicts the percentage of true positives in a block of solutions. The red line depicts the overall true positive rate (78%), while the green and blue lines depict the true positive rates of *caspo-ts* and *caspo-ts*^{D}, respectively.[3] We observe that for the *caspo-ts* system there are a lot of blocks with either a lot of true positives or very few. This suggests that true positives are clustered in the sequence of enumerated solutions. We observe that the diverse enumeration scheme does not show this behavior. This is especially important for enumerating true positive solutions of real world instances where only a limited number of solutions can be checked because of time constraints. With the original *caspo-ts* system, it can happen that the first cluster does not contain any true positives, making it impossible to find any true positive solution within a given time budget. The graph also shows that the diverse enumeration scheme does not sample over the full solution space. We see that before around 23000 solutions, the true positive rate is below the ideal 78% and then jumps up afterward. Thus, we conjecture that our enumeration scheme mainly breaks up local clusters of solutions with the artificial benchmark. Still, it is able to discover almost all clauses compared to the whole solution set (the frequency is 0 only once), while *caspo-ts* does not discover 20 clauses at all.

---

[3] For example, when x has value 3000, the y value in blue gives the true positive rate among the solutions 2001 to 3000 computed by *caspo-ts*^{D}.

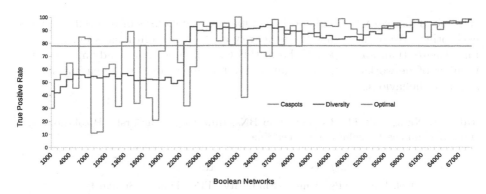

Fig. 3. True positive rate of BNs grouped in blocks of 1000 networks. (Color figure online)

3.2   HPN-DREAM Challenge Dataset

Next, we show the results of applying diverse solution enumeration to the HPN-DREAM challenge dataset [11]. We discuss the results according to three aspects: (1) time to compute the first true positive BN, (2) similarity among the family of solutions, and (3) Boolean functions computed by the original *caspo-ts* and the extended *caspo-ts^D* system. We start the analysis with four cell lines, and then we provide a detailed analysis of the Boolean functions of one cell line discovered by *caspo-ts* and *caspo-ts^D*.

Given that model checking is a computationally hard problem, we stop an experiment after a system verifies (using the model checker) 46 BNs per cell line[4]. The model checking task was performed on a server with 1.5 Tb of RAM. Table 1 shows the number of TP BNs obtained for each cell line. We see that the number of TP BNs differs comparing *caspo-ts* and *caspo-ts^D*. For MCF7 we obtain 0 TP BNs with *caspo-ts* and 4 TP BNs with *caspo-ts^D*, while for BT549 we obtain 2 and 14 TP BNs, respectively. For the other two cell lines BT20 and UACC812, we obtain a comparable number of BNs. We observe that we can get more TP solutions by checking the same number of BNs with *caspo-ts^D*. This is an important improvement given the fact that model checking the BNs is a computationally hard problem. Next, we consider the time column showing the time to compute the first TP BN for each cell line. We see that we are unable to get TP BN with *caspo-ts* in case of the MCF7 cell line, which shows that the *caspo-ts* system is stuck in a part of the search space where there are only FPs. Otherwise, for the other cell lines the time to get the first TP BN is comparable. We conclude that the difficulty to model check a BN depends on the cell line and not on the order in which solutions are found. Finally, the similarity column shows the similarity score among the set of TP BNs for each

---

[4] Note that the model checker could only verify 32 out of 46 solutions within one month for cell line BT20 in case of *caspo-ts^D*. There may exist more TPs for this cell line.

cell line. This score is calculated by comparing the clauses of one cell line with each other. We observe that the similarity among the solutions is much higher for *caspo-ts* than for *caspo-ts^D*. From this we conclude that, studying the same number of networks, *caspo-ts^D* can discover more clauses representing diverse signaling behaviors.

Table 1. Number of TP BNs out of 46 BNs, time to get the first TP solution, and similarity among TP solutions per cell line.

Cell Line	*caspo-ts*			*caspo-ts^D*		
	TPs	Time	Similarity	TPs	Time	Similarity
MCF7	0	—	—	4	6.7 h	0.51
BT549	2	8.4 min	0.92	14	7.9 min	0.44
UACC812	20	26 s	0.81	15	27 s	0.45
BT20	13	20 h	0.86	7+	20 h	0.32

Now, we analyze in more detail the UACC812 cell line using *caspo-ts*. Fig. 4 shows the union of 10 TP BNs obtained by *caspo-ts*. There are four different kinds of nodes in the graph: (1) stimuli shown in green, (2) inhibitors shown in red, (3) readouts shown in blue, and (4) unobserved nodes shown in white. Note that blue nodes with red borders are readouts, which are also inhibitors. There are two different kinds of edges shown in red and green color. Green edges are used to show a positive influence ($\leftarrow$), and red edges are used to show a negative influence ($\vdash$). We have discovered 25 clauses with *caspo-ts*, and we observe that the learned BNs only contain Boolean functions with clauses of size one. We also notice that the learned BNs are very similar to each other, as we see in Table 1 with the similarity score of 0.81. This relates to the fact that the ASP solver used by *caspo-ts* uses a backtracking algorithm to enumerate solutions and, hence, the solutions only change in small parts.

Next, we analyze the UACC812 cell line using *caspo-ts^D*. Fig. 5 shows the union of 10 TP BNs obtained by *caspo-ts^D*. Nodes, edges and colors have the same meaning as in Fig. 4. Unlike with *caspo-ts*, here we identified clauses with more than one element. They are represented by black rectangles where the nodes of incoming edges are their elements. Additionally, we use dashed edges to represent clauses that were also discovered by *caspo-ts*. In total, *caspo-ts^D* discovered 66 clauses, 41 more than *caspo-ts*. It identified 23 out of the 25 clauses discovered by the original system, and 43 additional clauses, studying the same number of BNs. It is important to note that even though for the UACC812 cell line *caspo-ts^D* learned 5 TP BNs less than *caspo-ts*, the number of clauses learned by *caspo-ts^D* is 3 times higher. Since we find much more clauses with *caspo-ts^D*, we can get an impression of the whole solution space by just inspecting a limited number of solutions. This analysis shows the efficacy of the extended *caspo-ts^D* system in a real case scenario, where it is difficult to study the complete solution space because of time constraints.

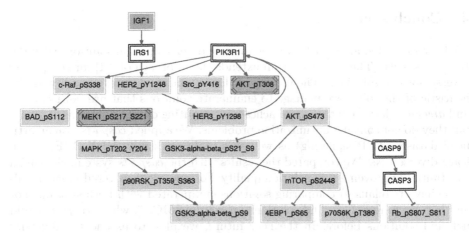

Fig. 4. *caspo-ts*: 10 optimal TPs BNs concatenated for cell line UACC812. All BNs are identically optimal. (Color figure online)

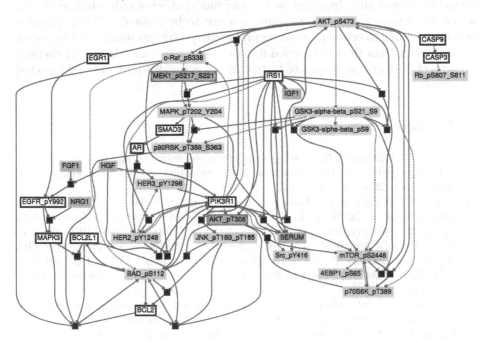

Fig. 5. *caspo-ts^D*: 10 optimal TP BNs concatenated for cell line UACC812. The dashed edges are used to represent clauses that were also discovered by *caspo-ts*. AND gates are represented by black boxes.

4   Conclusion

We have presented an algorithm to enumerate diverse optimal solutions with the *caspo-ts* system. The new algorithm extends the approach of [21] for computing diverse optimal solutions. The novelty of this extension is that by modifying the heuristic of the solver we manage to enumerate solutions that are both *optimal* and *diverse*. There are other approaches for computing diverse solutions [6,10,18] but they do not consider optimization problems. We applied *caspo-ts^D* on an artificial dataset (TCR signaling) as well as a real case study (HPN-DREAM) to learn diverse BNs. We compared the results with the *caspo-ts* system, showing a substantial improvement in solution quality. For one, we discovered more signaling behaviors (clauses) comparing solutions enumerated with both systems. For another, we were able to find solutions for cell line MCF7, where *caspo-ts* could not find solutions before. In the near future, we plan to extend the diversity algorithm in two directions. First, we are planning to experiment with solver parameters in order to introduce some randomness into the search. Second, we intend to extend the algorithm to call the model-checker only on answer sets which are diverse (according to some measure to be defined). This is possible because the time to enumerate over-approximated solutions using the ASP solver is much lower than the time needed to check solutions using the model-checker. We expect both enhancements to further improve the diversity of the discovered solutions.

References

1. Albert, R., Othmer, H.G.: The topology of the regulatory interactions predicts the expression pattern of the segment polarity genes in Drosophila melanogaster. J. Theor. Biol. **223**(1), 1–18 (2003)
2. Biane, C., Delaplace, F.: Abduction based drug target discovery using Boolean control network. In: Feret, J., Koeppl, H. (eds.) CMSB 2017. LNCS, vol. 10545, pp. 57–73. Springer, Cham (2017). https://doi.org/10.1007/978-3-319-67471-1_4
3. Brewka, G., Delgrande, J., Romero, J., Schaub, T.: Implementing preferences with *asprin*. In: Calimeri, F., Ianni, G., Truszczynski, M. (eds.) LPNMR 2015. LNCS (LNAI), vol. 9345, pp. 158–172. Springer, Cham (2015). https://doi.org/10.1007/978-3-319-23264-5_15
4. Calzone, L., et al.: Mathematical modelling of cell-fate decision in response to death receptor engagement. PLoS Comput. Biol. **6**(3), e1000702 (2010)
5. Carlin, D.E., et al.: Prophetic granger causality to infer gene regulatory networks. PloS one **12**(12), e0170340 (2017)
6. Eiter, T., Erdem, E., Erdoğan, H., Fink, M.: Finding similar or diverse solutions in answer set programming. In: Hill, P.M., Warren, D.S. (eds.) ICLP 2009. LNCS, vol. 5649, pp. 342–356. Springer, Heidelberg (2009). https://doi.org/10.1007/978-3-642-02846-5_29
7. Gebser, M., Kaminski, R., Kaufmann, B., Schaub, T.: Clingo = asp + control: preliminary report. arXiv preprint arXiv:1405.3694 (2014)

8. Gelfond, M., Lifschitz, V.: The stable model semantics for logic programming. In: Kowalski, R., Bowen, K., (eds.) Proceedings of the Fifth International Conference and Symposium of Logic Programming (ICLP 1988), pp. 1070–1080. MIT Press (1988)

9. Guziolowski, C., et al.: Exhaustively characterizing feasible logic models of a signaling network using answer set programming. Bioinformatics 29(18), 2320–2326 (2013)

10. Hebrard, E., Hnich, B., O'Sullivan, B., Walsh, T: Finding diverse and similar solutions in constraint programming. In: Proceedings of the Twentieth National Conference on Artificial Intelligence (AAAI 2005), pp. 372–377. AAAI Press (2005)

11. Hill, S.M., et al.: Inferring causal molecular networks: empirical assessment through a community-based effort. Nat. Methods 13(4), 310–318 (2016)

12. Hill, S.M., et al.: Context specificity in causal signaling networks revealed by phosphoprotein profiling. Cell Syst. 4(1), 73–83 (2017)

13. Kaminski, R., Schaub, T., Wanko, P.: A tutorial on hybrid answer set solving with *clingo*. In: Ianni, G., et al. (eds.) Reasoning Web 2017. LNCS, vol. 10370, pp. 167–203. Springer, Cham (2017). https://doi.org/10.1007/978-3-319-61033-7_6

14. Kauffman, S.A.: The Origins of Order: Self-Organization and Selection in Evolution. Oxford University Press, Oxford (1993)

15. Klamt, S.A., Saez-Rodriguez, J., Lindquist, J.A., Simeoni, L., Gilles, E.D.: A methodology for the structural and functional analysis of signaling and regulatory networks. BMC Bioinf. 7(1), 56 (2006)

16. MacNamara, A., Terfve, C., Henriques, D., Bernabé, B.P., Saez-Rodriguez, J.: State-time spectrum of signal transduction logic models. Phys. Biol. 9(4), 045003 (2012)

17. Mitsos, A., Melas, I.N., Siminelakis, P., Chairakaki, A.D., Saez-Rodriguez, J., Alexopoulos, L.G.: Identifying drug effects via pathway alterations using an integer linear programming optimization formulation on phosphoproteomic data. PLoS Comput. Biol. 5(12), e1000591 (2009)

18. Nadel, A.: Generating diverse solutions in SAT. In: Sakallah, K.A., Simon, L. (eds.) SAT 2011. LNCS, vol. 6695, pp. 287–301. Springer, Heidelberg (2011). https://doi.org/10.1007/978-3-642-21581-0_23

19. Ostrowski, M., Paulevé, L., Schaub, T., Siegel, A., Guziolowski, C.: Boolean network identification from perturbation time series data combining dynamics abstraction and logic programming. Biosystems 149, 139–153 (2016)

20. Rau, A., Jaffrézic, F., Foulley, J.-L., Doerge, R.W.: An empirical Bayesian method for estimating biological networks from temporal microarray data. Stat. Appl. Genet. Mol. Biol. 9(1) (2010)

21. Romero, J., Schaub, T., Wanko, P.: Computing diverse optimal stable models. In: OASIcs-OpenAccess Series in Informatics, vol. 52. Schloss Dagstuhl-Leibniz-Zentrum fuer Informatik (2016)

22. Rosenblueth, D.A., Muñoz, S., Carrillo, M., Azpeitia, E.: Inference of Boolean networks from gene interaction graphs using a SAT solver. In: Dediu, A.-H., Martín-Vide, C., Truthe, B. (eds.) AlCoB 2014. LNCS, vol. 8542, pp. 235–246. Springer, Cham (2014). https://doi.org/10.1007/978-3-319-07953-0_19

23. Sharan, R., Karp, M.: Reconstructing boolean models of signaling. J. Comput. Biol. 20(3), 249–257 (2013)

24. Shmulevich, I., Dougherty, E.R., Zhang, W.: Gene perturbation and intervention in probabilistic Boolean networks. Bioinformatics 18(10), 1319–1331 (2002)

25. Simons, P.: Extending the stable model semantics with more expressive rules. In: Gelfond, M., Leone, N., Pfeifer, G. (eds.) LPNMR 1999. LNCS (LNAI), vol. 1730, pp. 305–316. Springer, Heidelberg (1999). https://doi.org/10.1007/3-540-46767-X_22

26. Thakar, J., Albert, R.: Boolean models of within-host immune interactions. Curr. Opin. Microbiol. **13**(3), 377–381 (2010)

27. Videla, S., et al.: Revisiting the training of logic models of protein signaling networks with a ASP. In: Gilbert, D., Heiner, M. (eds.) CMSB 2012. LNCS, pp. 342–361. Springer, Heidelberg (2012). https://doi.org/10.1007/978-3-642-33636-2_20

28. Watterson, S., Marshall, S., Ghazal, P.: Logic models of pathway biology. Drug Discov. Today **13**(9), 447–456 (2008)

29. Wu, G., Dawson, E., Duong, A., Haw, R., Stein, L.: ReactomeFiviz: a Cytoscape app for pathway and network-based data analysis. F1000Research **3**, 146 (2014)

Characterization of the Experimentally Observed Clustering of VEGF Receptors

Emine Güven[1]([✉]), Michael J. Wester[2], Bridget S. Wilson[2,3],
Jeremy S. Edwards[2,4], and Ádám M. Halász[5]

[1] Department of Biomedical Engineering, Düzce University, Düzce, Turkey
emine.guven33@gmail.com
[2] N. M. Center for the SpatioTemporal Modeling of Cell Signaling,
University of New Mexico, Albuquerque, NM, USA
wester@math.unm.edu, {bwilson,jsedwards}@salud.unm.edu
[3] Comprehensive Cancer Center, University of New Mexico Health Sciences Center,
Albuquerque, NM, USA
[4] Departments of Chemical and Biological Engineering,
Chemistry and Chemical Biology, Molecular Genetics and Microbiology,
University of New Mexico, Albuquerque, NM, USA
[5] Department of Mathematics, West Virginia University, Morgantown, WV, USA
halasz@math.wvu.edu

Abstract. Cell membrane-bound receptors control signal initiation in
many important cellular signaling pathways. In many such systems,
receptor dimerization or cross-linking is a necessary step for activation, making signaling pathways sensitive to the distribution of receptors in the membrane. Microscopic imaging and modern labeling techniques reveal that certain receptor types tend to co-localize in clusters,
ranging from a few to tens, and sometimes hundreds of members. The
origin of these clusters is not well understood but they are likely not
the result of chemical binding. Our goal is to build a simple, descriptive
framework which provides quantitative measures that can be compared
across samples and systems, as groundwork for more ambitious modeling
aimed at uncovering specific biochemical mechanisms. Here we discuss a
method of defining clusters based on mutual distance, applying it to a
set of transmission microscopy images of VEGF receptors. Preliminary
analysis using standard measures such as the Hopkins' statistic reveals
a compelling difference between the observed distributions and random
placement. A key element to cluster identification is identifying an optimal length parameter L^*. Distance based clustering hinges on the separation between two length scales: the typical distance between neighboring
points within a cluster vs. the typical distance between clusters. This
provides a guiding principle to identify L^* from experimentally derived
cluster scaling functions. In addition, we assign a geometric shape to
each cluster, using a previously developed procedure that relates closely
to distance based clustering. We applied the cluster [support] identification procedure to the entire data set. The observed particle distribution
results are consistent with the random placement of receptors within the
clusters and, to a lesser extent, the random placement of the clusters

© Springer Nature Switzerland AG 2018
M. Češka and D. Šafránek (Eds.): CMSB 2018, LNBI 11095, pp. 75–92, 2018.
https://doi.org/10.1007/978-3-319-99429-1_5

on the cell membrane. Deviations from uniformity are typically due to large scale gradients in receptor density and/or the emergence of "mega-clusters" that are very likely the expression of a different biological function than the one behind the emergence of the quasi-ubiquitous small scale clusters.

Keywords: Membrane receptors · Clustering Hierarchical clustering · VEGF

1   Introduction

We introduce a novel method of analyzing clustering of molecular constituents on the cell membrane. This allows us to quantify the effects of clustering on cellular function. Figure 1 illustrates the experimental details and major quantitative analysis steps we perform in this study. Our first objective is to derive a natural length scale based on hierarchical distance-based clustering. The particles are grouped into naturally defined clusters via a clear scale separation between intra- and inter-cluster distances; this natural length scale, L^*, is interpreted as the largest intra-cluster distance between a particle and its nearest neighbor. We also test cluster centroid distributions to validate this choice for each image. To choose L^*, we consider the distribution of clusters at various length scales to observe the dependency on length when we apply distance based hierarchical clustering. We find that the short intra-cluster and long inter-cluster distances can be separated into two distinct binomial distributions, the inflection point of the sum of which marks the transition between the two types of distances and so makes for a natural choice of the optimum length. Empirically, this corresponds approximately to where the first derivative of the number of clusters per length separation is maximized. Once we identify the optimum length to perform clustering for each image, we use the domain reconstruction algorithm (DRA) to define shapes of clusters. The shape is defined by a contour around the cluster, following DRA developed previously for the study of live trajectories of labeled receptors [18].

The motivation for this project comes from considerable evidence that the cell membrane is separated into isolated compartments or regions, which we call micro-domains, such as lipid rafts [15] and protein islands [22]. The formation of clusters and their role in cell signaling has prompted computational studies of the roles of micro-domains in signal initiation [19]. There is general agreement that the composition of these micro-domains is heterogeneous. Previously, the distribution of receptors has been characterized using nearest neighbor distance distributions, and Hopkins' and Ripley's statistics [5,23]. Hopkins' and Ripley's statistics establish *clustering tendency* [10], which can then be followed by cluster identification.

The paper is organized as follows. In the remainder of this section, we provide some background on the biological context and importance of receptor clustering (Sect. 1.1), while in Sect. 1.2 we outline the mathematical ideas used to characterize clustering. Experimental details, mathematical definitions and method-

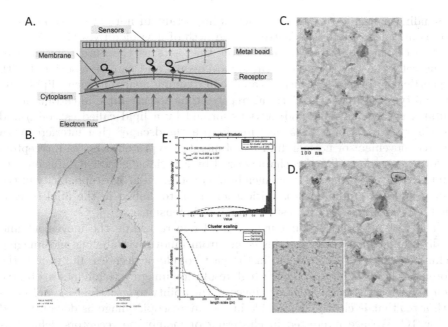

Fig. 1. (A) Principle of TEM: Immuno-gold nanoparticles conjugated to a specific antibody bind to membrane receptors (VEGFR). Cells (some incubated with VEGF) are exposed to the labeling preparation. Samples consist of cell membrane sheets fixed to a glass substrate; individual receptors on the membrane are labeled by the immuno-gold nanoparticles. (B) High resolution (25,000x) images (C) capture features of the membrane; the nano-particles used for labeling appear as dark spots whose location can be estimated with a precision of a few nanometers. Our goal (D) is to identify and analyze the clustering pattern. To achieve this, we follow an analysis procedure (E), that establishes the clustering tendency and independently aims to identify a natural length scale that gives rise to the clustering pattern.

ologies, and their justification are provided in Methods (Appendix A). We further described our proposed method for cluster identification, which relies crucially on the identification of an optimum clustering length scale in Methods (Appendix A). In Results (Sect. 2), we present the outcome of applying the clustering tendency measures as well as the cluster identification algorithm to the experimental data set. We first discuss the features of the overall distribution of particles in Sect. 2.1, cluster identification (Sect. 2.2) and then summarize the distribution of locations of the clusters in Sect. 2.3. We end with a Discussion (Sect. 3).

1.1 Biological Context

We investigate the spatial distribution of VEGF receptors in transmission electron microscopy (TEM) images. Vascular endothelial growth factor (VEGF) is a signaling molecule [1,8,11,16] which is important in normal development as well as in cancer because it controls the growth of new blood vessels [7,17,21].

The cell membrane is a lipid bilayer that acts as the outer containment of the cell (Fig. 1A). It is far from homogeneous, and can be associated with the cortical cytoskeleton which is capable of dynamic rearrangement, lipid rafts (special types of lipids that form aggregates that are insoluble in certain detergents), caveloae (cave-like indentations formed by a lipid called caveolin) and protein agglomerations. These features form a "landscape" that interferes with the free movement or normal diffusion of membrane proteins. Instead, receptors perform *anomalous diffusion* characterized by a variable effective diffusion constant. VEGF receptors are trans-membrane proteins. The hydrophobicity of the transmembrane amino acids is critical for them to embed in the lipid bilayer, but they can diffuse along its surface in two dimensions.

The vast majority of our current knowledge regarding the movement and localization of membrane proteins comes from innovative labeling and imaging techniques [20,22] that emerged in the past couple of decades. In general, the proteins or lipids of interest require high resolution imaging techniques to observe their behavior; therefore, imaging hinges on the ability to *label* these biomolecules with a *tag* that is clearly identifiable in the microscopic image as demonstrated in Fig. 1C. We are interested in clustering of membrane receptors defined as the accumulation of receptors in a fraction of the available area as shown in Figs. (1C, 4A).

1.2   Clustering

Clustering can have a significant impact on signaling in pathways that require receptor oligomerization, by increasing the likelihood of receptor-receptor binding. Therefore, quantitative measures of clustering are a crucial ingredient to predictive mathematical models of cell signaling.

Our analysis aims to (1) assess *clustering tendency* following an established methodology [5,23], (2) provide an improved *definition of clusters* leading to an identification as exemplified in Fig. 3, and (3) characterize the distribution of receptors in the micrographs in a way that *quantifies* the phenomenon of clustering. As shown in Figs. (1D, 4D), to establish clustering we would like to test for clustering tendency and characterize the clustering (Fig. 1E) using statistical tests.

It is useful to classify such images and to compare how the patterns change between different experimental conditions or cell types. A quantitative characterization is also a prerequisite for any mathematical modeling aimed at explaining the observed distributions. We first used nearest neighbor distance distributions (see Fig. 2) as well as Hopkins' and Ripley's statistics (see Fig. 5), which all rely on the mutual (Euclidean) distance between pairs of points. The methods are widely used in fields such as ecology [3] and network engineering [14]. We will generally refer to a set of N points in an area of size A. When focused on *establishing the fact* of clustering, we test against the hypothesis of random uniform placement of the points, also known as a spatial Poisson process with *density* $\lambda = N/A$.

Hopkins' and Ripley's Statistics. Ripley's functions are discussed in detail in [3] and the Hopkins' statistic in [10,23]. Methods (Appendix A) summarizes these ideas, both of which consider points that are within some radius of each sample point.

Nearest Neighbor Distance Distributions. While the Hopkins' test clustering tendency, the Ripley's test and the nearest neighbor distance (NND) are helpful in deriving a distance scale to characterize receptor clustering. The nearest-neighbor distance is obtained by selecting, for each point in a set, the point from the rest of the set that is the closest. The NND distribution gives a measure of the local density of points. If N points are distributed uniformly in an area A, then the typical nearest-neighbor distance should be close to the linear size b of the area available to each point: $b^2 = A/N$. For a uniform random configuration with density $\lambda = N/A = 1/b^2$, the probability density function (PDF) and the cumulative distribution (CDF) for the nearest neighbor distances are [14]:

$$f_{\text{NN}}(r) = 2\pi\lambda r \exp\left(-\pi\lambda r^2\right); \quad F_{\text{NN}}(r) = 1 - \exp\left(-\pi\lambda r^2\right). \tag{1}$$

We note that the cumulative distribution (CDF) only depends on the ratio r/b, therefore the expected value, median, and mode of the NND distribution scale linearly with b. [1]

In this study, we estimate the statistics for the first and second (next) nearest neighbor distances for all the images. The next nearest neighbor distance (Fig. 2) may be more relevant to clustering, since VEGF receptors may form dimers, and thus the nearest neighbor of a receptor may be its dimer partner. The corresponding distribution functions also scale linearly with b.

Hierarchical Clustering. Having assessed the clustering tendency for a given micrograph, we may turn to cluster identification. The method of distance based hierarchical clustering has been applied in various contexts [9]. Our starting point is the work of Espinoza and coworkers [5], who initially adapted the method to the analysis of nano-gold labeled membrane proteins. They investigated an optimal length parameter L^* for membrane bound receptors and proposed an *intrinsic clustering distance* d_I, the value of L that maximizes the number of clusters containing at least two points [2]. Hierarchical clustering relies on a single *length parameter* L, which induces a pattern of connections between the points of a given set. Two points are directly connected if their distance is less than L. Connected groups result by transitivity, similarly to connected graphs: a point that is directly connected to a member of a connected group is indirectly connected to all members of the group. The connected groups amount to a partition of the point set and define the clusters.

---

[1] For the nearest neighbor distance, the mean is $\langle r\rangle_{\text{NN}} = b/2$ and the mode (maximum probability) $r_{\text{NN}}^* = b/\sqrt{2\pi}$ corresponds to the radius of a circle of area A/N, $\pi r_{\text{NN}}^{*\,2} = b^2 = 1/\lambda$.

[2] We use "optimal length parameter" (L^*) to distinguish from the specific choice of [5].

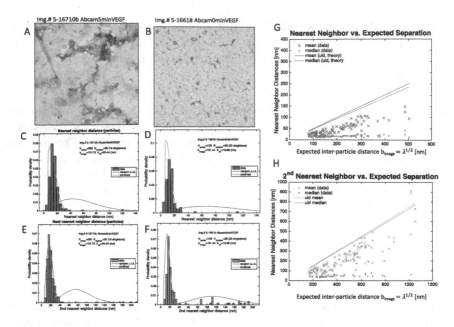

Fig. 2. Analysis of clustering in the resting cells using nearest (C, D) and next nearest neighbor (E, F) distance distributions for the images 5-16710b and 5-16618 (A, B). Compared to the plots (C, E), a larger fraction of the distances on (D, F) falls outside the high density mode, possibly forming a second mode that corresponds to inter-cluster distances. Particle distances are the actual values computed from the biological data. Random refers to the results for a random distribution. Confined are the results for a random distribution of the total number of particles in an image but at a higher density or a smaller area comparable to the total (cumulative) area of the clusters only. We use the confined curves as a check to determine the optimum length L. The scatter plots on the right (G, H) shows the nearest neighbor distributions over all the micrographs compared to expected inter-particle distance of nearest and next nearest neighbors.

Here, we refine this idea by developing criteria based on the global L-dependence of the number of clusters. The partition of a given set of points depends on the distance parameter; using a larger L for the same set will result in some of the clusters merging. Assuming there is a natural partition of the given set, this partition is recovered, at best, only for a bounded range of L values.

We note that clustering is a widely important topic, with many different algorithms implemented in both public domain and proprietary software. Applications of clustering include astronomy, forestry, geography, image analysis, microscopy, social science, statistics, and many more. No method works in all situations, and each application area has its own unique set of requirements. Clustering techniques come in several varieties: (1) k-means clustering, where the number of clusters is specified in advance, in which the distances to the centroids are minimized within each cluster. Variations exist that use Bayesian techniques to determine an optimum number of clusters from a pool of possibilities; (2) Gaussian (and other distributional) mixture models in which the data is assumed to come from a fixed number of distributions [2]; (3) density-based methods that examine the local density of points in a neighborhood to determine clusters—two examples are the Voronoi-based method of Levet et al. [12] and notably, DBSCAN (Density Based Spatial Clustering of Applications with Noise) [6], several different Matlab®implementations of which can be found online; and (4) hierarchical methods that cluster points based on their distance separation (in some metric). Matlab®comes with a standard implementation of this technique.

Each method depends on one or more parameters. For density-based and hierarchical algorithms, the minimum distance separating clusters and the minimal number of points that are needed to form a cluster are typically user-specified. Although there are implementations that can choose the distance on their own, this is done only on an example by example basis; no overall dataset technique seems to be available. A problem with many of the methods is that even for the same algorithm, each implementation is different and so often produces different results (sometimes just slightly, but large discrepancies can occur for a given example) as we saw in the Matlab®implementations of DBSCAN. Moreover, some methods (in particular, density-based methods) may be sensitive to the order of points fed into them. Therefore, we chose to use the standard hierarchical clustering algorithm to underlie our own clustering methodology in order to achieve simple and reproducible results pertinent to quantifying clustering images of cell membrane-bound receptors. We wanted a clustering technique that was faithful to the science that presumably lies beneath the clustering mechanism of the labeled biological particles as we have described them above. Our goal was to create a hypothesis based model where we consider some membrane regions to be more attractive for clustering than others, which will allow us to develop a numerical model to test against the biological data.

2   Results

The data set discussed here consists of 81 high resolution (2500x) TEM micrographs of labeled VEGF receptors on PAE-KDR cells. Labeled receptors appear as dark spots (points), whose coordinates are extracted for subsequent analysis. The same type of cells were used for all micrographs; cells were stimulated with varying amounts of VEGF, and labeling was performed with two different commercial sources of primary antibodies to VEGFR, in order to validate results. The principle of sample acquisition and imaging is illustrated in Fig. 1; for more details refer to the Methods section.

2.1   Non-uniform Spatial Distribution of Receptors

The number of detected receptors and their spatial distribution pattern vary among the images. Figure 3 shows several images, illustrating the different dis-

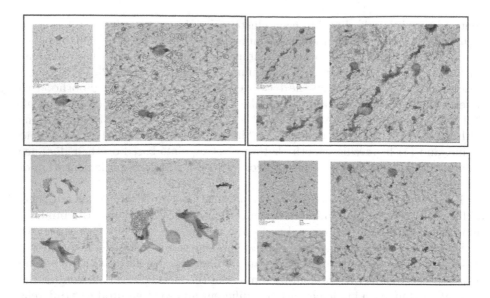

Fig. 3. TEM images of labeled VEGF receptors on the membrane of PAE-KDR cells and maps of particles. The gold particles appear as dark spots, whose coordinates are extracted in a semi-automatic procedure. The image numbers starting from the top left grouping are as follows: 5-16622, 5-16610, 5-16712, 5-16616. For each grouping, **Top left column:** high resolution image; **Bottom left column:** details of image; **Right column:** clustering pattern of dark spots using the clustering analysis procedure for the particle coordinates.

tribution patterns. Other than the tendency of points to accumulate in small areas (clustering), there is no obvious *prima facie* correlation between the different patterns and experimental conditions.

The concept of clustering we are interested in is illustrated by the image in Fig. 4. The points surrounded by contours are "naturally" grouped together. We refer to a set of points like these as a "cluster". Distance is the central element that defines clusters – points within a cluster are clearly closer to each other than they are to points in any other cluster; the typical distance within a cluster is much smaller than the typical distance between clusters, $L_P \ll L_C$. *Cluster identification* algorithms provide a partition of a set of points into clusters, and generally return such a partition *for any set of points*. The other necessary element is establishing *clustering tendency*, which employs *quantitative measures* comparing the distribution of interest to random distributions.

Clustering Tendency. We used several statistical tests to establish clustering tendency. We calculated Ripley's function and obtained the Hopkins' statistic for all our images. Typical results obtained for one image are shown in Fig. 5. Both of these tests are commonly used to determine the likelihood of clustering in a given sample to the expectation from a comparable random, uniform independent

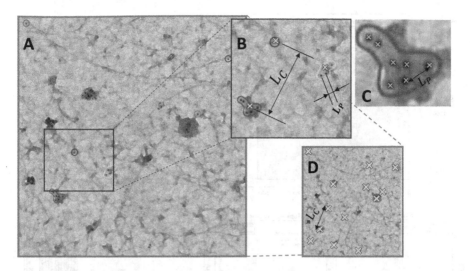

Fig. 4. (A) Image 5-16610 illustrates the idea of clustering based on the complete separation of two distance scales (or ranges). Clusters are identified by colored lines surrounding the points. This identification was performed as described in the subsequent sections; however, in this example, most observers would probably agree that the identification of clusters as indicated is natural or obvious. (B) On the zoomed in detail, we outline the distance between two clusters (L_C) and between two points inside the same cluster (L_P). We naturally identify the clusters as such because the distribution of these two types of distances is separated in the sense that the smallest inter-cluster distance is still larger than the largest intra-cluster distance. (C) The points within a given cluster appear to follow a random distribution (on the condition that the points stay within the footprint of the cluster) (D) The location of clusters (or cluster centroids used to represent the location of a cluster) also appears to follow an unconstrained, uniform, independent (UID) random distribution. (Color figure online)

distribution. We show in Fig. 5B (blue line) Ripley's K statistics for the image under consideration. The blue line compares the probability of finding other particles at a given distance from a particle in the sample; the likelihood in the range of ≤ 600 nm is larger than the expectation for u.i.d. (uniformly and independently distributed).

As stated earlier, the Hopkins' statistic H is a random variable, $0 \leq H \leq 1$. A distribution symmetric around $H = 0.5$ is consistent with u.i.d., whereas values above 0.5 indicate clustering. In this particular image, the distribution is strongly skewed (red bars in Fig. 5C), with close to 100% between $[0.9, 1]$.

The general clustering tendency in our images is illustrated in Fig. 5E showing the distribution of the mean of the Hopkins statistics obtained from all the images in the sample. The distribution of the Hopkins values for all our images falls between 0.6 and 1 and most of the images have the Hopkins' values very close to 1.

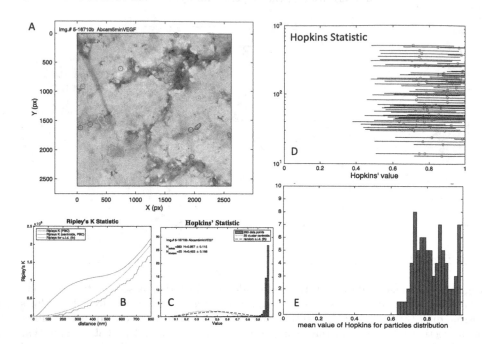

Fig. 5. (B) is the Ripley's K statistic and (C) is the Hopkins' statistic for the image 5-16616 (A). (D) is the mean Hopkins' statistic values (red circles) for all the images where the black lines represent error bars of ±1 standard deviation from the mean, and (E) is the histogram of the mean of the Hopkins' value of all the images. (D) and (E) shows the consistency of almost all the images in that they have Hopkins' value very close to 1 which is a sign of clustering tendency. (Color figure online)

Nearest Neighbor Distances. We extracted the first and second nearest neighbor distances (NND and NND2) for each image, and compared the distributions to the expected uniform random distribution, Eqs. (1) and (S4), using the density $\lambda = N_p/A_{image}$ (N_p and $A_{image} = 7 \cdot 10^6$ px^2 represent the number of points and area of each image).

For typical examples of the biological data (Fig. 2), the majority of the NN distances fall in a single mode, similar in shape, but shifted to the left compared to the random curve. This *high density mode* has an average in the range of ≈20–30 px and its location varies little with the number of points in the image. This is contrast with the uniform random case, when the average NN distance should scale with[3] $N_p^{-\frac{1}{2}}$. The local density, as shown by the nearest and next nearest neighbor distance scatter plots on the right of Fig. 2, is independent of the average density (especially noticeable for the median data). Another consistent feature is that the smallest observed distance never falls below 5 nm ≈ 7 px. This *exclusion distance* is consistent with the fact that the particle coordinates

---

[3] From Eqs. (1) and (S4): $\langle r \rangle_{NN} = \frac{1}{2}b$, $\langle r \rangle_{NN2} = \frac{3}{4}b$ where $b^2 = \frac{A}{N}$; all images have the same area.

represent the centroids of the 6 nm $\approx$ 8.7 px diameter nano-gold probes, and so have to be taken into account when comparing to a model distribution.

We found that the high density modes in the NND and NND2 distributions were well approximated by the distribution functions (1) and (S4), corresponding to uniform random placement of the same number of points in a *smaller area* (i.e., the actual cluster area), resulting in a higher density parameter $\lambda_{\text{eff}} > \lambda = N_p/A_{\text{image}}$. The dashed curves shown on the NND plots in Fig. 2 correspond to point particles distributed uniformly but with an *effective spacing* parameter $D_{\text{eff}} < b_{\text{image}}$, and set to zero for $r < D_{\text{excl}}$. At this initial stage, the values of D_{eff} and the corresponding density $\lambda_{\text{eff}} = 1/D_{\text{eff}}^2 = N_p/A_{\text{eff}}$ were obtained in an ad hoc manner.

2.2 Cluster Identification

Similarly to Espinoza and coworkers [5], we rely on distance based clustering. This approach requires a choice for the value of the length parameter L. Increasing the value of L used to cluster the same set of points will generally result in the merger of some clusters. Using a very small or very large L will result a trivial partition into single point clusters ($N_C(0) = N$) or a single cluster containing all the points ($N_C(\infty) = 1$).

The idea of [5] was to identify a special value of L by looking at the number of clusters of size larger than 1, $N_C^{(2+)}(L)$, as a function of L. As L is increased, the number of clusters starts with $N_C^{(2+)}(0) = 0$ (each point is a singleton cluster); it increases as singleton clusters merge into larger ones, but eventually declines as the overall number of "nontrivial" clusters declines; eventually it reaches $N_C^{(2+)}(\infty) = 1$ as L exceeds the largest pairwise distance.

Each distance based cluster corresponds to a connected graph, whose vertices are the members of the cluster, and the distance between two adjacent points is at most equal to L. We can think of the points as vertices of a graph, where two points are linked if and only if their distance is less than L; the clusters correspond to the *connected* subgraphs. Conversely, the distance between any two points assigned to *different* clusters must be strictly larger than L. This is consistent with the intuition that clustering means that entities separate *"naturally"* into groups, in a way that members of a group are closer to each other than they are to any member of another group.

2.3 Analysis of the Distribution Within Clusters

We constructed clusters for values of the distance parameter ranging from 0.698 nm (1 pixel) to a few hundred in order to obtain the full $N_C(L)$ dependence. We can visualize the distribution of clusters for various length scales. The dependence of the number of clusters on the length scale is plotted in Fig. 6 for a single image as well as for the entire set analyzed. We first scaled the theoretical (random) distributions to correspond to a higher particle density D_{eff} based on the area of the computed clusters, which is the confined particle curve.

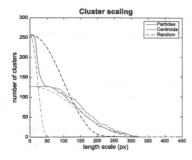
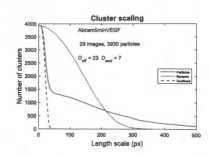

Fig. 6. (*left*) number of clusters as a function of the length parameter (*red lines*) in a single TEM image; (*right*) we summed the number of clusters from a set of 29 images, merging the images into a single larger one. These results are compared with the average number of clusters expected from a random distribution using the same number of points in the same area (*blue lines*). (Color figure online)

We then added a correction to take into account a minimal separation of points by setting the corresponding pdf $p(x)$ to zero for $x \leq D_{\text{exclusion}}$ in the case of nearest neighbor distance distributions. Based on the definition of the exclusion distance (i.e., the physical size of the probes), we use the confined curves as a check when determining the optimum length L. The behavior for $L > D_{\text{exclusion}}$ should be close to that of random particles, but the perfect $N_C(L)$ curve will take values for $L < D_{\text{exclusion}}$, which is unphysical, so they are removed. Therefore, instead of having a discontinuity, we decided to shift the ideal curve by that small value (10 nm) which does not modify the large value behavior too much. Then, the curves were scaled vertically to match the integral of the high density mode (NND). Comparison with the random distribution shows clear and consistent deviations. Both the individual image and the cumulative plot exhibit a sharp initial decrease, followed by a significantly slower variation. The initial fast decrease corresponds to the fact that particles are much closer to their neighbors than the average distance. In an ideal clustered scenario, where the intra-cluster separation between particles is smaller than the shortest distance between clusters, the number of clusters would decrease until the largest intra-cluster distance, and one would then observe a plateau until L becomes comparable to the inter-cluster distance. While we do not observe a perfect scale separation, both the single image and the cumulative curves exhibit a clear change in behavior around 16 nm (23 pixels), and a shoulder that extends to approximately 49 nm (70 pixels).

2.4   Analysis of the Spatial Distribution of the Clusters

After analyzing the distribution of particles within the clusters, we considered the distribution and localization of centroids, that is, taking each cluster on a given image as a particle in the given area. We found the centroid of each cluster by taking the average of the coordinates of the particles composing each

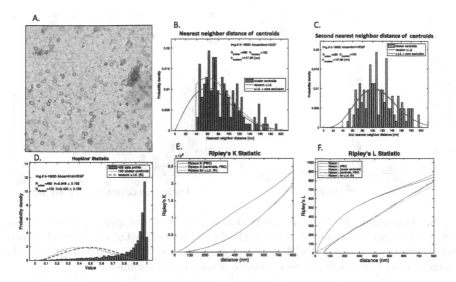

Fig. 7. Statistical measures of clustering applied to the cluster centroids for image 5-16620. Centroids are defined as the average of the position vectors of all particles in the cluster. NND distributions (B, C) of the 150 centroids are much closer to random placement than the positions of the particles themselves. The Hopkins' and Ripley's statistics (D, E, F) essentially indicate no clustering tendency. (Color figure online)

Table 1. Experimental data sets: column 1 is the type of antibody with related time t in minutes at which the cells were fixed, columns 2 through 4 gives the number of images in total, total number of particles and the average number of particles per image for corresponding data sets, respectively.

Experimental data sets			
Antibody type and time	Total number of images	Total number of receptors	Average number of receptors
Abcam 0 min	21	2625	125.0
Abcam 2 min	14	2457	175.0
Abcam 5 min	29	3930	135.5
CS 0 min	8	1015	126.8
CS 2 min	10	935	93.5

one. We then analyzed the distribution of centroids as the same way we did for particles within the clusters. We used the nearest neighbor distribution as well as Hopkins' and Ripley's tests to analyze the clustering formation. With very few exceptions, the tests confirmed that the centroids do not exhibit clustering, and their distribution is consistent with random uniform placement. The reason for the need of a new and simple predictive model is to provide quantitative measurements by quantifying both clustering and the effect of micro-domains

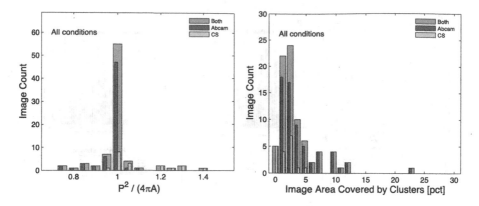

Fig. 8. Histograms of all the data sets that describes the properties of the clusters such as area, perimeter and shape. (L) The form factor distribution shows that most clusters are close to circular, with some outliers. (R) The fraction of area occupied by clusters, the closest possible direct measure of our model parameter f.

on the amount of clustering. As in shown by Fig. 6 on the left, the green curve corresponds to the distribution of centroids at the given length scales. The result of the centroid distribution is consistent with the distribution of particles within the clusters as is shown in Fig. 7.

Number of Particles, Clusters and the Optimal Clustering Length of Images. By using distance based hierarchical clustering, we identified the number of clusters and characteristic length of images with a given number of particles. One way to approach the clustering phenomenon is analyzing the data image by image, then comparing the set of data for resting and activated cells. We also analyzed the characteristic properties of each data set by comparing two data subsets, **Abcam** and **CS**, for different stimulus time durations as can be seen in Table 1. Grouping somewhat similar objects based on those characteristics is another approach. Therefore, we produced histograms for resting and stimulated cells in terms of the number of particles, characteristic lengths, and number of clusters in order to understand the characteristics of each data set. Even though there are a number of outliers in the histograms of number of particles, length scales and number of clusters versus the number of images, an instant observation is that the majority of the images spread around an average with respect to the corresponding parameter scales. See Appendix A for further details.

Area and Perimeter of Clusters. We calculated the area (A) and perimeter (P) of the polygons by using the domain reconstruction algorithm. We observed that $P^2/(4\pi A)$, which we call the form factor, is important because this value gives an idea about the shape of the clusters in a given micrograph. As it is shown in Fig. 8, the majority of clusters are circularly shaped, but there are outliers that are not close to a circle in shape. The analyses outlined above were performed for each of the images in the dataset. We generated several plots for each image file for each type of analysis. It is not possible or useful to include all of these results here.

3   Discussion

Membrane proteins are distributed unevenly and non-randomly on plasma membranes. Special features of cell membrane subdomains and compartments support evidence of heterogeneity. These features may be lipid rafts, protein islands and/or cytoskeletal corrals [13]. When cells are stimulated, protein distributions on the cell membrane change. We analyzed results derived from PAE-KDR cells that were exposed for 2–5 min to increasing concentrations of VEGF that binds the high affinity receptor KDR. VEGF receptors were labeled with gold particles for visualization by TEM imaging. The resulting micrographs showed the receptors localized in singletons, small clusters and large clusters. These patterns changed between groups of cells.

We analyzed the distribution of receptors using nearest neighbor distance distributions, Hopkins' and Ripley's tests [23]. The Hopkins' and Ripley's tests are not to *quantify* the clusters in this type of experiment; they only give a qualitative indication of clustering. Therefore, there is a need for a simple and predictive model which can identify clusters and compare clustering between and within experimental conditions. Previously, Espinoza and co-workers [5] developed an approach based on the hierarchical clustering algorithm to define the characteristic distance which enables quantifying the density of the clustering in TEM images. They compared experimental conditions between activated mast cells with increasing amount of stimulus.

Here we apply a similar approach, with one important difference. We use the cluster scaling curve $N(L)$, and try to identify the optimal length scale L^* consistent with the idea of separation between the intra-cluster and inter-cluster distances. After identifying the optimum length and cluster identification for each image, we used the domain reconstruction algorithm to define the footprints of clusters. This is defined by a contour around the cluster, following work developed previously for the study of live trajectories of labeled receptors [18].

Based on cluster identification, we were able to identify features of the nearest and next-nearest neighbor distance distributions that reflected the distribution of particles within clusters along with those resulting from inter-cluster distances. This analysis pointed us to a second hypothesis, that the particles within a cluster are distributed randomly within the cluster's footprint, and the locations of clusters are also distributed randomly within the cell membrane. This model is consistent with much of our data; deviations are due to either "mega-clusters" containing hundreds of receptors, or large scale gradients in receptor density (or perhaps labeling efficiency) within individual cells.

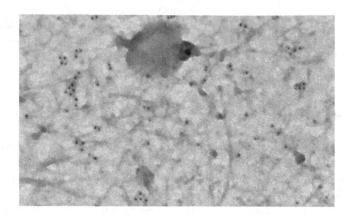

Fig. 9. The clusters of labeled receptors appear to be located in regions of the membrane that have different local properties. The interpretation may be somewhat misleading in that the shades on a transmission electron micrograph do not indicate depth, but rather the ability to absorb electrons. However, the clusters seen in the image occupy small areas of a different apparent consistency; there are very similar domains that do not contain labeled receptors.

A Hypothesis for The Origin of Clusters. The reason for receptor clustering is not completely understood, but it is generally accepted that it is a consequence of the physical features of the cell membrane, which influence the movement of receptors. All of our results support a simple hypothesis, that *the clusters form in specific, pre-existing regions (small domains) on the cell membrane* due to special physical properties of those regions. Figure 9 illustrates the idea. Some of the domains are empty, but the domains altogether represent a small fraction of the total area.

Direct detection of membrane micro-domains is challenging. Available methods provide a range of temporal, spatial and chemical resolutions. Thus, different pictures can arise from different experimental modalities for the cell/receptor combination, and it may not be possible simultaneously to detect several different features of membrane domain properties [4].

Static images of receptors (even in the absence of ligand) typically reveal a clustering pattern, where receptors tend to accumulate in groups ranging from a few to a few tens of receptors. Our results indicate that VEGF receptor clustering is a robust feature in this particular experimental system. The proposed method of identifying the optimal L and the identification of a geometric support for each cluster are well suited for automatic analysis and were successfully applied to a data set of moderate size. The size and number of VEGFR clusters varies among images; except for instances of "mega-clusters", the majority of observed receptors are in clusters of size 1–10, and confined to a fraction of less than 10% of the total area.

The main insight emerging from our analysis is that both the distribution of receptors within the geometric area of each cluster, as well as the distribution of the locations of the clusters, are consistent with random placement. This points to the hypothesis that some of the micro-domains found on the cell membrane have a specific molecular composition that results in an affinity for the receptors; receptors may diffuse in and out of the domain boundaries, but the crossing probability is asymmetric. The micro-domains are distributed randomly, and receptors trapped inside these domains are also free to move randomly within the confining area.

A   Methods

Experimental details, mathematical definitions and methodologies, and their justification are provided as supplementary materials available at https://www.dropbox.com/s/8bggjrzhzr1vsne/Supplementary-Materials-Clustering-Paper.pdf?dl=0.

References

1. Birk, D.A., Barbato, J., Mureebe, L., Chaer, R.A.: Current insights on the biology and clinical aspects of VEGF regulation. Vasc Endovasc. Surg **42**, 517–530 (2010)
2. Day, N.E.: Estimating the components of a mixture of normal distributions. Biometrika **56**(3), 463–474 (1969)
3. Dixon, P.M.: Encyclopedia of Environmetrics, vol. 3, pp. 1796–1803. Wiley, Chichester (2002). Chap. Ripley's K function
4. Edidin, M.: Lipid microdomains in cell surface membranes. Curr. Opin. Cell Biol. **7**(4), 528–532 (1997)
5. Espinoza, F.A., Oliver, J.M., Wilson, B.S.: Using hierarchical clustering and dendrograms to quantify the clustering of membrane proteins. Bull. Math. Biol. **74**(1), 190–211 (2011)
6. Ester, M., Kriegel, H.-P., Sander, J., Xu, X.: A density-based algorithm for discovering clusters in large spatial databases with noise. In: Simoudis, E., Han, J., Fayyad, U.M. (eds.) Proceedings of 2nd International Conference on Knowledge Discovery and Data Mining (KDD-1996), pp. 226–231. AAAI Press (1996). ISBN 1-57735-004-9
7. Ferrara, N., Hilla, K.J., Gerber, H.P., Novotny, W.: Discovery and development of bevazicumab, an anti-VEGF antibody for treating cancer. Net Rev. Drug Discov. **3**, 391–400 (2004)
8. Hanahan, D., Folkman, J.: Patterns and emerging mechanisms of the angiogenetic switch during tumorigenesis. Cell **86**, 353–364 (1996)
9. Jain, A., Murty, M., Flynn, P.: Data clustering: a review. ACM Comput. Surv. **31**(3), 264–323 (1999)
10. Jain, A.K., Dubes, R.C.: Algorithms for Clustering Data, p. 218. Prentice-Hall, Englewood Cliffs (1988)
11. Karamysheva, A.F.: Mechanisms of angiogenesis. Biochemistry (Mosc) **73**, 751–762 (2008)
12. Levet, F., et al.: SR-Tesseler: a method to segment and quantify localization-based super-resolution microscopy data. Nat. Methods **12**(11), 1065–1071 (2015)
13. Lingwood, D., Simons, K.: Lipid rafts as a membrane-organizing principle. Science **327**(5961), 46–50 (2010)

14. Moltchanov, D.: Distance distributions in random networks. Ad Hoc Netw. **10**(6), 1146–1166 (2012)

15. Nagy, P., et al.: Lipid rafts and the local density of ErbB proteins influence the biological role of homo-and heteroassociations of ErbB2. J. Cell Sci. **115**(22), 4251–4262 (2002)

16. Olsson, D.A., Kreuger, J., Claesson-Welsh, L.: VEGF receptor signaling - in control of vascular function. Nat. Rev. Mol. Cell Biol. **7**, 359–371 (2006)

17. Plouet, J., Schilling, J., Gospodarowicz, D.: Isolation and characterization of a newly identified endothelial cell mitogen produced by AtT-20 cells. EMBO J. **8**, 3801–3806 (1989)

18. Pryor, M.M.: Orchestration of ErbB3 signaling through heterointeractions and homointeractions. Mol. Biol. Cell **26**(22), 4109–4123 (2015)

19. Radhakrishnan, K., Halász, Á., McCabe, M.M., Edwards, J.S., Wilson, B.S.: Mathematical simulation of membrane protein clustering for efficient signal transduction. Ann. Biomed. Eng. **40**(11), 2307–2318 (2012)

20. Ritchie, K., Kusumi, A.: Single-particle tracking image microscopy. Methods Enzymol. **360**, 618–634 (2003)

21. Senger, D.R., Galli, S.J., Dvorak, A.M., Perruzzi, C.A., Harvey, V.S., Dvorak, H.F.: Tumor cells secrete a vascular permeability factor that promotes accumulation of ascites fluid. Science **219**, 983–985 (1983)

22. Wilson, S., et al.: Exploring membrane domains using native membrane sheets and transmission electron microscopy. In: McIntosh, T.J. (ed.) Lipid Rafts. Humana Press, Totowa (2007). https://doi.org/10.1007/978-1-59745-513-8_17

23. Zhang, J., Leiderman, K., Pfeiffer, J.R., Wilson, B.S., Oliver, J.M., Steinberg, S.L.: Characterizing the topography of membrane receptors and signaling molecules from spatial patterns obtained using nanometer-scale electron-dense probes and electron microscopy. Micron **37**, 14–34 (2006)

Synthesis for Vesicle Traffic Systems

Ashutosh Gupta[1]([✉]), Somya Mani[2]([✉]), and Ankit Shukla[3]([✉])

[1] IIT Bombay, Mumbai, India
akg@iitb.ac.in
[2] IBS-CSLM, Ulsan, South Korea
somyamn@gmail.com
[3] NCBS, Bangalore, India
ankitk@ncbs.res.in

Abstract. Vesicle Traffic Systems (VTSs) are the material transport mechanisms among the compartments inside the biological cells. The compartments are viewed as nodes that are labeled with the containing chemicals and the transport channels are similarly viewed as labeled edges between the nodes. Understanding VTSs is an ongoing area of research and for many cells they are partially known. For example, there may be undiscovered edges, nodes, or their labels in a VTS of a cell. It has been speculated that there are properties that the VTSs must satisfy. For example, stability, i.e., every chemical that is leaving a compartment comes back. Many synthesis questions may arise in this scenario, where we want to complete a partially known VTS under a given property. In the paper, we present novel encodings of the above questions into the QBF (quantified Boolean formula) satisfiability problems. We have implemented the encodings in a highly configurable tool and applied to a couple of found-in-nature VTSs and several synthetic graphs. Our results demonstrate that our method can scale up to the graphs of interest.

1 Introduction

Eukaryotic cells, including human cells, consist of multiple membrane-bound compartments. Material is transported among these compartments by the vesicle transport system (VTS). Briefly, the source compartment produces a membrane-bound packet of molecules called a vesicle. After release, this vesicle specifically recognizes the correct target compartment within the cell, and fuses with it [1]. A lot of information about the molecules that form the machinery of the VTS has been discovered, including their regulatory interaction with each other [2]. In spite of this detailed knowledge at the level of the molecules, the structure of the VTS network, or the road-map of the eukaryotic cell, is far from complete. For example, although the localization of various SNAREs – a class of molecules that participate in the control of VTS – in the cell is known, and also their site of action [3], for most SNAREs, how they first reached the compartments they reside in is not known. The current knowledge of the network is put together from a patchwork of biological experiments and is scattered across

© Springer Nature Switzerland AG 2018
M. Češka and D. Šafránek (Eds.): CMSB 2018, LNBI 11095, pp. 93–110, 2018.
https://doi.org/10.1007/978-3-319-99429-1_6

several publications. Even after this information is collected and put together, we find that the network obtained is still not complete; new vesicles and new contents in previously known vesicles are constantly being discovered (some new discoveries include [4–7]). The synthesis for the unknown pieces may be assisted by computation on the graph model of VTSs. In this paper, we are looking at the computational questions arising from the VTSs.

VTSs are regulated by the same molecules that they transport. For the purpose of this paper, the VTS molecules we focus on are the transmembrane SNARE proteins. SNAREs drive the recognition of the target compartment by vesicles and their subsequent fusion. The SNAREs can be divided into v-SNAREs (which are present on vesicles) and t-SNAREs (which are present on compartments). A vesicle fuses with a compartment if its v-SNARE can form a complex with the t-SNARE present on that compartment. Not all v- and t-SNARE combinations can form complexes; this constraint forms part of the basis for the specificity of vesicle traffic [8].

We use the model of VTSs that has been presented in [9]. Please look at Appendix A for a detailed discussion on pros and cons of the model. We model the system as a labeled graph, where compartments are nodes and transport vesicles are edges. The molecular compositions of the compartments and vesicles are the node and edge labels respectively. The molecules can be active or inactive on any a compartment or vesicle. The activity states of molecules are also included in the labels. Due to the biology of SNAREs of the VTSs our interest, a vesicle is enabled by a set of *four* molecules such that one part of the set occurs in the vesicle and the other part occurs in the target of the vesicle compartment. The partition always divides the set in the set of three and one molecules. The enabling molecules must be active in the vesicle and target compartment respectively. The pairs are called *fusing* sets and analogously the vesicle is considered to be *fused* with the destination compartment. Not all sets of molecules can participate in the fusion; in the biological cells, fusogenic SNARE complexes are discovered through experiments. Generally, the fusing pairs are found to be distinct for distinct vesicle-compartment fusions. To ensure that a molecule that has participated in a fusion does not interfere with fusion at compartments and vesicles, in the model, we require that the molecule is inactive on appropriate compartments. The activity of molecules is regulated by the other molecules, i.e., the presence and absence of the other molecules in a compartment or vesicle may make the molecule active or inactive. We call this regulation as *activity functions*. The regulation controls are defined by a fusion pairing relation containing pairs of molecules and activity Boolean functions.

In the model, we assume that the system is in steady state and the concentrations of the molecules in compartments do not change over time. Since our system is in steady state, we expect that any molecule that leaves a compartment must come back via some path on the graph. We call this property of VTS as *stability*.

As we have discussed earlier, our understanding of VTSs is partial. The synthesis of the unknown pieces may be *assisted* by computation on the graph

model of VTSs. In this paper, we consider several versions of the synthesis problem involving different parts of VTSs that can be synthesized, such as modifying
labels, adding/deleting edges, and learning activity function. We also consider
variations on the properties against which we do synthesis, namely stability, and
k-connectedness that states that the VTS remains connected after removing any
$k-1$ edges. We have assumed that the given partial VTS is always well-fused
whereas properties like stability and k-connectedness may not hold in the partial
VTS. In order to synthesize the parts of a VTS such that it satisfies the constraints, we encode the synthesis problem into one of satisfiability of quantified
Boolean formulas (QBFs).

We have implemented the encoding in a flexible tool, which can handle a wide
range of synthesis queries. We have applied our tool on several VTSs including
two found-in-nature VTSs.

Our experiments suggest that some of the synthesis problems are solvable
by modern solvers and the synthesis technology may be useful for biological
research.

The rest of the paper is organized as follows. In Sect. 2, we present the graph
model of VTSs and encoding of several constraints on VTSs. In Sect. 3, we
present the synthesis problems and their encoding into QBF satisfiability. In
Sect. 4, we present our implementation and experimental results. We discuss
related work in Sect. 5 and conclude in Sect. 6.

2   Preliminaries

In this section, we will present the model of VTS from [10]. We will also present
the constraints and properties on the VTSs, and their encoding as a QBF formula. We model a VTS as a labelled graph along with assisting pairing matrices
and activating functions.

Definition 1. *A VTS G is a tuple $(N, M, E, L, \mathcal{P}, g, f)$, where*

- N *is a finite set of nodes representing compartments in the VTS,*
- M *is the finite set of molecules flowing in the system,*
- $E \subseteq N \times (2^M - \emptyset) \times N$ *is the set of edges with molecule sets as labels,*
- $L : N \to 2^M$ *defines the molecules present in the nodes,*
- $\mathcal{P} \subseteq 2^M$ *is pairing relation,*
- $f : M \to 2^M \to \mathbb{B}$ *is activity maps for nodes, and*
- $g : M \to 2^M \to \mathbb{B}$ *is activity maps for edges.*

N, M, E, and L define a labelled graph. Additionally, $\mathcal{P}$ defines which molecules
can fuse with which molecules, and f and g are the activity functions for
molecules on nodes and edges respectively. The model captures the steady state
of a VTS. The analysis of the model will inform us about the network/graph
properties of VTSs.

A molecule k is *active* at node n if $k \in L(n)$ and $f(k, L(n))$ is true. A molecule
k is *active* at edges (n, M', n') if $k \in M'$ and $g(k, M')$ is true. We call G *well-
structured* if molecules M is divided into two partitions Q and R such that for

each $P \in \mathcal{P}, |P \cap Q| = 3 \wedge |P \cap R| = 1$, and for each $(n, M', n') \in E$, $n \neq n'$ and $M' \subseteq L(n) \cap L(n')$. In other words, molecules are of two types Q and R, pairing relations have sets of four molecules such that three are of one type and one is of another type (motivated by the biochemistry of the fusion), there are no self loops, and each edge carry only those molecules that are present in its source and destination nodes. An edge $(n, M', \_) \in E$ *fuses* with a node n' if there are non-empty set of molecules $M'' \subseteq M'$ and $M''' \subseteq L(n')$ such that M'' are active in the edge, M''' are active in n', and $M'' \cup M''' \in \mathcal{P}$. We call G *well-fused* if each edge $(n, M', n') \in E$ fuses with its destination node n' and can not fuse with any other node.

A *path* in G is a sequence $n_1, \ldots, n_\ell$ of nodes such that $(n_i, \_, n_{i+1}) \in E$ for each $0 < i < \ell$. For a molecule $m \in M$, an *m-path* in G is a sequence $n_1, \ldots, n_\ell$ of nodes such that $(n_i, M', n_{i+1}) \in E$ and $m \in M'$ for each $0 < i < \ell$. A node n' is *(m-)reachable* from node n in G if there is a $(m$-$)$path $n, \ldots, n'$ in G. We call G *stable* if for each $(n, M', n') \in E$ and $m \in M'$, n is m-reachable from n'. We call G *connected* if for each $n, n' \in N$, n' is reachable from n in G. We call G k-connected if for each $E' \subseteq E$ and $|E'| < k$, VTS $(N, M, E - E', L, \mathcal{P}, g, f)$ is connected.

2.1   Encoding VTS

The conditions on the VTSs for a given size can be encoded as a QBF formula with uninterpreted functions. To encode the constraints, we need variables for each aspect of VTS. Let us suppose that the size of the graph is ν and a number of molecules are μ. To fully finitize the problem, we also limit the maximum number π of edges present between two nodes. Here, we list the Boolean variables and uninterpreted function symbols that encode parts of VTSs.

1. Boolean variable $n_{i,m}$ indicates if $m \in L(i)$
2. Boolean variable $e_{i,j,q}$ indicates if qth edge exists between i and j.
3. Boolean variable $e_{i,j,q,m}$ indicates if qth edge between i and j contains m.
4. Boolean variable $p_{\{m_1, m_2, m_3, m_4\}}$ indicates if $\{m_1, m_2, m_3, m_4\} \in \mathcal{P}$
5. uninterpreted Boolean functions $f_m : \mathbb{B}^\mu \to \mathbb{B}$ encoding $f(m)$ map
6. uninterpreted Boolean functions $g_m : \mathbb{B}^\mu \to \mathbb{B}$ encoding $g(m)$ map

We also have auxiliary Boolean variables that will help us encode the well-fused property.

1. $a_{i,m}$ indicates that molecule m is active at node i, i.e., $f(m, L(i))$ holds
2. $b_{i,j,q,m}$ indicates that molecule m is active at qth edge (i, M', j) between i and j, i.e., $g(m, M')$ holds.

We will describe several constraints that encode VTSs in this section. In the next section, we will extend the encoding for the synthesis problem. To avoid cumbersome notation, we will not explicitly write the ranges of the indexing in the constraints. i and j will range over nodes, i.e., from 1 to ν. m will range over molecules, i.e., from 1 to μ. q will range over edges between two nodes, i.e., from 1 to π.

The following constraints encode the basic consistancy of VTSs.

$$\text{EdgeC} = \bigwedge_{i,j,q} (\bigvee_m e_{i,j,q,m}) \Rightarrow e_{i,j,q} \wedge \bigwedge_{i,q} \neg e_{i,i,q} \wedge \bigwedge_{i,j,q,m} e_{i,j,q,m} \Rightarrow (n_{i,m} \wedge n_{j,m})$$

$$\text{ActivityC} = \bigwedge_{i,j,q,m} b_{i,j,q,m} \Rightarrow e_{i,j,q,m} \quad \wedge \quad \bigwedge_{i,m} a_{i,m} \Rightarrow n_{i,m}$$

$$\text{PairingC} = \exists qr. \bigwedge_{m_1,m_2,m_3,m_4} (p_{\{m_1,m_2,m_3,m_4\}} \Rightarrow qr_{m_1} + qr_{m_2} + qr_{m_3} + qr_{m_4} = 3)$$

$$\text{Fusion1} = \bigwedge_{i,j,q} e_{i,j,q} \Rightarrow \bigvee_{m_1,m_2,m_3,m_4} (\bigwedge_{l=1}^4 (b_{i,j,q,m_l} \vee a_{j,m_l}) \wedge \bigvee_{l=1}^4 b_{i,j,q,m_l} \wedge$$
$$\bigvee_{l=1}^4 a_{j,m_l} \wedge p_{\{m_1,m_2,m_3,m_4\}})$$

$$\text{Fusion2} = \bigwedge_{i,j,q,m_1,m_2,m_3,l \in \{1,..,3\}} b_{i,j,q,m_1} \wedge .. \wedge b_{i,j,q,m_l} \Rightarrow$$
$$\neg \bigvee_{j \neq j', m'_{l+1},..,m'_4} (a_{j',m'_{l+1}} \wedge .. \wedge a_{j',m'_4} \wedge p_{\{m_1,..,m_l,m'_{l+1},..,m'_4\}})$$

$\text{Consistancy} = \text{EdgeC} \wedge \text{ActivityC} \wedge \text{PairingC} \wedge \text{Fusion1} \wedge \text{Fusion2}$

EdgeC states that each edge has at least one molecule, there are no self loops, and edge labels are consistent with node labels. ActivityC states that active molecule are present. PairingC states that all molecules are divided into two types using qr_m bit, which encodes if m belongs to one type or another, and any fusing set of molecules must have three molecules involved from one type and one molecule from the other. Fusion1, and Fusion2 states the well-fused condition. Consistancy is the conjunction of all of the above.

Activity Functions. We also need to encode that the activity of the molecules are controlled by activity functions. The input VTS may include concrete activity functions for some molecules, and for the others the functions may be unknown and to be synthesized. The concrete functions can be given to us in many different ways, for example as a lookup table, or a concise Boolean formula. In the following section, we will assume the appropriate encoding is used for the concrete functions and represent them by NodeFun_m and EdgeFun_m for node and edge regulations respectively. We will use f_m and g_m to represent functions that are unknown in a VTS. Later we will be synthesizing the unknown activity functions and replace f_m and g_m with parameterized constraints that encode a space of candidate functions.

2.2 VTS Properties

For the synthesis of incomplete systems, we need properties against which we synthesize the missing parts. Here we will discuss two such properties proposed in earlier works [10], namely stability and k-connectedness.

Stability Property. We use Boolean variable $r_{i,j,m,p}$ to indicate if there is an m-path from i to j of length less than or equal to p. We use m-reachability to encode the stability condition in VTSs. The following constraint recursively encodes that node j is m-reachable from node i in less than p steps. Subsequently, we encode stability condition using the reachability variables.

$$\texttt{Paths}(r) = \bigwedge_{i,j,m,p} r_{i,j,m,p} \Rightarrow (\bigvee_q e_{i,j,q,m} \vee \bigvee_{i \neq i'} (\bigvee_q e_{i,i',q,m}) \wedge r_{i',j,m,p-1})$$

$$\texttt{Loop}(r) = \bigwedge_{i,j,m} (\bigvee_q e_{i,j,q,m}) \Rightarrow r_{j,i,m,\nu}$$

$$\texttt{Stability} = \exists r. \texttt{Paths}(r) \wedge \texttt{Loop}(r)$$

k-connected Property. k-connectedness expresses robustness against failure of few edges. Let us use $d_{i,j,q}$ to indicate qth edge between i and j is failed and $r'_{i,j}$ to indicate if there is a path from i to j in the modified VTS. In the following, $\texttt{Fail}(d,k)$ encodes that only existing edges can be failed and exactly $k-1$ edges are failed. $\texttt{FReach}(d,r')$ defines reachability in the modified VTS. We use a new variable $r'_{i,j,p}$ to encode reachability from i to j in at most p steps. $\texttt{Connected}(r')$ says that all nodes are reachable from any other node.

$$\texttt{Fail}(d,k) = \bigwedge_{i,j,q} d_{i,j,q} \Rightarrow e_{i,j,q} \wedge \sum_{i,j,q} d_{i,j,q} = k-1$$

$$\texttt{FReach}(d,r') = \bigwedge_{i,j,p} r'_{i,j,p} \Rightarrow [\bigvee_q (e_{i,j,q} \wedge \neg d_{i,j,q}) \vee (\bigvee_{i' \neq i} r'_{i',j,p-1} \wedge \bigvee_q (e_{i,i',q} \wedge \neg d_{i,i',q})]$$

$$\texttt{Connected}(r') = \bigwedge_{i,j} (r'_{i,j,\nu} \vee r'_{j,i,\nu})$$

We will be synthesizing k-connected graphs. We define $\texttt{Connected}(k)$ that says for all possible valid failures the graph remains reachable.

$$\texttt{Connected}(k) = \forall d. (\texttt{Fail}(d,k) \Rightarrow \exists r'. \texttt{FReach}(d,r') \wedge \texttt{Connected}(r'))$$

Since d variables in $\texttt{Connected}(k)$ are universally quantified, $\texttt{Connected}(k)$ introduces quantifier alternations. Therefore, synthesis against this property will require QBF reasoning. We may make the formula quantifier free by considering all possible failures separately and introducing a vector of reachability variables for each failure. However, this will blow up the size of the formula and may not be solvable by a SAT solver.

3   Synthesis for VTS

In this section, we will present a list of synthesis problems that may arise from the partially available information about a VTS and our synthesis method for the problems.

3.1   Problem Statements

We will assume that we are given a VTS, whose all components are not specified. Our objective is to find the missing parts. The missing parts can be in any of the components of VTS. For example, some undiscovered edges or nodes, or insufficient knowledge about the presence of molecules in some part of the VTS. To cover most of the likely variations of this missing information, we have encoded the following variants of VTS synthesis problem.

1. Fixing VTS by adding edges
2. Fixing VTS by adding molecules to the labels
3. Fixing VTS by learning activity functions
4. Fixing VTS by both adding/deleting parts.

3.2   Encoding Incomplete VTS

In our synthesis method, we take a VTS $G = (N, M, E, L, \mathcal{P}, g, f)$ as input. We allow activity functions not to be specified. We construct the following constraints to encode the available information about G. We encode both the present and the absent components in G. Later, the constraints will help us encode the synthesis problems.

$$\texttt{PresentE} = \wedge \{ e_{i,j,q,m} | (i, M_1, j), \ldots, (i, M_{q'}, j) \in E \wedge q \leq q' \wedge m \in M_q \}$$

$$\texttt{PresentN} = \wedge \{ n_{i,m} | m \in L(i) \wedge i \in N \}$$

$$\texttt{PresentP} = \wedge \{ p_{\{m_1, m_2, m_3, m_4\}} | \{m_1, m_2, m_3, m_4\} \in \mathcal{P} \}$$

$$\texttt{KnownActiveN} = \wedge \{ a_{i,m} = \texttt{NodeFun}_m(n_{i,1}, \ldots, n_{i,\mu}) | f_m \text{ is defined.} \}$$

$$\texttt{KnownActiveE} = \wedge \{ b_{i,j,q,m} = \texttt{EdgeFun}_m(e_{i,j,q,1}, .., e_{i,j,q,\mu}) | g_m \text{ is defined.} \}$$

$$\texttt{PresentCons} = \texttt{PresentE} \wedge \texttt{PresentN} \wedge \texttt{PresentP} \wedge \texttt{KnownActiveE} \wedge$$
$$\texttt{KnownActiveN}$$

We also collect the variables that are not set to true in `PresentCons`.

$$\texttt{AbsentELabel} = \{ e_{i,j,q,m} | (i, M_1, j), \ldots, (i, M_{q'}, j) \in E \wedge 0 < q \leq q' \wedge m \notin M_q \}$$

$$\texttt{AbsentE} = \{ e_{i,j,q} | (i, M_1, j), \ldots, (i, M_{q'}, j) \in E \wedge q' < q \leq \pi \}$$

$$\texttt{AbsentNLabel} = \{ n_{i,m} | m \notin L(i) \wedge i \in N \}$$

$$\texttt{AbsentP} = \{ p_{\{m_1, m_2, m_3, m_4\}} | \{m_1, m_2, m_3, m_4\} \notin \mathcal{P} \}$$

$$\texttt{UnknownActive} = \bigwedge \{ a_{i,m} = f_m(n_{i,1}, \ldots, n_{i,\mu}) | f_m \text{ is undefined.} \} \cup$$
$$\{ b_{i,j,q,m} = g_m(e_{i,j,q,1}, .., e_{i,j,q,\mu}) | g_m \text{ is undefined.} \}$$

We have defined `AbsentELabel`, `AbsentE`, `AbsentN`, and `AbsentP` as sets. They will be converted into formulas depending on the different usage in the synthesis problems.

3.3 Encoding Synthesis Property

We will do synthesis against the following property that says the VTS is stable and 3-connected.

$$\text{Property} = \text{Stability} \wedge \text{Connected}(3)$$

The property was proposed in [9]. However, the biological relevance of the property is debatable and open for change. Our tool is easily modifiable to support any other property that may be deemed interesting by the biologists.

3.4 Encoding Synthesis Constraints

Now we will consider the encodings for the listed synthesis problems. The presented variations represent the encodings supported by our tool. Additionally, the combinations of the variation are also possible and our tool easily supports them. For simplicity of the presentation, we assume that if we are synthesizing an aspect of VTS, then all other aspects are fully given. Therefore, we will describe two kinds of constraints for synthesis problems. One will encode the variable part in the synthesis problem and the other encodes the fixed parts. Subsequently, the two constraints will be put together with Consistancy and Property to construct the constraints for synthesis.

Fixing VTS by Adding Edges: Now we will consider the case when we add new edges to VTS to satisfy the properties. In the following, the pseudo-Boolean formula AddE encodes that at most *slimit* new undeclared edges may be added in the VTS. FixedForEdge encodes the parts of the VTS that are not allowed to change.

$$\text{AddE}(slimit) = \sum \text{AbsentE} \leq slimit$$

$$\text{FixedForEdge} = \text{PresentCons} \wedge \text{UnknownActive} \wedge$$

$$\neg \bigvee \text{AbsentELabel} \cup \text{AbsentNLabel} \cup \text{AbsentP}$$

We put together the constraints and obtain the following formula.

$$\text{SynthE}(slimit) = \text{Consistancy} \wedge \text{Property} \wedge \text{FixedForEdge} \wedge \text{AddE}(slimit)$$

Similar to what we have seen Consistancy encodes the basic constraints about VTS, Property encodes the goal, and the rest two are defined just above. A satisfying model of SynthE will make some of the edges in AbsentE true such that Property is satisfied. We limit the addition of the edges, since we look for a fix that require minimum number of changes in the given VTS. We start with $slimit = 1$ and grow one by one until SynthE($slimit$) becomes satisfiable.

In the later synthesis problems, we will construct a similar QBF formula with same first two parts and the last two are due the requirements of the synthesis problem.

Fixing VTS by Adding Molecules to the Labels: The system may also be fixed only by modifying labels on the edges or the nodes instead of adding edges. Here let us consider only adding molecules to the labels of edges. In the following, the formula encodes that only *slimit* edge labels may be added.

AddLabelEdge(*slimit*) = $\sum$ UnknwonEdgeLabel $\leq$ *slimit*

FixedForLabel = PresentCons $\wedge$ UnknownActive$\wedge$

$\neg \bigvee$ AbsentE $\cup$ AbsentNLabel $\cup$ AbsentP

SynthLabel(*slimit*) =

(Consistancy $\wedge$ Property $\wedge$ FixedForLabel $\wedge$ AddLabelEdge(*slimit*))

Similar to the previous encoding, we solve the satisfiability of the above formula to obtain additional molecules that may be added to the edge labels to satisfy the properties.

Fixing VTS by Learning Activity Functions: Now we consider a scenario where some of the activity functions for some of the molecules are missing. The activity functions are μ-input Boolean functions. First, we choose a class of formulas for the candidate functions. We encode the candidates in a formula with parameters. By assigning different values for the parameters, a solver may select different candidates for the activity functions. We will illustrate only one class of formulas. However, we support other classes of formulas, for example, k-CNF.

In the following, the formula NNFTemplate encodes a set of negation normal form functions that take $y_1, .., y_\mu$ as input and contain λ literals. We use Gate to encode a gate that takes a parameter integer x to encode various gates. We use Leaf to encode the literal at some position. Both are stitched to define NNFTemplate. To encode the set of NNF formulas with λ literals, it has finite-range integer variables $z_1, .., z_{2\lambda}$ as parameters.

Gate$(x, w_1, w_2) = (x = 1 \Rightarrow w_1 \wedge w_2) \wedge (x = 2 \Rightarrow w_1 \vee w_2)$

Leaf$(x, [y_1, .., y_\mu]) = \bigwedge_{l=1}^{\mu} (x = 2l - 1 \Rightarrow y_l) \wedge (x = 2l \Rightarrow \neg y_l)$

NNFTemplate$([z_1, .., z_{2\lambda}], [y_1, .., y_\mu]) =$

$\exists w_1, .., w_{2\lambda}. \; w_1 \wedge \bigwedge_{l=1}^{\lambda} w_{\lambda+l} = \text{Leaf}(z_{\lambda+l}, [y_1, .., y_\mu]) \wedge w_l = \text{Gate}(z_l, w_{2l}, w_{2l+1})$

Using the template we define the constraints $\texttt{FindFunctions}(z, \lambda)$ that encodes the candidate functions that satisfy the activity requirements, where z is the vectors of parameters for encoding parameters for each molecule, and λ limits the size of the candidate functions. We fix the all other aspects of the VTS to be fixed via constraints $\texttt{FixedForFunctions}$.

$$\texttt{FindFunctions}(z, \lambda) =$$

$$\bigwedge \{ \bigwedge_{i} n_{i,m} \Rightarrow a_{i,m} = \texttt{NNFTemplate}([z_{m,1}, .., z_{m,2\lambda}], [n_{i,1}, \ldots, n_{i,\mu}])$$

$$|f_m \text{ is undefined}\}$$

$$\bigwedge \{ \bigwedge_{i,j,q} e_{i,j,q,m} \Rightarrow b_{i,j,q,m} = \texttt{NNFTemplate}([z_{i,j,q,1}, .., z_{i,j,q,2\lambda}], [e_{i,j,q,1}, .., e_{i,j,q,\mu}])$$

$$|g_m \text{ is undefined}\}$$

$$\texttt{FixedForFunctions} = \texttt{PresentCons} \wedge$$

$$\neg \bigvee \texttt{AbsentE} \cup \texttt{AbsentELabel} \cup \texttt{AbsentNLabel} \cup \texttt{AbsentP}$$

$$\texttt{SynthFunction}(z, \lambda) =$$

$$(\texttt{Consistancy} \wedge \texttt{Property} \wedge \texttt{FixedForFunctions} \wedge \texttt{FindFunctions}(slimit))$$

We construct $\texttt{SynthFunction}(z, \lambda)$ similar to the earlier variations. By reading of the values of z in a satisfying model of the formula, we learn the synthesized function.

3.5 Fixing VTS by Both Adding/Deleting Parts

Now we will consider repairing of VTS by allowing not only addition but also deletion of the molecules, edges, functions, or pairing matrix. We have encoded the repairing in our tool by introducing flip bits for each variable that is modifiable in the VTS. We illustrate the repairing on one class of variables and rest can be easily extended. Let us consider repairing of node labels. For each bit $n_{i,m}$, we create a bit $flip_{i,m}$. We add constraints that take xor of VTS assigned values for $n_{i,m}$ and $flip_{i,m}$. We also limit the number of $flip_{i,m}$ that can be true, therefore limiting the number of flips. The above constraints are encoded in $\texttt{FlipN}(slimit)$.

$$\texttt{FlipN}(slimit, flip) = \bigwedge \{ n_{i,m} \oplus flip_{i,m} | m \in L(i) \wedge i \in N \} \wedge$$

$$\bigwedge \{ \neg n_{i,m} \oplus flip_{i,m} | m \notin L(i) \wedge i \in N \} \wedge \sum_{i,m} flip_{i,m} \leq slimit$$

Similar to the earlier variations, we construct `SynthRepairNode`($slimit$) for the repair. In that, `FixedForNodeRepair` encodes all the parts of VTS that do not change.

`FixedForNodeRepair` $=$ `PresentE` $\wedge$ `PresentP` $\wedge$ `KnownActiveE`$\wedge$

$\quad$ `UnknownActive` $\wedge$ `KnownActiveN` $\wedge \neg \bigvee$ `AbsentE` $\cup$ `AbsentNLabel` $\cup$ `AbsentP`

`SynthRepairNode`($slimit, flip$) $=$

$\quad$ (`Consistancy` $\wedge$ `Property` $\wedge$ `FixedForNodeRepair` $\wedge$ `FlipN`($slimit, flip$))

A satisfying model of `SynthRepairNode`($slimit, flip$) will assign some $flip$ bits to true. We will learn from the assignments the needed modifications in the VTS.

4 Implementation and Experiments

We have implemented the encodings in a tool called VTSSYNTH[1]. The tool takes a partially defined VTS as input in a custom designed input language. The input is then converted to the constraints over VTS. The tool can not only synthesize the above-discussed queries, but also their combinations. For example, our tool can modify labels of nodes or edges while learning activity functions. Our tool is developed in C++ and uses Z3 [11] infrastructure for processing formulas. Since some of the formulas involve alternation of quantifiers over Boolean variables Z3 is not a suitable choice for those examples. We translate the formulas created by Z3 tool into a standard QDIMACS [12] format and use as an input for QBF solvers. We use DEPQBF [13] for solving of QBF formulas. Our tool includes about 7000 lines of code.

We have applied VTSSYNTH on six partially defined VTSs. The results are presented in Fig. 3 for both the solvers DEPQBFand Z3. To use Z3, we remove `Connected` constraints, such that the queries becomes quantifier-free. The experiments were done on a machine with Intel(R) Core(TM) i3-4030U CPU @ 1.90 GHz processor and 4 GB RAM with 30 min (1800 s) timeout. The first four VTSs are synthetic but inspire from literature for typical motifs in VTSs. The third VTS is a subgraph of the last VTS. The fifth VTS is taken from [14]. The last VTS represent mammalian SNARE map created by studying the literature references.

The table shows timing for various synthesis queries. For each synthesis query, we have two columns. One column reports the timing and the other reports the minimum changes needed to obtain a valid VTS. ∞ indicates that any number of changes with the synthesis query search space can obtain the VTS. In the table, we are reporting five synthesis queries The first one only adds new labelled edges to the graph. We have ranked the all possible graph edits with the simple rank

---

[1] https://github.com/arey0pushpa/pyZ3.

of minimum updates. The second query adds new labels to the edge. The third query synthesizes NNF Boolean functions only containing ∧ and ∨ gates for activity functions, while allowing more edges to be added. The result shows the basic template of 4 leaves and 3 gates. To illustrate the versatility of our tool, the fourth query synthesizes 3-CNF functions (encoding not presented). Finally, we report queries that allows both addition and deletion of edges, and labels of node and labels.

5   Related Work

In recent years, there has been a wide range of methods developed for the similar synthesis problems [15–17]. They range from filling gaps an implementation of C programs from the pool of template predicates to learn a program from example runs of the program. In the course of developing such methods, the background technology, i.e. solving of quantified constraints has been evolving rapidly [13,18].

There has been some work in applying synthesis technique in biology especially in gene regulatory networks [19,20]. A very recent work [20] synthesize executable gene regulatory networks from single-cell gene expression data. Synthesis technique is also used in optimal synthesis for chemical reaction networks [21]. The [20] uses constraint (satisfiability) solving techniques for the synthesis whereas [19] uses SMT for synthesis. The paper [21] in addition to using SMT over ODE, uses a template-guided approach. In our case queries contain quantifiers so we have employed QBF solving with Z3 for the solving the synthesis problem. To our best knowledge, this is the first application of synthesis in VTS.

6   Conclusion

In this paper, we presented encodings of the synthesis problems that may arise from VTSs. We demonstrated that our tool based on the encodings scale up to the relevant sizes of the VTSs for some synthesis queries. Our tool timed out on larger examples. We are working to improve the performance of our tool. We will take this tool to the biologists and develop wet experiments that may validate some synthesis results from the tool. Our model of VTSs is static graphs. In future, we will study the dynamic behaviors of VTSs. It will allow us to predict behaviors after the perturbations in the VTSs and more ways to test the predicted synthesis results.

Appendix

A   Discussion on the Choice of VTS Model

The molecules transported by the VTS are themselves its regulators. The molecules in a compartment/vesicle may be active or inactive. The molecules that are responsible for vesicle fusion are called SNARE proteins [8, 22]. Active SNAREs present on vesicles (v-SNAREs) bind with their cognate active SNAREs on the target compartment (t-SNAREs) to enable vesicle fusion. A cell contains multiple kinds of v- and t-SNAREs. Only specific pairs of v and t SNAREs can bind to each other and participate in fusion. Fusion compatible v- and t- SNAREs are determined by biological experiments. Different vesicle-compartment fusions in the cell are brought about by different v- and t-SNARE pairs. A molecule that participates in a given fusion reaction must not interfere with fusion at different compartments or vesicles. Therefore, SNAREs must be kept in an inactive form in appropriate compartments/vesicles. The activity of molecules is regulated by the other molecules, i.e., the presence and absence of the other molecules in a compartment or vesicle may make the molecule active or inactive. We call this regulation as activity functions. In the VTS model, we assume that the system is in steady state and the concentrations of the molecules in the compartments do not change over time. We define SNARE pairing specificity by a fusion pairing relation containing pairs of SNAREs and molecular regulation by activity Boolean functions. Since the system is in steady state, we expect that any molecule that leaves a compartment must come back via some path on the graph. We call this property of VTS as stability.

Our model is inspired by [9]. On the timescales of minutes, our following assumptions reasonably capture the important aspects of the Rothman-Schekman-Sudhof (RSS) model [23] of vesicle traffic system.

1. A cell is a set of compartments exchanging vesicles.
2. Compartments are neither created nor destroyed.
3. Each compartment is in steady state, gain and loss balance.
4. Molecules are neither created nor destroyed.
5. Molecules move via vesicles of uniform size.
6. Identical vesicles have identical target compartments.
7. Fusion of vesicles to compartments is driven by specific SNARE pairing.
8. The activity of a SNARE can be regulated by other molecules present on the same compartment or vesicle.
9. An active SNARE pair is necessary and sufficient for fusion.

SNARE proteins are the agents of vesicle fusion in eukaryotic cells. When SNAREs on vesicles (v-SNAREs) encounter their cognate SNAREs on target compartments (t-SNAREs), they form SNARE complexes [8], and a single SNARE complex releases enough energy to enable membrane fusion [24]. SNAREs are identified by the presence of a conserved 60–70 stretch of amino acids called the SNARE motif. Based on amino acid sequence, SNARE motifs fall into 4 classes: Qa, Qb, Qc, and R [8]. Across all intracellular vesicle fusion reactions, the associated SNARE complexes contain one of each of the four kinds of SNARE motifs; the v-SNARE contributes a single SNARE motif, usually it is an R-SNARE (although, exceptions are known: Sec22b and Ykt6 are both R-SNAREs which form parts of t-SNAREs [25]) and the rest of the three SNARE motifs are contributed by the t-SNARE. In the cell, different vesicle fusion reactions are associated with distinct v- and t-SNARE pairs.

The paper [9] consider three Q SNARES as a single molecule, we have extended this model by considering each complex molecule as distinct. In contrast to the [9], we allow Q and R-SNARE type distribution across the whole system to be uneven. In our model fusion is driven by an active combination of three Q SNARE and one R SNARE molecule. We have relaxed the pairing matrix constraint to comply with this fact. For biological efficiency and optimality reasons, we do not allow self-edges to be present in the VTS.

B   The Natural VTSs

Here we will present the two VTS collected from the literature.

B.1   Mammalian VTS

The Fig. 1 represent mammalian SNARE map created by studying the wide array of literature. To construct the map, we have assumed that vesicles only contain a single active v-SNARE, and we have attributed t-SNAREs and inactive v-SNAREs that travel between compartments to one of the known vesicles that go between the same source and target compartments. In order to identify the active SNARE complex involved in any particular vesicle fusion, we used two criteria. The SNARE complex is formed *in vivo*. In most papers, this is determined by immunoprecipitation of the SNARE complex from the relevant cell fraction. Blocking SNARE complex formation (for example, using antibodies against these

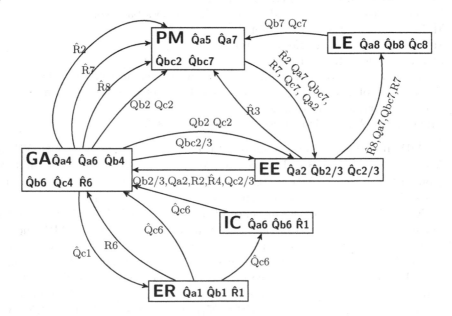

Fig. 1. A found-in-nature VTS. Nodes and edges are labelled with sets of molecules. ˆindicates that the molecule is active.

SNAREs, or using cytosolic forms of these SNAREs) blocks the specific transport step. Note that these vesicles have been collected from multiple cell types, and any given cell type is likely to contain only a subset of the vesicles in the map.

In this figure, the rectangles represent compartments, the identities of compartments are written within ER = endoplasmic reticulum, ERGIC = ER-Golgi intermediate compartment, RE = recycling endosome, EE = early endosome, LE=late endosome, LYS = lysosome, PM = plasma membrane. The arrows represent vesicle edges.

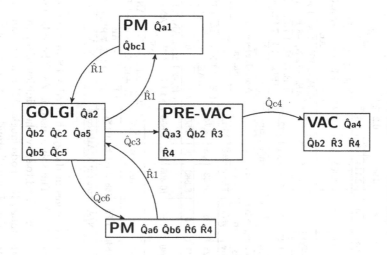

Fig. 2. Yeast VTS

B.2   Yeast VTS

In Fig. 2, we present the yeast VTS. We have borrowed the VTS from [14]. It has been adapted from the paper by separating the v and the t SNAREs. It is clear that it is an incomplete description of the VTS. For example, the inactive molecules were not reported in the reference. We are currently searching for more literature that can help us complete all known information about the VTS.

Table a

Table a	Add edge		Add molecules		Learning NNF (only ∧ and ∨)		Learning k-CNF		Add/Delete parts	
	Time	#C	Time	#C	Time	#C	Time	#C	Time	#C
plos1-dia[3C]	0.326	∞	0.312	∞	0.669	∞	0.966	∞	0.277	-1 E, -1 AE, -1 AN. +1 E, +1 N.
plos2-dia[4C]	0.266	0	0.322	0	1.409	0	2.114	0	0.337	0
sub-mammal[3C]	0.767	1 E	1.049	5 PE	3.523	1E	4.961	1E	1.172	-1 E, -2 PE, -1 AN. +1 E, +4 PE, +4 N, +2 AN, +2 AE.
node4[3C]	1.554	1 E	3.859	12 PE	5.286	∞	4.502	∞	2.194	-2 E, -2 PE, -1 N, -1 AN, -1 AE. +12 N, +8 E, +1 PE.
yeast-graph[3C]	95.016	1 E	timeout	N/A	1571.42	1 E	530.210	1 E	72.316	-1 E, -1 N, -1 AE, -1 AN, -1PE. +1 E, 7 PE, 8 N.
mammal-graph[3C]	timeout	N/A	timeout	N/A	timeout	N/A	timeout	N/A	timeout	N/A

Table b

Table b	Add edge		Add molecules		Learning NNF (only ∧ and ∨)		Learning k-CNF		Add/Delete parts	
	Time	#C	Time	#C	Time	#C	Time	#C	Time	#C
plos1-dia	0.041	∞	0.320	∞	0.225	∞	0.33	∞	3.74	-1E, -1PE, - 1 N, -1 PE. +1 AE, +1 PE, +1 N
plos2-dia	3.97	0	2.647	0	5.941	0	5.680	0	3.56	0
sub-mammal	3.483	1 E	4.379	5 PE	29.980	1 E	10.405	1 E	3.650	-1 E, -2 PE, -1 AN. +1 E, +4 PE, +4 N, +2 AN, +2 AE
node4	4.150	1 E	10.562	12 PE	3.401	∞	4.760	∞	5.05	-2 E, -2 PE, -1 N, -1 AN, -1 AE. +12 N, +8 E, +1 PE
yeast-graph	40.225	1 E	timeout	N/A	1393.84	1 E	468.161	1 E	69.81	-1 E, -1 N, -1 AE, -1 AN, -1PE. +1 E, 7 PE, 8 N.
mammal-graph	timeout	N/A	timeout	N/A	timeout	N/A	timeout	N/A	timeout	N/A

Fig. 3. Run-times for synthesis queries. #C stands for minimum changes in the synthesized VTS in comparison with the given partial VTS. Time is reported in seconds. (a) The solver used is DepQBF (b) The solver used is Z3. The sub-mammal is a subgraph of the complete mammal-graph. In the Add/Delete parts column, '+'n sign is used to show the addition of n number of the molecules, similarly '-'n is used to show the removal of n number of molecules. In the table, N is node labels, AN is active node molecules, E is edges, PE is molecule presence on the edge and AE is active molecules on the edge. The [kC] stands for k graph connectedness which is part of only DepQBF experiments.

References

1. Alberts, B., et al.: Essential Cell Biology. Garland Science, Auerbach (2013)
2. Bonifacino, J.S., Glick, B.S.: The mechanisms of vesicle budding and fusion. Cell **116**(2), 153–166 (2004)
3. Hong, W.J., Lev, S.: Tethering the assembly of snare complexes. Trends Cell Biol. **24**(1), 35–43 (2014)
4. Chanaday, N.L., Kavalali, E.T.: How do you recognize and reconstitute a synaptic vesicle after fusion? F1000Research **6**, 1734 (2017)
5. D'Agostino, M., Risselada, H.J., Lürick, A., Ungermann, C., Mayer, A.: A tethering complex drives the terminal stage of snare-dependent membrane fusion. Nature **551**(7682), 634 (2017)
6. Rodepeter, F.R.: Indication for differential sorting of the rat v-SNARE splice isoforms VAMP-1a and -1b. Biochem. Cell Biol. **95**(4), 500–509 (2017)
7. Zhao, Y., Holmgren, B.T., Hinas, A.: The conserved SNARE SEC-22 localizes to late endosomes and negatively regulates RNA interference in Caenorhabditis elegans. RNA **23**(3), 297–307 (2017)
8. Jahn, R., Scheller, R.H.: SNAREs-engines for membrane fusion. Nat. Rev. Mol. Cell Biol. **7**(9), 631 (2006)
9. Shukla, A., Bhattacharyya, A., Kuppusamy, L., Srivas, M., Thattai, M.: Discovering vesicle traffic network constraints by model checking. PloS ONE **12**(7), e0180692 (2017)
10. Gupta, A., Shukla, A., Srivas, M., Thattai, M.: SMT solving for vesicle traffic systems in cells. In: SASB (2017)
11. de Moura, L., Bjørner, N.: Z3: an efficient SMT solver. In: Ramakrishnan, C.R., Rehof, J. (eds.) TACAS 2008. LNCS, vol. 4963, pp. 337–340. Springer, Heidelberg (2008). https://doi.org/10.1007/978-3-540-78800-3_24
12. Qbflib.org: QDIMACS Standard Version 1.1 (2018)
13. Lonsing, F., Biere, A.: DepQBF: a dependency-aware QBF solver. J. Satisf. Boolean Model. Comput. **7**, 71–76 (2010)
14. Burri, L., Lithgow, T.: A complete set of snares in yeast. Traffic **5**(1), 45–52 (2004)
15. Solar-Lezama, A.: The sketching approach to program synthesis. In: Hu, Z. (ed.) APLAS 2009. LNCS, vol. 5904, pp. 4–13. Springer, Heidelberg (2009). https://doi.org/10.1007/978-3-642-10672-9_3
16. Alur, R., et al.: Syntax-guided synthesis. In: Formal Methods in Computer-Aided Design, FMCAD 2013, Portland, OR, USA, 20–23 October 2013, pp. 1–8 (2013)
17. Gulwani, S.: Automating string processing in spreadsheets using input-output examples. In: Proceedings of the 38th ACM SIGPLAN-SIGACT Symposium on Principles of Programming Languages, POPL 2011, Austin, TX, USA, 26–28 January 2011, pp. 317–330. ACM (2011)
18. de Moura, L., Bjørner, N.: Efficient E-matching for SMT solvers. In: Pfenning, F. (ed.) CADE 2007. LNCS (LNAI), vol. 4603, pp. 183–198. Springer, Heidelberg (2007). https://doi.org/10.1007/978-3-540-73595-3_13
19. Shavit, Y.: Automated synthesis and analysis of switching gene regulatory networks. Biosystems **146**, 26–34 (2016)
20. Fisher, J., Köksal, A.S., Piterman, N., Woodhouse, S.: Synthesising executable gene regulatory networks from single-cell gene expression data. In: Kroening, D., Păsăreanu, C.S. (eds.) CAV 2015. LNCS, vol. 9206, pp. 544–560. Springer, Cham (2015). https://doi.org/10.1007/978-3-319-21690-4_38

21. Cardelli, L., et al.: Syntax-guided optimal synthesis for chemical reaction networks. In: Majumdar, R., Kunčak, V. (eds.) CAV 2017. LNCS, vol. 10427, pp. 375–395. Springer, Cham (2017). https://doi.org/10.1007/978-3-319-63390-9_20

22. Wickner, W., Schekman, R.: Membrane fusion. Nat. Struct. Mol. Biol. **15**(n7), 658 (2008)

23. Rothman, J.E.: The machinery and principles of vesicle transport in the cell. Nat. Med. **8**(10), 1059–1063 (2002)

24. Van Den Bogaart, G., Holt, M.G., Bunt, G., Riedel, D., Wouters, F.S., Jahn, R.: One snare complex is sufficient for membrane fusion. N. Struct. Mol. Biol. **17**(3), 358 (2010)

25. Hong, W.: Snares and traffic. Biochimica et Biophysica Acta (BBA)-Mol. Cell Res. **1744**(2), 120–144 (2005)

Formal Analysis of Network Motifs

Hillel Kugler[1]($\boxtimes$), Sara-Jane Dunn[2,3], and Boyan Yordanov[2]($\boxtimes$)

[1] Bar-Ilan University, Ramat Gan, Israel
hillelk@biu.ac.il
[2] Microsoft Research, Cambridge, UK
{sara-jane.dunn,yordanov}@microsoft.com
[3] Wellcome Trust-Medical Research Council Stem Cell Institute,
University of Cambridge, Cambridge, UK

Abstract. A recurring set of small sub-networks have been identified as the building blocks of biological networks across diverse organisms. These network motifs have been associated with certain dynamical behaviors and define key modules that are important for understanding complex biological programs. Besides studying the properties of motifs in isolation, existing algorithms often evaluate the occurrence frequency of a specific motif in a given biological network compared to that in random networks of similar structure. However, it remains challenging to relate the structure of motifs to the observed and expected behavior of the larger network. Indeed, even the precise structure of these biological networks remains largely unknown. Previously, we developed a formal reasoning approach enabling the synthesis of biological networks capable of reproducing some experimentally observed behavior. Here, we extend this approach to allow reasoning about the requirement for specific network motifs as a way of explaining how these behaviors arise. We illustrate the approach by analyzing the motifs involved in sign-sensitive delay and pulse generation. We demonstrate the scalability and biological relevance of the approach by revealing the requirement for certain motifs in the network governing stem cell pluripotency.

Keywords: Biological programs · Formal analysis · Network motifs

1 Introduction

Network motifs [2,20] are basic interaction patterns that recur throughout biological networks, where they are observed more frequently than in random networks with similar properties (e.g. a comparable number of components and interactions). The same small set of network motifs appears to serve as the building blocks of biological networks for diverse organisms [3,17,32]. Each network motif can operate as an elementary circuit with a well-defined function, which is integrated within a larger network and has a role in performing the required information processing [22]. Since the introduction of the concept of network motifs and the identification and experimental validation of initial instances [2],

© Springer Nature Switzerland AG 2018
M. Češka and D. Šafránek (Eds.): CMSB 2018, LNBI 11095, pp. 111–128, 2018.
https://doi.org/10.1007/978-3-319-99429-1_7

a wide range of additional motifs with new roles have been uncovered. Network motifs have been identified within transcriptional networks [16,27], signaling networks [3], neuronal networks [25], and metabolic networks [24]. In addition to biological networks, recurring motifs have been identified in engineered systems, including electronic circuits and the world wide web [20].

The study of network motifs provides an attractive research direction towards understanding complex biological programs and uncovering modularity and reusable patterns of computation in the design of biological circuits. While many of the associated problems are challenging, especially when dealing with large biological networks, a wide range of computational methods have been developed [12,29]. For example, novel motifs have been identified by comparing the occurrence frequency of a sub-network in a known biological network to that in random networks of similar structure, while various graph methods have been applied to algorithmically scan a network for specific motifs. Substantial research effort has been devoted to dealing with the algorithmic challenges of network motif identification and the related graph algorithms [1,18,23,29]. This has led to the development of a number of computational tools, including mfinder [12], MAVisto [26], NeMoFinder [6], FANMOD [30], Grochow-Kellis [9], Kavosh [11], MODA [31], NetMODE [15], Acc-MOTIF [19] and QuateXelero [13] (see also [29] for a review and detailed comparisons). Verification techniques have also been applied to study network motifs. In [5], certain motifs and their dynamic properties were characterized using temporal logic, and parallel model checking was used to verify properties of networks with around ten components. In [10], approximate methods for analyzing gene regulatory networks were developed utilizing network motifs.

Following the identification of biologically-relevant motifs and the exploration of their dynamical properties in isolation, understanding how their presence or absence within a larger biological network defines that network's behavior becomes a central problem. This problem is compounded by the fact that the precise structure of such biological networks often remains largely unknown, due to noisy and sometimes irreproducible experimental data. This makes it challenging to search for motifs within the network or to explore the connections between a network's structure and its behavior.

Previously, we developed a formal reasoning approach enabling the synthesis and analysis of biological networks (e.g. incorporating gene regulation, signaling, etc.) that were only partially known [8,33]. The method, summarized briefly in Sect. 2.1, introduced the concept of an Abstract Boolean Network as a formalism for describing discrete dynamical models of biological networks, where the precise interactions or update rules were unknown. These models could then be constrained with specifications of some required behaviors, thereby providing a characterization of the set of all networks capable of reproducing some experimental observations.

In this paper, we extend the approach from [8,33] to enable automated reasoning about the requirement for specific network motifs as part of a biological network that is only partially known. This allows us to incorporate within the same framework constraints relating to the structure of the network, represented as logical formulas over the presence or absence of different motifs, together with constraints about the network's dynamic behavior. Our reasoning approach then allows us to draw conclusions about certain motifs being essential or disallowed for reproducing the required behavior, thus helping explaining how the observed behaviors arise from various motifs.

We illustrate the approach by analyzing the motifs involved in sign-sensitive delay and pulse generation – two distinct behaviors that have been associated with certain network structures and observed biological properties [16,17]. We consider a generic 3-layered network topology that serves as a prototype for a variety of biological programs and find that under the qualitative, Boolean modeling formalism of [8,33], positive feedback is required to implement both behaviors.

To demonstrate the scalability and biological relevance of the approach, we apply the proposed method to identify the motifs requirements in the network governing stem cell pluripotency [7,8]. This reveals that positive feedback, as well as a particular incoherent feed-forward motif, is also essential for maintaining pluripotency in the qualitative model.

We envision that the method proposed in this paper will provide a powerful tool for researchers interested in exploring the structural properties of biological networks and understanding how different motifs lead to various biological behaviors. In the future, this tool could support theoretical studies, where the connections between network structure and function are explored, as well as experimental studies, for example allowing researches to focus on the core, essential modules of biological networks.

2   Methods

In the following, we introduce some notation and summarize the approach from [33], which is implemented in the computational tool RE:IN (Sect. 2.1) and serves as a foundation for the extensions we propose in this paper (Sects. 2.2 and 2.3).

2.1   Abstract Boolean Network Analysis

Following the notation from [33], an *Abstract Boolean Network (ABN)* is a tuple $\mathcal{A} = (C, I, I^?, r)$, where

- C is the finite set of components,
- I is the set of definite (positive and negative) interactions between the components from C,
- $I^?$ is the set of possible (positive and negative) interactions, and
- r assigns a subset of regulation conditions (possible update functions) to each component from C.

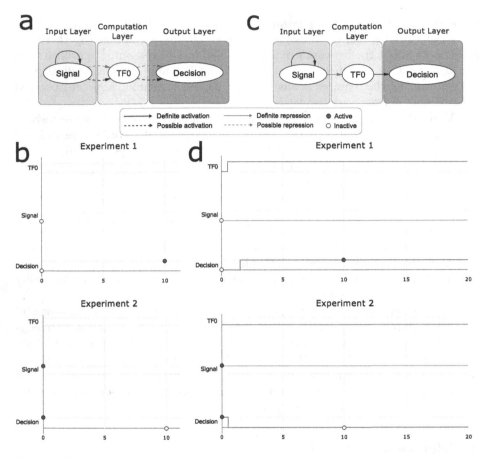

Fig. 1. Abstract Boolean Networks constrained against experimental observations. *(a)* The generic network architecture we consider is comprised of three layers (Input/Computation/Output). The input and output layers each include a single component, while the number of computation components n can be varied (in this case $n = 1$). This simple ABN includes 4 optional interactions and 1 definite interaction. Formally, the ABN is defined as $\mathcal{A} = (C, I, I^?, r)$, where $C = \{\text{Signal}, \text{TF0}, \text{Decision}\}$, $I = \{(\text{Signal}, \text{Signal}, +)\}$, $I^? = \{(\text{Signal}, \text{TF0}, +), (\text{Signal}, \text{TF0}, -), (\text{TF0}, \text{Decision}, +), (\text{TF0}, \text{Decision}, -)\}$, and r allows all regulation conditions for each component. *(b)* Experimental constraints encode expected states along different network trajectories. Here, two experimental constraints are illustrated, which specify the initial state of the signal and decision components, and their state at step 10. *(c)* A single, concrete network that is consistent with the constraints in (b) is generated using our SMT-based approach. *(d)* Experiment trajectories from the concrete model in (c) illustrate how the experimental constraints are satisfied by this network.

ABNs are discrete, dynamic models suitable for studying biological systems, when often the existence of interactions between components are hypothesized, but not definitively known [8,33]. An example of an ABN is illustrated in Fig. 1a.

Each component $c \in C$ describes a different chemical signal, protein, gene, etc. that can exist in one of two states: active or inactive. The dynamics of the system are defined by the regulation condition assigned to each component, which serves as an 'update function' that specifies the state of the component at step $k + 1$, given the state of all of its regulators (other components $c' \in C$ with interactions to c) at step k. Here we consider synchronous updates, such that deterministic trajectories emerge from each initial state, though RE:IN also allows the exploration of asynchronous trajectories.

ABNs are abstract models because of the uncertainty in the precise network topology and regulation rules for each component. An ABN is transformed into a concrete Boolean Network by instantiating a subset of the possible interactions, discarding all other optional interactions, and assigning a specific regulation condition for each gene. By virtue of the unique combination of interactions and regulation conditions, different concrete models derived from the same ABN can have different dynamical behaviors.

The concept of a *Constrained Abstract Boolean Network (cABN)* was introduced in [33] as a formalism for describing a set of Boolean Network models that are consistent with some experimentally observed biological behaviors. A cABN is defined in terms of an ABN, together with a set of constraints over the states of the components from C. These constraints encode experimental observations, where separate executions of the system correspond to different biological 'experiments'. For example, the observations encoded in Fig. 1b specify a biological program where cells make a particular decision only in the absence of some signal. Experiment 1 requires that there exists a trajectory, where both the 'Signal' and 'Decision' components are initially inactive and 'Decision' becomes active at step 10. Similarly, Experiment 2 requires that there exists a trajectory, where both the 'Signal' and 'Decision' components are initially active and 'Decision' becomes inactive at step 10. These constraints limit the feasible assignments of regulation conditions and possible interactions, such that all concrete networks from the cABN are guaranteed to reproduce all experimental observations.

cABN analysis was solved in [33] by encoding it as a Satisfiability Modulo Theories (SMT) problem. This enables the enumeration of individual concrete models that are consistent with the experimental observations. For example, the concrete model from Fig. 1c is generated from the ABN in Fig. 1a and is consistent with the constraints from Fig. 1b. This is demonstrated using the trajectories visualized in Fig. 1d. The SMT-based approach from [33] also allows reasoning about hypotheses describing unknown biological behaviors to make novel predictions from all consistent models collectively, without the need to enumerate individual concrete networks.

In addition to the above, the analysis can reveal both required and disallowed interactions of the cABN. An interaction $i \in I^?$ is required if the experimentally observed behavior cannot be reproduced without it (i.e. all concrete models of the cABN include the interaction i). Similarly, an interaction is disallowed if including it in a concrete model means that the observed behavior can no longer be reproduced. The analysis of required and disallowed interactions yields insight into how network structures lead to certain dynamic behaviors, and is the starting point for the extensions we propose here.

In the following, we use set notation to denote the existence or non-existence of interactions in an ABN or a cABN. For example, $(c, c', +) \in I$ denotes that a definite, positive interaction exists in $\mathcal{A}$, $(c, c', -) \in I^?$ denotes that an optional, negative interaction exists, and $(c, c', *) \notin I$ denotes that no definite interactions (i.e. the wild card $*$ stands for either $+$ or $-$) exist between c and c'.

2.2 Motif Assignment

Definition 1 (Motif). *A motif is a tuple $\mathcal{M} = \{C, I, I^?\}$, where C is the finite set of components, I is the set of definite and $I^?$ the set of possible interactions (similarly to the definition of ABNs).*

Examples of several different motifs are illustrated in Fig. 3. In contrast to ABNs, motifs (Definition 1) are static networks, without regulation conditions (update functions) to make them dynamical systems. However, because interactions from $I^?$ are uncertain, motifs are abstract – a motif defined as in Definition 1 with a non-empty $I^?$ describes a set of $2^{|I^?|}$ concrete, static networks.

Definition 2 (Motif Assignment). *Given an ABN $\mathcal{A} = (C_{\mathcal{A}}, I_{\mathcal{A}}, I^?_{\mathcal{A}}, r)$ and a motif $\mathcal{M} = \{C_{\mathcal{M}}, I_{\mathcal{M}}, I^?_{\mathcal{M}}\}$ a motif assignment is a map $\theta : C_{\mathcal{M}} \to C_{\mathcal{A}}$.*

Note that Definition 2 can also be applied to cABNs instead of ABNs by omitting the set of experimental observations from the cABN.

Given an ABN $\mathcal{A} = (C_{\mathcal{A}}, I_{\mathcal{A}}, I^?_{\mathcal{A}}, r)$ and a motif $\mathcal{M} = \{C_{\mathcal{M}}, I_{\mathcal{M}}, I^?_{\mathcal{M}}\}$, let $\bar{I}_{\mathcal{A}} = I_{\mathcal{A}} \cup I^?_{\mathcal{A}}$ and $\bar{I}_{\mathcal{M}} = I_{\mathcal{M}} \cup I^?_{\mathcal{M}}$ denote the set of all interactions (definite and optional) in the ABN and the motif. Given a motif assignment $\theta : C_{\mathcal{M}} \to C_{\mathcal{A}}$, let $I_{\mathcal{A}, \theta, \mathcal{M}} = \{(\theta(c), \theta(c'), *) \in I_{\mathcal{A}} \mid c \in C_{\mathcal{M}} \wedge c' \in C_{\mathcal{M}}\}$ denote the set of definite interactions from the ABN between components that the motif maps to. Similarly, let $I^?_{\mathcal{A}, \theta, \mathcal{M}} = \{(\theta(c), \theta(c'), *) \in I^?_{\mathcal{A}} \mid c \in C_{\mathcal{M}} \wedge c' \in C_{\mathcal{M}}\}$ denote the set of optional interactions from the ABN between components that $\mathcal{M}$ maps to[1].

Definition 3 (Valid Motif Assignment). *A given motif assignment θ between ABN $\mathcal{A}$ and motif $\mathcal{M}$ is valid if and only if*

1. $\forall (c, c', +) \in I_{\mathcal{A}, \theta, \mathcal{M}} . (\theta^{-1}(c), \theta^{-1}(c'), +) \in \bar{I}_{\mathcal{M}}$,
2. $\forall (c, c', -) \in I_{\mathcal{A}, \theta, \mathcal{M}} . (\theta^{-1}(c), \theta^{-1}(c'), -) \in \bar{I}_{\mathcal{M}}$,
3. $\forall (c, c', +) \in I_{\mathcal{M}} . (\theta(c), \theta(c'), +) \in \bar{I}_{\mathcal{A}}$, and
4. $\forall (c, c', -) \in I_{\mathcal{M}} . (\theta(c), \theta(c'), -) \in \bar{I}_{\mathcal{A}}$

The conditions from Definition 3.1–3.4 ensure that the motif components are assigned to ABN components in such a way that the interactions match. In other words, each definite (positive or negative) interaction in the ABN (between components that the motif maps to) matches an interaction (definite or optional) in

---

[1] While, in general, the motif assignment θ is not invertible, $\theta^{-1}(c)$ and $\theta^{-1}(c')$ can be defined for the interactions $(c, c', *) \in I_{\mathcal{A}, \theta, \mathcal{M}}$ and $(c, c', *) \in I^?_{\mathcal{A}, \theta, \mathcal{M}}$.

the motif (Definition 3.1–3.2) and each definite interaction in the motif matches an interaction in the ABN (Definition 3.3–3.4).

Given an optional interaction $(c, c', *) \in I_\mathcal{A}^?$, let $\mathcal{I}_{c,c'}^* \in \mathbb{B}$ denote the Boolean choice variable representing whether the interaction is included in a concrete model (see [33] for details of the SMT encoding of ABNs and cABNs). Asserting that $\mathcal{I}_{c,c'}^*$ is true can be interpreted as modifying the ABN such that $(c, c', *) \notin I_\mathcal{A}^?$ and $(c, c', *) \in I_\mathcal{A}$ (i.e. ensuring that the interaction is definitely present). Similarly, asserting that $\mathcal{I}_{c,c'}^*$ is false can be interpreted as modifying the ABN such that $(c, c', *) \notin I_\mathcal{A}^?$ but $(c, c', *) \notin I_\mathcal{A}$ (i.e. ensuring that the interaction is definitely absent).

The notion of a valid motif assignment (Definition 3) is sufficient to guarantee that the components of the motif are mapped to components of the ABN in such a way that all definite interactions are matched. However, it is possible that optional interactions of the ABN map to definite interactions of the motif or do not match any motif interactions. Therefore, while the interactions of the ABN match that of the motif, it is not possible to guarantee that every concrete network represented by the ABN matches the motif. The additional constraints defined in the following ensure that this is indeed the case.

Definition 4 (Motif Assignment Constraints). *Given a motif assignment θ between ABN $\mathcal{A}$ and motif $\mathcal{M}$, the motif assignment constraints are*

$$\begin{aligned}
\mathcal{C}_\theta = \{\mathcal{I}_{c,c'}^+ \mid (c, c', +) \in I_M \wedge (\theta(c), \theta(c'), +) \in I_\mathcal{A}^?\} \cup \\
\{\mathcal{I}_{c,c'}^- \mid (c, c', -) \in I_M \wedge (\theta(c), \theta(c'), -) \in I_\mathcal{A}^?\} \cup \\
\{\neg\mathcal{I}_{c,c'}^+ \mid (c, c', +) \notin \bar{I}_M \wedge (\theta(c), \theta(c'), +) \in I_\mathcal{A}^?\} \cup \\
\{\neg\mathcal{I}_{c,c'}^- \mid (c, c', -) \notin \bar{I}_M \wedge (\theta(c), \theta(c'), -) \in I_\mathcal{A}^?\}.
\end{aligned}$$

The additional constraints from (Definition 4) assert that an optional interaction in the ABN that matches a definite interaction of the motif is always included. Similarly, an optional ABN interaction that does not match any motif interaction is never included. These additional constraints guarantee that the interactions of all concrete networks of the ABN match those of the motif, under the given motif assignment.

2.3   Motif Constraints

The motif assignment constraints (Definition 4) ensure that a given motif $\mathcal{M}$ is implemented in all concrete networks of an ABN $\mathcal{A}$ between the specific components defined by the motif assignment θ. In general, however, we are interested in guaranteeing that motif $\mathcal{M}$ is implemented in the ABN $\mathcal{A}$ by any of its components rather than the the specific set of components specified by θ.

Definition 5 (Motif Constraints). *Given an ABN $\mathcal{A}$ and a motif $\mathcal{M}$, the motif constraints $\mathcal{C}_{\mathcal{M},\mathcal{A}}$ are defined in terms of the motif assignment constraints (Definition 4) as $\mathcal{C}_{\mathcal{M},\mathcal{A}} = \bigvee_{\theta \in \hat{\Theta}} \mathcal{C}_\theta$, where $\hat{\Theta}$ is the set of valid motif assignments between $\mathcal{A}$ and $\mathcal{M}$ (Definition 2).*

The motif constraints from Definition 5 guarantee that motif $\mathcal{M}$ is implemented by every concrete network of ABN $\mathcal{A}$, even though the precise components used to implement $\mathcal{M}$ might differ.

Given an ABN $\mathcal{A}$ and a set of motifs, logical formulas (e.g. $\neg\mathcal{M}, \mathcal{M} \vee \mathcal{M}', \mathcal{M} \wedge \mathcal{M}'$, etc) could be constructed and interpreted by replacing each motif $\mathcal{M}$ with its corresponding motif constraints $\mathcal{C}_{\mathcal{M},\mathcal{A}}$.

2.4   Implementation

As part of the RE:IN framework [8,33], a high-level, domain specific language was proposed for describing cABNs by defining the sets of components and interactions, as well as associated experimental observations. We implement the methods described in Sects. 2.2 and 2.3 as an extension of RE:IN that enables the reasoning about cABNs with additional structural constraints about the presence or absence of various motifs.

Currently, the generation of motif constraints (Definition 5) is implemented as a pre-processing step using a straightforward, exhaustive algorithm, where all motif assignments are first generated and then filtered to preserve only the valid ones using the conditions from Definition 3. Various cABN analysis problems are then encoded and solved using an SMT solver as shown previously [8,33], while the additional motif constraints are also incorporated.

Two notable modifications are introduced to the method described in Sects. 2.2 and 2.3 for improved usability. First, when the name of a motif component matches the name of a cABN component, no other assignments are considered for that component. This enables the specification of partially known motifs, where the mapping of some of the motif components to the cABN components is given. Second, a dummy 'Context' component is always included within the set of motif components. Given a motif assignment, the 'Context' component matches any cABN component that is not already mapped to by the motif. This provides additional control in specifying how a motif could be implemented as part of the cABN's network. For example, including optional positive and negative interactions from 'Context' to every motif component and vice versa does not impose additional constraints on the motif's implementation. Without any additional 'Context' interactions, on the other hand, the motif can only be fully isolated and disconnected from all other components of the cABN.

2.5   Reasoning About Motifs

Combining the previously-developed SMT-based reasoning strategies [8,33] with an encoding of the motif constraints from Sect. 2.3 enables the automated reasoning about structural (motif) properties of a network, together with the requirements about reproducing certain dynamical behaviors. Among the different analysis questions this method could support, in this work we focus specifically on identifying required (essential) and disallowed motifs. Similarly to the identification of required and disallowed interactions (Sect. 2.1), a motif $\mathcal{M}$ is required if the experimentally observed behavior could not be reproduced without it, while it is disallowed if enforcing that the motif is present in the network guarantees that the observed behavior can no longer be reproduced. These hypotheses can be tested as follows. Applying the SMT analysis of an ABN, if no concrete models are identified with the constraint $\mathcal{M}$ (the motif is present in the cABN) then the motif is disallowed. If, on the other hand, no concrete models are identified with the constraint $\neg\mathcal{M}$ (the motif is not present in the network), then the motif is essential.

3 Results

To illustrate the analysis method proposed in Sect. 2, we apply it to study the importance of certain motifs for biological networks. First, we consider a generic network architecture (Fig. 2a) composed of an input, computation, and output layer, which serves as a prototype for many biological programs (Sect. 3.1). We study the models consistent with this network topology that could give rise to two distinct dynamical behaviors (Fig. 2b) and identify the motifs from a given set (Fig. 3) that are required or disallowed for producing this behavior as described in Sect. 2.5. Then, in Sect. 3.2 we apply our motif analysis method to the recently identified biological program governing stem cell decision making [7], demonstrating the scalability and biological relevance of the approach.

3.1 Biological Program Prototype

We construct a simple abstract network topology in order to explore how various motifs give rise to different dynamical behaviors. The network has a single input component that represents some biochemical signal and a single output (readout) component. The output component might represent a biochemical signal that affects a downstream process or, as in this example, could represent a particular cellular decision (e.g. to differentiate, divide, etc). Information processing is performed in the 'computation' layer, which includes a number of components. While all interactions in the network are unknown, we assume that information flows from the input layer, through the computation layer, and into the output layer. As a result, we consider a network with a densely connected computation layer (possible positive and negative interactions between each pair of computation components). The input (signal) component might affect any of the computation components, so possible positive and negative interactions from the input (signal) to all computation components are included. Similarly, possible positive and negative interactions from each computation component to the output (decision) component are included. A definite self-activation is included for the signal to guarantee that once set at the beginning of computation, its value does not change, but no other self-regulation interactions are allowed. The resulting network architecture with $n = 3$ computation components is visualized in Fig. 2a. In all subsequent analysis, we impose the additional constraint that a positive and a negative interaction between the same components are never included together in concrete models.

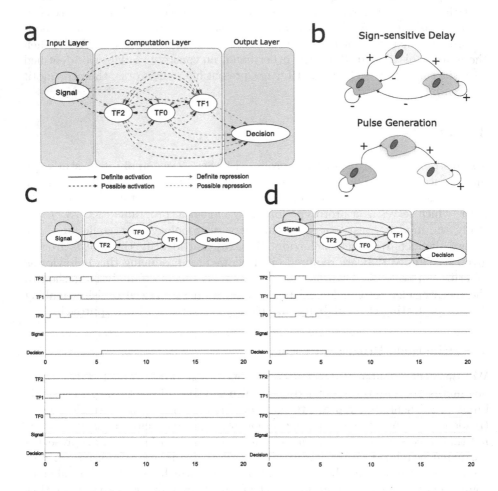

Fig. 2. A generic network architecture generates biological programs implementing either sign-sensitive delay or pulse generation. *(a)* The ABN we consider is comprised of three computation components, with optional positive and negative interactions between them. *(b)* The sign-sensitive delay and pulse generation behaviors are represented graphically as transitions between different cellular states. Signs on the edges indicate the presence (+) or absence (−) of input signal, while the output is active only in the cell type shown in green (right-most cell). *(c)* An example of a biological program implementing sign-sensitive delay has the characteristic delay during activation (top trajectories) but responds faster during deactivation (bottom trajectories). *(d)* An example of a biological program implementing pulse generation produces a single pulse of the output when the input signal is present (top trajectories), while the output remains inactive when the signal is not present (bottom trajectories). (Color figure online)

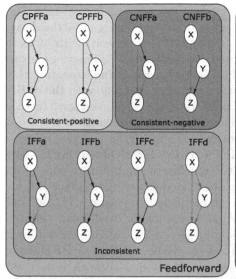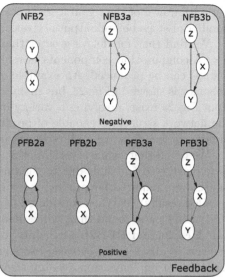

Fig. 3. The set of 15 network motifs we define and use to analyze biological programs. These are sub-divided into a set of feedforward and feedback motifs. An optional positive and negative interaction from the 'Context' component to every motif component, and from every motif component to 'Context', is also included but not shown in the figure. This ensures that the motif of interest can be identified in the ABN being explored regardless of how the motif components are connected within the network external to the motif (see Sect. 2.4).

Sign-Sensitive Delay. The first dynamical behavior we consider requires that a system produce an output in response to some input (e.g. the 'Decision' component becomes active if and only if 'Signal' is present in Fig. 2a). However, while the effect on the output is immediate when the signal is withdrawn, there is a delay on the output's activation when the signal is supplied[2]. Due to the asymmetric response to changes in the input signal, the behavior is called sign-sensitive delay [16,17] and was shown to have a potential role for making decisions based on noisy inputs by filtering out fluctuations in input stimuli [16,17].

We encode the requirement for a sign-sensitive delay as depicted in Fig. 2b. When no signal is present, the system can stabilize in a state where the output is inactive (shown in gray). When the signal is supplied, a transition to an intermediate state occurs (light green), although the output is still not activated. Withdrawing the signal at this point resets the system to the initial state, while continuous application of the signal leads to a transition to a state where the output is active (green). This active state is stable as long as the signal is present,

---

[2] Depending on the exact implementation, the delay can be observed when the signal switches from active to inactive instead, but this variation of a sign-sensitive delay is not considered here.

but withdrawal of the signal leads quickly back to the initial, inactive state. The number of steps before output activation upon supplying the signal is the delay.

We find that for delays greater than a single step, a network with at least $n = 3$ computation components is required, and with $n = 3$, delays of up to 4 steps can be produced. An example of a concrete network implementing this behavior is shown in Fig. 2c, but many such networks consistent with the cABN from Fig. 2a exist, involving a variety of network motifs. To investigate further the network structures capable of producing sign-sensitive delays, we defined a number of feed-forward and feed-back motifs shown in Fig. 3.

We found that, while several motifs (**IFFa, IFFc-d, PFB3a-b, NFB3a-b**) were disallowed (possible due to the limited number of computation nodes), no motifs were required for producing a sign-sensitive delay of 4 steps (see [14] for detailed results). We then considered pairs of motifs (e.g. by testing $\mathcal{M} \wedge \mathcal{M}'$ and $\neg \mathcal{M} \wedge \neg \mathcal{M}'$) and found that, while many pairs were jointly disallowed, only the pair **PFB2a** and **PFB2b** was jointly required (see [14]). This indicates that a positive feedback motif between two components (either **PFB2a** or **PFB2b**) is essential for implementing a 4-step sign-sensitive delay in a network with $n = 3$ computation components, but one of these motifs can be substituted for the other.

Pulse Generation. The second dynamical behavior we consider requires that a system produce a transient output pulse in response to some input. We encode this behavior as depicted in Fig. 2b. While no signal is present, the system remains stably in a state where the output is inactive (shown in gray). When the signal is supplied, a transition to an intermediate state occurs (green) where the output is activated. Further application of the signal causes a transition to a state where the output is no longer active (light green). Currently, no 'resetting' (i.e. withdrawal of the signal causing a transition to the initial state) is considered. The number of steps during which the output is active upon supplying the signal is the pulse width.

We find that for pulse widths greater than a single step, a network with at least $n = 3$ computation components is required, and with $n = 3$, pulse widths up to 4 steps can be produced. An example of a concrete network implementing this behavior is shown in Fig. 2d.

As in the previous case study, we were interested in exploring how the motifs from Fig. 3 affect the capacity for pulse generation. We found that, several motifs (**CPFFa, IFFc-d, PFB3a-b, NFB3a-b**) were disallowed and **PFB2b** was required. This indicates that a positive feedback motif between two components, implemented specifically through two negative interactions, is essential for implementing a generator for pulses of width 4 in a network with $n = 3$ computation components.

3.2   Stem Cell Pluripotency Program

The sign-sensitive delay and pulse generation examples demonstrate the utility of our approach in revealing how different network motifs give rise to different dynamic behaviors within a relatively simple network. Here we apply our analysis to a realistic biological network. Previously, we used RE:IN to study the biological program governing maintenance of naïve pluripotency: the property uniquely exhibited by embryonic stem cells (ESCs) to generate all adult cell types, as well as the germline. This capacity is lost as cells begin to differentiate into the different adult lineages throughout embryogenesis. While the critical transcription factors (TFs) that regulate pluripotency had been identified, and the culture environments required to sustain the cells progressively refined [21], it was not previously understood how environmental signals were processed by the core TFs to govern the pluripotent state. We investigated the biological program governing pluripotency by defining an ABN based on gene expression profiles of mouse ESCs, which was subsequently constrained against a rich set of behaviors derived from experiments in which ESCs are subject to different culture conditions and molecular perturbations [8]. This network has since been further refined with additional data [7] (Fig. 4a). It is of interest to learn from this complex starting set of networks whether there are specific motifs that are required to generate, and thereby explain, the experimental observations of pluripotency.

Our analysis revealed that three motifs are required in all concrete models, **IFFa, PFB2a, PFB3a**, and four are disallowed, **CNFFa, IFFc, NFB2, NFB3a**. The set of disallowed motifs is a trivial result, as none of these are present in the ABN, regardless of how the optional interactions are instantiated. In contrast, there are a number of possible configurations of components in the ABN that are equivalent to the three motifs identified as required. We subsequently tested the requirement for specific component assignments of these motifs, and found that there are four cases present in all concrete models that are consistent with the experimental constraints. The four required motifs are shown in Fig. 4b, revealing that these models all require both feed-forward and feed-back motifs between specific network components.

This example demonstrates how our approach scales to complex networks that explain critical aspects of cellular decision-making, and can reveal essential elements of these biological programs required to explain observed behavior. Further analysis could reveal whether the motifs we have identified serve to determine the robustness of the pluripotency network, which is known to vary between different culture conditions [8]. Furthermore, it would be of interest to study whether these required motifs are repeated throughout networks governing similar biological behaviors, and in particular, whether they arise in pluripotency networks from other mammals, revealing elements of pluripotency that are conserved between species.

4   Discussion

The SMT-based formal reasoning approach from [8,33] allows us to encode dynamic Boolean models of biological networks, where the precise set of interactions and regulation conditions for each component are unknown. These models can be constrained against specifications of experimental observations in order

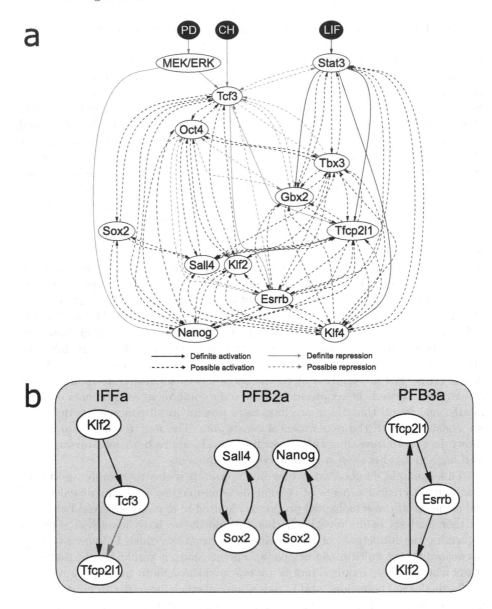

Fig. 4. Network motifs required for stem cell pluripotency. (a) The ABN defined for three input signals provided as culture conditions to mouse ESCs - LIF, CH and PD - and 13 downstream components that have been functionally validated as critical pluripotency factors [7]. (b) Four instantiations of the three motifs required to satisfy the experimental constraints.

to identify concrete models capable of reproducing the required behavior, or to test novel hypotheses without selecting a particular concrete model from the set of models consistent with the experiments.

Certain structural constraints can also be handled directly by the previous method. For example, assigning an interaction as definite guarantees that only concrete models that incorporate that interaction are considered. Similarly, removing an interaction as either optional or definite guarantees that none of the considered models incorporate this interaction. However, expressing more complex structural properties, such as those required for reasoning about motifs, is challenging using the approach from [8,33]. The extensions we propose in this work allow for complex properties describing the presence or absence of arbitrary motifs to be specified and tested, even when the network being studied is partially unknown. This leads to a natural framework for incorporating rich structural constraints and jointly reasoning about motifs and dynamical properties.

While the problem of identifying motifs does not scale favorably to large networks [4,28], in [8] it was demonstrated that relatively small networks of core components can explain a rich set of biologically-relevant behaviors and cellular decisions. Therefore, even the straightforward, exhaustive algorithm implemented currently to generate motif constraints (Sect. 2.4) proves suitable for examining networks of biological significance. Indeed, the computation times for generating motif constraints are negligible compared to the task of verifying whether any concrete models exist that satisfy all constraints (see detailed results in [14]). All analysis reported in this paper was accomplished on the order of seconds to minutes for simple networks (Sect. 3.1) and tens of minutes for the biological network governing stem cell pluripotency on a 3.6 GHz Intel Xeon (E5-1620) computer with 32 GB RAM (Sect. 3.2). Even so, many of the algorithmic advances towards more efficient identification of motifs in large networks [12,18,29] could be adapted as part of the pre-processing step of motif constraint generation.

An obvious limitation of the proposed method is that the cABN models being analyzed are qualitative (Boolean), discrete, and also deterministic, due to the synchronous update semantics we assume (although asynchronous updates are also supported by the method from [8,33]). Therefore, dynamical behaviors associated with motifs or networks that require more detailed modeling assumptions cannot be handled directly. Thus, certain properties, for example relating to noise propagation and attenuation, or precise timing of signals, are not currently supported by our motif analysis.

Still, a number of interesting biological questions can be framed in terms of the analysis of structural, motif-based properties of partially known networks with respect to the dynamical behaviors they produce. In this work, we focused specifically on identifying motifs that are essential or disallowed for producing certain behaviors. The proposed approach can also be used for more in-depth studies, for example in order to identify whether an essential motif must always involve specific components in the network, as illustrated in the stem cell case study from Sect. 3.2. This is achieved by testing whether concrete models without a given motif involving a particular network component exist. Further detailed studies could also explore not just the presence of absence of motifs, but also

how these motifs must be connected to the rest of the network (e.g. for a particular flow of information between components) in order to achieve the required
behavior by exploring different combinations of optional and definite interactions
between motif components and the 'Context'.

Another interesting question is what additional insights the motif analysis
proposed in this work can provide beyond the identification of required and disallowed interactions, already possible with the methods from [8,33]. Intuitively,
if a given motif must be implemented by specific components of a network,
then all the motif's interactions would be identified as required, which is the
case for some of the motifs from the stem cell case study (Sect. 3.2). However,
identifying motifs that are essential but appear in different places in different
concrete networks that are consistent with experimental observations, reveals
deeper connections between structure and behavior. The degree to which a network is constrained, either by reducing the number of optional interactions or by
specifying additional behaviors that must be reproduced, limits the possibility of
implementing motifs between different components and could lead to more predictions about essential or disallowed motifs. In contrast, larger, less-constrained
biological networks can achieve the same behavior in different ways and certain
motifs would no longer be required.

The case studies presented here illustrate how the proposed method can
be used for theoretical studies of the properties and requirements for different
motifs (Sect. 3.1) and demonstrate that the approach scales up to and provides
useful insights about realistic biological networks (Sect. 3.2). Extending these
studies and experimentally validating the motif requirements in these networks
is a direction for future research.

5   Summary

To deal with the challenge of studying the connections between the structure
and function of biological networks, even when these networks are only partially
understood, we extended the SMT-based reasoning approach RE:IN from [8,33].
The proposed methods involve the algorithmic generation of motif constraints,
encoding the requirement that a given motif is present in some cABN. Rich
structural requirements can then be incorporated in addition to the functional
properties encoded as part of the cABN through logical formulas over such motif
constraints. We illustrated the method by predicting that certain motifs are
essential or disallowed for producing sign-sensitive delay and pulse generation in
a network representing a prototype of biological programs, and in the biological
network governing stem cell pluripotency [7]. The proposed methods enable the
study of network motifs in the context of partially unknown, abstract networks
and can support future theoretical and experimental studies, where reasoning
about network structure and function in the same framework is essential.

References

1. Alon, N., Dao, P., Hajirasouliha, I., Hormozdiari, F., Sahinalp, S.C.: Biomolecular network motif counting and discovery by color coding. Bioinformatics **24**(13), i241–i249 (2008)
2. Alon, U.: An Introduction to Systems Biology: Design Principles of Biological Circuits. CRC Press, Boca Raton (2006)
3. Amit, I., et al.: A module of negative feedback regulators defines growth factor signaling. Nat. Genet. **39**(4), 503 (2007)
4. Babai, L., Luks, E.M.: Canonical labeling of graphs. In: Proceedings of the Fifteenth Annual ACM Symposium on Theory of Computing, pp. 171–183. ACM (1983)
5. Barnat, J., Brim, L., Cerna, I., et al.: From simple regulatory motifs to parallel model checking of complex transcriptional networks. Pre-proceedings of Parallel and Distributed Methods in Verification (PDMC 2008), Budapest, pp. 83–96 (2008)
6. Chen, J., Hsu, W., Lee, M.L., Ng, S.K.: NeMoFinder: dissecting genome-wide protein-protein interactions with meso-scale network motifs. In: Proceedings of the 12th ACM SIGKDD International Conference on Knowledge Discovery and Data Mining, pp. 106–115. ACM (2006)
7. Dunn, S.J., Li, M.A., Carbognin, E., Smith, A.G., Martello, G.: A common molecular logic determines embryonic stem cell self-renewal and reprogramming. bioRxiv, p. 200501 (2017)
8. Dunn, S.J., Martello, G., Yordanov, B., Emmott, S., Smith, A.: Defining an essential transcription factor program for naïve pluripotency. Science **344**(6188), 1156–1160 (2014)
9. Grochow, J.A., Kellis, M.: Network motif discovery using subgraph enumeration and symmetry-breaking. In: Speed, T., Huang, H. (eds.) RECOMB 2007. LNCS, vol. 4453, pp. 92–106. Springer, Heidelberg (2007). https://doi.org/10.1007/978-3-540-71681-5_7
10. Ito, S., Ichinose, T., Shimakawa, M., Izumi, N., Hagihara, S., Yonezaki, N.: Formal analysis of gene networks using network motifs. In: Fernández-Chimeno, M., et al. (eds.) BIOSTEC 2013. CCIS, vol. 452, pp. 131–146. Springer, Heidelberg (2014). https://doi.org/10.1007/978-3-662-44485-6_10
11. Kashani, Z.R.M., et al.: Kavosh: a new algorithm for finding network motifs. BMC Bioinform. **10**(1), 318 (2009)
12. Kashtan, N., Itzkovitz, S., Milo, R., Alon, U.: Efficient sampling algorithm for estimating subgraph concentrations and detecting network motifs. Bioinformatics **20**(11), 1746–1758 (2004)
13. Khakabimamaghani, S., Sharafuddin, I., Dichter, N., Koch, I., Masoudi-Nejad, A.: QuateXelero: an accelerated exact network motif detection algorithm. PLoS One **8**(7), e68073 (2013)
14. Kugler, H., Dunn, S.J., Yordanov, B.: Formal analysis of network motifs. bioRxiv (2018)
15. Li, X., Stones, D.S., Wang, H., Deng, H., Liu, X., Wang, G.: NetMODE: network motif detection without nauty. PLoS One **7**(12), e50093 (2012)
16. Mangan, S., Alon, U.: Structure and function of the feed-forward loop network motif. Proc. Nat. Acad. Sci. **100**(21), 11980–11985 (2003)
17. Mangan, S., Zaslaver, A., Alon, U.: The coherent feedforward loop serves as a sign-sensitive delay element in transcription networks. J. Mol. Biol. **334**(2), 197–204 (2003)

18. McKay, B.: Practical graph isomorphism. Congr. Numerantium **30**, 45–87 (1981)
19. Meira, L.A., Máximo, V.R., Fazenda, Á.L., Da Conceição, A.F.: acc-Motif: accelerated network motif detection. IEEE/ACM Trans. Comput. Biol. Bioinform. (TCBB) **11**(5), 853–862 (2014)
20. Milo, R., Shen-Orr, S., Itzkovitz, S., Kashtan, N., Chklovskii, D., Alon, U.: Network motifs: simple building blocks of complex networks. Science **298**(5594), 824–827 (2002)
21. Nichols, J., Smith, A.: Pluripotency in the embryo and in culture. Cold Spring Harb. Perspect. Biol. **4**(8), a008128 (2012)
22. Nurse, P.: Life, logic and information. Nature **454**(7203), 424–426 (2008)
23. Pržulj, N.: Biological network comparison using graphlet degree distribution. Bioinformatics **23**(2), e177–e183 (2007)
24. Ravasz, E., Somera, A.L., Mongru, D.A., Oltvai, Z.N., Barabási, A.L.: Hierarchical organization of modularity in metabolic networks. Science **297**(5586), 1551–1555 (2002)
25. Reigl, M., Alon, U., Chklovskii, D.B.: Search for computational modules in the C. elegans brain. BMC Biol. **2**(1), 25 (2004)
26. Schreiber, F., Schwöbbermeyer, H.: MAVisto: a tool for the exploration of network motifs. Bioinformatics **21**(17), 3572–3574 (2005)
27. Shen-Orr, S.S., Milo, R., Mangan, S., Alon, U.: Network motifs in the transcriptional regulation network of Escherichia coli. Nat. Genet. **31**(1), 64 (2002)
28. Shervashidze, N., Vishwanathan, S., Petri, T., Mehlhorn, K., Borgwardt, K.: Efficient graphlet kernels for large graph comparison. In: Artificial Intelligence and Statistics, pp. 488–495 (2009)
29. Tran, N.T.L., Mohan, S., Xu, Z., Huang, C.H.: Current innovations and future challenges of network motif detection. Brief. Bioinform. **16**(3), 497–525 (2015)
30. Wernicke, S., Rasche, F.: FANMOD: a tool for fast network motif detection. Bioinformatics **22**(9), 1152–1153 (2006)
31. Wong, E., Baur, B., Quader, S., Huang, C.H.: Biological network motif detection: principles and practice. Brief. Bioinform. **13**(2), 202–215 (2011)
32. Yeger-Lotem, E., et al.: Network motifs in integrated cellular networks of transcription-regulation and protein-protein interaction. Proc. Natl. Acad. Sci. U.S.A. **101**(16), 5934–5939 (2004)
33. Yordanov, B., Dunn, S.J., Kugler, H., Smith, A., Martello, G., Emmott, S.: A method to identify and analyze biological programs through automated reasoning. NPJ Syst. Biol. Appl. **2**(16010) (2016)

Buffering Gene Expression Noise by MicroRNA Based Feedforward Regulation

Pavol Bokes[1,2]([✉]), Michal Hojcka[1], and Abhyudai Singh[3]

[1] Department of Applied Mathematics and Statistics,
Comenius University, 84248 Bratislava, Slovakia
pavol.bokes@fmph.uniba.sk
[2] Mathematical Institute, Slovak Academy of Sciences, 81473 Bratislava, Slovakia
[3] Department of Electrical and Computer Engineering, University of Delaware,
Newark, DE 19716, USA

Abstract. Cells use various regulatory motifs, including feedforward loops, to control the intrinsic noise that arises in gene expression at low copy numbers. Here we study one such system, which is broadly inspired by the interaction between an mRNA molecule and an antagonistic microRNA molecule encoded by the same gene. The two reaction species are synchronously produced, individually degraded, and the second species (microRNA) exerts an antagonistic pressure on the first species (mRNA). Using linear-noise approximation, we show that the noise in the first species, which we quantify by the Fano factor, is sub-Poissonian, and exhibits a nonmonotonic response both to the species lifetime ratio and to the strength of the antagonistic interaction. Additionally, we use the Chemical Reaction Network Theory to prove that the first species distribution is Poissonian if the first species is much more stable than the second. Finally, we identify a special parametric regime, supporting a broad range of behaviour, in which the distribution can be analytically described in terms of the confluent hypergeometric limit function. We verify our analysis against large-scale kinetic Monte Carlo simulations. Our results indicate that, subject to specific physiological constraints, optimal parameter values can be found within the mRNA–microRNA motif that can benefit the cell by lowering the gene-expression noise.

1 Introduction

Gene regulatory circuits encode diverse mechanisms to counter stochasticity arising from low-copy numbers of circuit constituents. Perhaps the most well-known example of this is negative feedback realized via gene autoregulation, where an expressed protein inhibits its own transcription/translation [8,10,32,38,42,48]. While such negative autoregulation is quite ubiquitous for *E. coli* transcription factors [3], it is surprisingly rare for eukaryotic transcription factors [45]. It is

© Springer Nature Switzerland AG 2018
M. Češka and D. Šafránek (Eds.): CMSB 2018, LNBI 11095, pp. 129–145, 2018.
https://doi.org/10.1007/978-3-319-99429-1_8

possible that the time delays associated with transporting the protein from the cytoplasm to nucleus compromise the noise buffering properties of negative feedback.

An alternative option is an incoherent feedforward loop that has been shown to be effective in maintaining a desired expression level in spite of changes in gene dosage [6], or upstream fluctuations in transcription factor levels [29,33,43,46]. Interestingly, increasing evidence shows that many eukaryotic genes are regulated by a specific feedforward architecture – the transcribed intronic regions of a gene that are removed during splicing are further processed to make a microRNA that targets the same gene's mRNA [11]. This creates a strong coupling between the two species both in the sense of stoichiometry, and also timing of production events. We systematically study how such coupling in a feedforward loop alters noise in mRNA copy numbers, and identify parameter regimes which provide the most (and least) effective noise suppression.

Cellular regulatory circuits can be represented, up to a suitable level of detail, by systems of chemical kinetics. Unfortunately, exact characterisations of the copy-number distributions in a reaction system are often unavailable. Systems operating at a complex-balanced equilibrium are a notable exception in that they admit tractable product-form distributions [4,5,24]. Steady-state distributions have also been characterised in terms of generating functions in a growing collection of simple models that are not complex-balanced [19,20,40,49]. Such representations typically involve the use of special mathematical functions [2].

Approximative methods often provide a viable alternative in analysing a reaction system if exact results are unavailable or intractable. The linear-noise approximation (LNA) and moment-closure methods can reveal useful insights into the noise behaviour even for relatively complex reaction networks [12–14,16, 18,41]. The quasi-steady state (QSS) approximation often leads to formulation of simplified models that can be more amenable to exact characterisation [7,25, 36,47].

Here we apply these methodologies to analyse a reaction-kinetics model of a feedforward loop. Section 2 formally introduces the model and reviews some of its essential features as identified by an LNA analysis. Section 3 contains the derivation of the LNA results and cross-validates them with large-scale kinetic Monte Carlo simulations. Section 4 focuses on a special case in which the model admits a product-form distribution predicted by the Chemical Reaction Network Theory [4,5]. Section 5 introduces a QSS approximation and shows that it can outperform the LNA in a specific parametric regime. The paper ends with a discussion of the current results and sketches lines of future inquiry.

2   The Statement of the Model and Main Results

We consider a discrete stochastic chemical kinetics system composed of two species X (mRNA) and Y (microRNA) which are subject to reaction channels

$$R_1 : \emptyset \xrightarrow{\ k\ } X + Y, \qquad R_2 : X + Y \xrightarrow{\ \frac{\delta(1-q)}{k}\ } \emptyset,$$

$$R_3 : X + Y \xrightarrow{\ \frac{\delta q}{k}\ } Y, \qquad R_4 : Y \xrightarrow{\ 1\ } \emptyset, \qquad R_5 : X \xrightarrow{\ \varepsilon\ } \emptyset. \tag{1}$$

The molecules X and Y are produced synchronously with rate constant k through the reaction channel R_1. In the reaction channels R_2 and R_3, a pair of X and Y molecules interact, which leads to the elimination of X; the Y molecule survives its antagonistic action on X with probability q, where $0 \le q \le 1$ are allowed. For studies with a decoupled production of X and Y and $q = 0$ we refer the reader e.g. to [26, 35, 37].

The strength of the antagonistic interaction is measured by the parameter δ. The reaction rates of both second-order reactions are divided by k in order to achieve a classical scaling [27] of the reaction system (1) with respect to the parameter k. Doing so makes the ensuing analysis more transparent.

The reaction channels R_4 and R_5 describe the spontaneous degradation of Y and X. By measuring time in units of the expected lifetime of Y, we are able to fix the reaction constant of R_4 to one. The parameter ε gives the ratio of Y to X lifetimes. In particular, small values of ε pertain to the assumption that X be much more stable than Y.

In this paper we aim to examine how the choice of parameter values affects the stochastic noise in the X copy number at steady state. We use the Fano factor, which is defined as the ratio of the variance to the mean, as our chosen noise metric. It is well known that the Fano factor is equal to one for the Poisson distribution. We refer to a distribution with Fano factor lower than one as sub-Poissonian.

Using the linear-noise approximation (LNA), we obtain for the Fano factor

$$F_{\mathrm{LNA}} = 1 - \frac{\varepsilon\delta}{(\varepsilon + \delta)^2(1 + \varepsilon + \delta)} \quad \text{for } q = 1. \tag{2}$$

The details of the derivation follow in Sect. 3. Note that the LNA result (2) is independent of the parameter k. The function (2) is visualised as a heat map in the upper-left panel of Fig. 1.

Elementary analysis of (2) reveals that

1. The steady-state distribution of X is sub-Poissonian if $\varepsilon > 0$ and $\delta > 0$.
2. If there is no interaction ($\delta = 0$) or if X is stable ($\varepsilon = 0$), the Fano factor is equal to one.
3. For a fixed positive value of ε, the Fano factor is a nonmonotonic function of δ, initially decreasing before reaching a minimum, and slowly increasing back to one as $\delta \to \infty$.
4. The analogous holds if the roles of ε and δ are reversed in Property 3.
5. The function $F_{\mathrm{LNA}}(\varepsilon, \delta)$ is discontinuous at $(\varepsilon, \delta) = (0, 0)$. A range of limiting values can be achieved depending on along which ray the origin is approached.

6. The function $F_{\mathrm{LNA}}(\varepsilon, \delta)$ does not have an unconstrained minimum. An infimum of 0.75 is approached if $\varepsilon = \delta \to 0$.

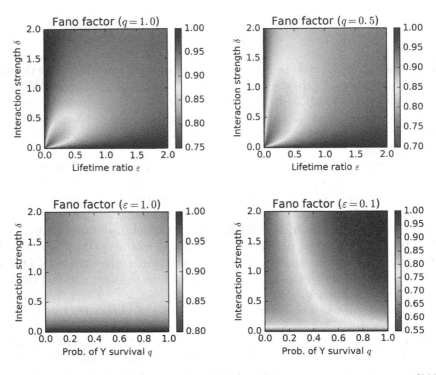

Fig. 1. The Fano factor of species X (mRNA) by linear-noise approximation (LNA) depending on the model parameters.

The function F_{LNA} can be efficiently evaluated also for $0 \leq q < 1$, but the algebra reveals little. Graphical examination of F_{LNA} indicates that all the above properties hold for $0 < q \leq 1$ (see the upper-right panel of Fig. 1 for $q = 0.5$). The infimum of the Fano factor in Property 6 is lower than 0.75 if $q < 1$ and is approached along a different ray emanating from the origin. In the exceptional case $q = 0$ the dependence of F_{LNA} on ε and δ is monotonous (the lower panels of Fig. 1).

Property 2 suggests, but does not provide a definitive proof, that the distribution of X is Poissonian if $\varepsilon = 0$ or $\delta = 0$. In the non-interaction case ($\delta = 0$), the proof is straightforward: the dynamics of X is that of a simple immigration-and-death process, which is known to generate a Poisson distribution [9,23]. If X is stable ($\varepsilon = 0$), the distribution is again Poisson, but the proof requires a more subtle reasoning based on the Chemical Reaction Network Theory. We present the details in Sect. 4.

The discontinuity of $F_{\mathrm{LNA}}(\varepsilon, \delta)$ at the origin indicates that a surprisingly rich behaviour, in terms of the chosen noise metric, can be recovered by focusing solely on the $\varepsilon, \delta \ll 1$ parametric region. We pursue this line of inquiry in Sect. 5, where we identify a family of discrete distributions, which describe the limit behaviour of (1) in this parametric region. For reasons made explicit later, we refer to this description as the quasi-steady-state (QSS) model. We demonstrate that the QSS model can be superior to the LNA in predicting simulation results.

3   Linear-Noise Approximation

In linear-noise approximation, the mean behaviour is given by the law-of-mass-action formulation of the reaction system (1), which is

$$\dot{x} = k - \frac{\delta xy}{k} - \varepsilon x, \quad \dot{y} = k - \frac{\delta(1-q)xy}{k} - y. \tag{3}$$

Setting the derivatives in (3) to zero and solving the resulting algebraic system in x and y yield the stationary mean values

$$x = k\tilde{x}, \quad y = k\tilde{y}, \tag{4}$$

where

$$\tilde{x} = \frac{2}{\delta q + \varepsilon + \sqrt{(\delta q + \varepsilon)^2 + 4\delta\varepsilon(1-q)}}, \quad \tilde{y} = q + \varepsilon(1-q)\tilde{x}. \tag{5}$$

Note that the means (4) scale with the production rate constant k.

In order to obtain the LNA of the variance, we need to determine the steady-state fluctuation and dissipation matrices of the reaction system (1). The dissipation matrix A is equal to the linearisation matrix of the system (3), i.e.

$$A = -\begin{pmatrix} \delta\tilde{y} + \varepsilon & \delta\tilde{x} \\ \delta(1-q)\tilde{y} & 1 + \delta(1-q)\tilde{x} \end{pmatrix}. \tag{6}$$

The fluctuation matrix B is obtained in the following manner [28]: for each reaction channel in the system (1), we calculate the outer product of the reaction vector [1] with itself, and multiply it by the (steady-state) reaction rate; then we sum the results over all reaction channels. In our particular example this leads to

$$B = k\tilde{B}, \quad \text{where} \quad \tilde{B} = \begin{pmatrix} 2 & 2 - \tilde{y} \\ 2 - \tilde{y} & 2 \end{pmatrix}. \tag{7}$$

The fluctuation–dissipation theorem [34] states that the covariance matrix Σ of the random vector of steady-state X and Y copy numbers satisfies

$$A\Sigma + \Sigma A^\mathsf{T} + B = 0,$$

i.e.

$$\Sigma = k\tilde{\Sigma}, \quad \text{where} \quad A\tilde{\Sigma} + \tilde{\Sigma}A^\mathsf{T} + \tilde{B} = 0. \tag{8}$$

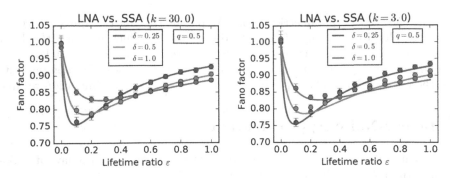

Fig. 2. The Fano factor of species X (mRNA) as function of model parameters as given by the linear-noise approximation (LNA, solid lines) and stochastic simulation algorithm (SSA, discrete markers). Two values of k are used, one large ($k = 30$, left panel) and one moderate ($k = 3$, right panel). The LNA is independent of k. Error bars indicate 99.9% confidence intervals for the simulation-based Fano factors.

Note that the covariance matrix (8) scales with the production rate constant k. The linear algebraic system (8) can be written in a flattened form as

$$\begin{pmatrix} a_{11} & a_{12} & 0 \\ a_{21} & a_{11} + a_{22} & a_{12} \\ 0 & a_{21} & a_{22} \end{pmatrix} \begin{pmatrix} \tilde{\Sigma}_{11} \\ \tilde{\Sigma}_{12} \\ \tilde{\Sigma}_{22} \end{pmatrix} + \begin{pmatrix} 1 \\ 2 - \tilde{y} \\ 1 \end{pmatrix} = \mathbf{0}. \tag{9}$$

Our chosen metric of the noise in species X, the Fano factor, is given by

$$F_{\text{LNA}} = \frac{\Sigma_{11}}{x} = \frac{\tilde{\Sigma}_{11}}{\tilde{x}}.$$

Since both the mean x and the variance Σ_{11} scale linearly with k, the Fano factor is independent of the parameter.

For $0 \leq q < 1$, we solve (9) for every combination of parametric values numerically by a fast linear-algebra solver. For $q = 1$, Eqs. (5) and (6) simplify to

$$\tilde{x} = \frac{1}{\delta + \varepsilon}, \quad \tilde{y} = 1, \quad A = -\begin{pmatrix} \delta + \varepsilon & \frac{\delta}{\delta + \varepsilon} \\ 0 & 1 \end{pmatrix}.$$

The linear system (9) becomes upper triangular, and the formula (2) is obtained after few elementary steps.

A classical system-size-expansion argument [22] guarantees that the LNA accurately describes the reaction system (1) as k tends to infinity. We demonstrate the asymptotics in Fig. 2, in which we compare the LNA of the Fano factor to the value obtained by the application of Gillespie's stochastic simulation algorithm (SSA) [17]. We observe a perfect agreement between the two if $k = 30$ (Fig. 2, left panel). Although the agreement remains satisfactory for $k = 3$, the SSA results are now seen to deviate systematically from the LNA prediction for moderate values of ε and δ (Fig. 2, right panel).

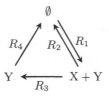

Fig. 3. Graphical representation of a reduced reaction system obtained from (1) by disabling the spontaneous degradation of X (reaction channel R_5). The reduced reaction system is seen to be weakly reversible (unless $q = 0$). Additionally, it has zero deficiency and admits no conservation laws. Therefore, the copy numbers of X and Y are independent and Poissonian.

The reaction species mean values and standard deviations were calculated using Stochpy's [30] implementation of Gillespie's direct method [17]. We skipped over the first 30 units of time to avoid the influence of an initial transient; we estimated the moments from the next 10^5 iterations (for $k = 3$) or 10^6 iterations (for $k = 30$) of the algorithm. The Fano factor was calculated as the ratio of the squared standard deviation to the mean value. The procedure was repeated to obtain 25 independent Fano factor estimates. The repetition increased accuracy and facilitated the construction of confidence intervals.

4   Stability of X Implies Poisson Distribution

In Sect. 3, we reported that if Y has a chance of surviving the antagonism with X ($q > 0$) and if X is stable ($\varepsilon = 0$), then the LNA of the Fano factor is equal to one. Here we expand on this observation by providing an actual proof that the steady-state copy number of X (and also that of Y) follows the Poisson distribution. The argument is based on the application of the Chemical Reaction Network Theory (CRNT) [5].

Setting $\varepsilon = 0$ in the reaction system (1) is tantamount to removing the reaction channel R_5 for spontaneous degradation of X. The four remaining reaction channels involve $N = 3$ complexes (meant in the CRNT sense [15]), namely the empty set $\emptyset$, the pair X + Y, and the singleton Y. The three complexes are represented as vertices of the reaction graph (Fig. 3). The reaction graph has a single linkage class ($l = 1$). Given that the linkage class is strongly connected, the chemical system is weakly reversible in the sense of the CRNT.

The reaction vectors span the entire two-dimensional ($s = 2$) space of X and Y copy number pairs. The deficiency of the reaction network is obtained by the well-known formula $\delta = N - l - s = 0$. According to CRNT [5], weakly reversible networks of zero deficiency admit a product-form steady-state distribution

$$P[X = m, Y = n] = C \frac{x^m y^n}{m! n!}, \tag{10}$$

where $x = k/\delta q$ and $y = kq$ are obtained by solving the law-of-mass action kinetics (3) at steady state, i.e. by setting $\varepsilon = 0$ in (4)–(5).

Since the reaction system does not admit any conservation laws, the support of the distribution (10) includes all pairs of nonnegative integers $m, n \geq 0$. The normalisation constant C is then readily determined as $C = e^{-x-y}$. In other words, the joint distribution of X and Y is the product of the marginal distributions, either of which is Poissonian with mean x and y, respectively.

5 Quasi-Steady-State Approximation

The LNA results presented in Sect. 2 suggested that the reaction system (1) exhibits a wide range of noise behaviour in the small $\varepsilon, \delta \ll 1$ region. In this section we develop an approximative description to the reaction system (1) that is applicable for such parametric choices. Specifically, we assume that

$$k = \varepsilon\kappa, \quad \delta = \varepsilon\kappa\alpha, \quad \varepsilon \ll 1, \tag{11}$$

where κ and α are the rescaled production and interaction rate constants. The scaling (11) guarantees that ε and δ are both small and that they approach the origin along a ray with slope given by $\kappa\alpha$. Note that taking ε and δ to zero whilst keeping the production rate k fixed would have led to a divergence in the level of X. By making the production also scale with ε we are able to approach a distinguished limiting distribution as ε tends to zero. Also note that (11) removes the classical scaling with respect to the production rate constant from the system (1). It turns out that this choice keeps the ensuing analysis more tractable.

The steady-state probability distribution $p_{m,n}$ of observing m molecules of X and n molecules of Y in the system satisfies the master equation

$$\alpha(1-q)(m+1)(n+1)p_{m+1,n+1} + \alpha q(m+1)np_{m+1,n} + (n+1)p_{m,n+1}$$
$$+ \varepsilon\kappa p_{m-1,n-1} + \varepsilon(m+1)p_{m+1,n} - (\alpha mn + n + \varepsilon\kappa + \varepsilon m)p_{m,n} = 0. \tag{12}$$

The first three terms in (12) give the probability influx into the reference state (m, n) due to the reaction channels R_2, R_3 (interactions) and R_4 (decay of Y). These terms are not multiplied by ε, i.e. the three reactions are considered to be fast. The next two terms give the probability influx due to the channels R_1 (production) and R_5 (decay of X). These terms are of order ε, i.e. the two reactions are slow. The final, negative, term in (12) gives the probability efflux from the reference state (m, n). The master equation states that the influx and the efflux balance out.

The separation between fast and slow reactions suggests that we should use quasi-steady-state (QSS) reduction techniques to study (12). The molecule Y plays the role of the transient, highly reactive, species (or the QSS species [31, 44]). Each time a molecule of Y is produced, a brief period of interaction with X ensues, with ends with the elimination of Y (whether through natural degradation or through the interaction with X if $q < 1$).

We seek a power-series solution

$$p_{m,n}(\varepsilon) = p_{m,n}^{(0)} + \varepsilon p_{m,n}^{(1)} + O(\varepsilon^2). \tag{13}$$

Inserting (13) into (12) and collecting $O(1)$ terms yields

$$p_{m,n}^{(0)} = 0 \quad \text{for} \quad n > 0, \tag{14}$$

which states that the probability of observing a non-zero number of the QSS species Y is $O(\varepsilon)$ small. Further analysis helps establish additional relations

$$p_{m,n}^{(1)} = 0 \quad \text{for} \quad n > 1, \tag{15}$$

which state that the probability of observing two or more molecules of Y is $O(\varepsilon^2)$ small. Although relations (14) and (15) are sufficient for our present analysis, they can be generalised to $p_{m,n}^{(k)} = 0$ for $n > k$, which state that the probability of having more than k molecules of Y is $O(\varepsilon^{k+1})$.

In light of (14), it remains to determine the terms $p_{m,0}^{(0)}$ in order to obtain the limiting probability distribution. To this end, it is sufficient to use the master equation (12) for $n = 0$ and $n = 1$, i.e.

$$
\begin{aligned}
n = 0: \quad & \alpha(1 - q)(m + 1)p_{m+1,1} + p_{m,1} \\
& + \varepsilon(m + 1)p_{m+1,0} - \varepsilon(\kappa + m)p_{m,0} = 0, \tag{16}
\end{aligned}
$$

$$
\begin{aligned}
n = 1: \quad & 2\alpha(1 - q)(m + 1)p_{m+1,2} + \alpha q(m + 1)p_{m+1,1} + 2p_{m,2} \\
& + \varepsilon\kappa p_{m-1,0} + \varepsilon(m + 1)p_{m+1,1} - (\alpha m + 1 + \varepsilon\kappa + \varepsilon m)p_{m,1} = 0. \tag{17}
\end{aligned}
$$

Inserting the power-series ansatz (13) into (16)–(17) and collecting $O(\varepsilon)$ terms yields

$$
\begin{aligned}
n = 0: \quad & \alpha(1 - q)(m + 1)p_{m+1,1}^{(1)} + p_{m,1}^{(1)} \\
& + (m + 1)p_{m+1,0}^{(0)} - (\kappa + m)p_{m,0}^{(0)} = 0, \tag{18}
\end{aligned}
$$

$$
n = 1: \quad \alpha q(m + 1)p_{m+1,1}^{(1)} + \kappa p_{m-1,0}^{(0)} - (\alpha m + 1)p_{m,1}^{(1)} = 0, \tag{19}
$$

whereby we made use of the relations (14) and (15). In particular, the relations (15) guarantee that (18)–(19) forms a closed system of difference equations in the unknown series $p_{m,0}^{(0)}$ and $p_{m,1}^{(1)}$.

In order to solve (18)–(19), we introduce the generating functions

$$f(x) = \sum_{m=0}^{\infty} x^m p_{m,0}^{(0)}, \quad g(x) = \sum_{m=0}^{\infty} x^m p_{m,1}^{(1)}. \tag{20}$$

Multiplying (18)–(19) by x^m and summing over $m \geq 0$ yield a system of ordinary differential equations

$$\alpha(1 - q)\frac{dg}{dx} + g = (x - 1)\frac{df}{dx} + \kappa f = 0, \tag{21}$$

$$\alpha(q - x)\frac{dg}{dx} - g = -\kappa x f. \tag{22}$$

Next, we turn the system (21)–(22) of two first-order ordinary differential equations into a single second-order ordinary differential equation.

First, we eliminate g by adding the Eqs. (21) and (22) up, and dividing the result by $1 - x$, which gives

$$\alpha \frac{dg}{dx} = \kappa f - \frac{df}{dx}. \tag{23}$$

Second, we eliminate dg/dx by adding up the $(x - q)$-multiple of (21) and the $(1 - q)$-multiple of (22) before dividing the result by $x - 1$, whereby we obtain

$$g = (x - q) \frac{df}{dx} + \kappa q f. \tag{24}$$

Differentiating (24), we find

$$\frac{dg}{dx} = \frac{d}{dx} \left((x - q) \frac{df}{dx} \right) + \kappa q \frac{df}{dx}. \tag{25}$$

Combining (23) and (25), we arrive at an ordinary differential equation of the second order for f, which reads

$$\frac{d}{dx} \left((x - q) \frac{df}{dx} \right) + \left(\kappa q + \frac{1}{\alpha} \right) \frac{df}{dx} - \frac{\kappa}{\alpha} f = 0. \tag{26}$$

We look for a solution to (26) in the form of a power series

$$f(x) = \sum_{m=0}^{\infty} c_m (x - q)^m. \tag{27}$$

Inserting (27) into (26), we have

$$\sum_{m=0}^{\infty} \left(c_m m^2 (x - q)^{m-1} + \left(\kappa q + \frac{1}{\alpha} \right) c_m m (x - q)^{m-1} - \frac{\kappa}{\alpha} c_m (x - q)^m \right) = 0.$$

Equating like powers of $(x - q)$ yields a recursion

$$m \left(m + \kappa q + \frac{1}{\alpha} \right) c_m = \frac{\kappa}{\alpha} c_{m-1}, \quad m = 1, 2, \ldots,$$

solving which yields

$$c_m = \frac{c_0 \left(\frac{\kappa}{\alpha} \right)^m}{\left(\kappa q + \frac{1}{\alpha} + 1 \right)_m m!}, \tag{28}$$

where

$$(a)_m = a(a + 1) \cdot \ldots \cdot (a + m - 1), \quad (a)_0 = 1,$$

represents the m-th rising factorial from a number $a > 0$.

Substituting (28) into (27) we find that

$$f(x) = c_0 \times {}_0F_1 \left(\kappa q + \frac{1}{\alpha} + 1, \frac{\kappa}{\alpha} (x - q) \right), \tag{29}$$

in which the confluent hypergeometric limit function $_0F_1$ is defined by the convergent series

$$_0F_1(a, z) = \sum_{m=0}^{\infty} \frac{z^m}{(a)_m m!}. \tag{30}$$

Imposing the normalisation condition $f(1) = 1$, we determine the prefactor c_0 in (29) and obtain

$$f(x) = \frac{_0F_1\left(\kappa q + \frac{1}{\alpha} + 1, \frac{\kappa}{\alpha}(x - q)\right)}{_0F_1\left(\kappa q + \frac{1}{\alpha} + 1, \frac{\kappa}{\alpha}(1 - q)\right)}. \tag{31}$$

Basic properties of the confluent hypergeometric limit function can be established using its power-series representation (30). Repeatedly differentiating the series (30) term by term yields

$$\frac{\mathrm{d}^m}{\mathrm{d}z^m}\,_0F_1(a, z) = \frac{_0F_1(a + m, z)}{(a)_m}. \tag{32}$$

Comparing (30) with the power-series expansions of the normal and modified Bessel functions [2], we obtain

$$_0F_1(c, z) = \Gamma(c)z^{\frac{1-c}{2}}I_{c-1}(2\sqrt{z}), \quad _0F_1(c, -z) = \Gamma(c)z^{\frac{1-c}{2}}J_{c-1}(2\sqrt{z}), \quad z > 0, \tag{33}$$

where $\Gamma(z)$ is the gamma function, $J_\nu(z)$ is the Bessel function, and $I_\nu(z)$ is the modified Bessel function of order ν.

Repeatedly differentiating (31) yields

$$\frac{\mathrm{d}^m f(x)}{\mathrm{d}x^m} = \frac{\left(\frac{\kappa}{\alpha}\right)^m}{\left(\kappa q + \frac{1}{\alpha} + 1\right)_m} \times \frac{_0F_1\left(\kappa q + \frac{1}{\alpha} + 1 + m, \frac{\kappa}{\alpha}(x - q)\right)}{_0F_1\left(\kappa q + \frac{1}{\alpha} + 1, \frac{\kappa}{\alpha}(1 - q)\right)}, \tag{34}$$

which provides an approximation

$$p_{m,n}^{(0)} = \frac{\delta_{n,0}\left(\frac{\kappa}{\alpha}\right)^m}{m!\left(\kappa q + \frac{1}{\alpha} + 1\right)_m} \times \frac{_0F_1\left(\kappa q + \frac{1}{\alpha} + 1 + m, -\frac{\kappa q}{\alpha}\right)}{_0F_1\left(\kappa q + \frac{1}{\alpha} + 1, \frac{\kappa}{\alpha}(1 - q)\right)} \tag{35}$$

for the desired solution to the master equation (12). Evaluating the derivatives of $f(x)$ at $x = 1$, we obtain the factorial moments [21]

$$\mu_{(m)} = \langle X(X - 1) \cdot \ldots \cdot (X - m + 1)\rangle = \left.\frac{\mathrm{d}^m f(x)}{\mathrm{d}x^m}\right|_{x=1}$$

$$= \frac{\left(\frac{\kappa}{\alpha}\right)^m}{\left(\kappa q + \frac{1}{\alpha} + 1\right)_m} \times \frac{_0F_1\left(\kappa q + \frac{1}{\alpha} + 1 + m, \frac{\kappa}{\alpha}(1 - q)\right)}{_0F_1\left(\kappa q + \frac{1}{\alpha} + 1, \frac{\kappa}{\alpha}(1 - q)\right)}. \tag{36}$$

At the same time as noting that the mean $\langle X \rangle$ trivially coincides with the first factorial moment $\mu_{(1)}$, we also point out that the main characteristic of interest here, the Fano factor, can be expressed in terms of the first two factorial moments as

$$F_{\mathrm{QSS}} = 1 + \frac{\mu_{(2)}}{\mu_{(1)}} - \mu_{(1)}. \tag{37}$$

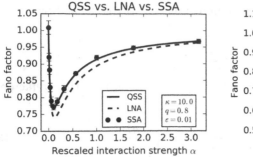 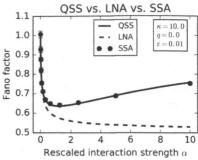

Fig. 4. The Fano factor of species X (mRNA) as function of rescaled model parameters as given by the quasi-steady state (QSS) model, the linear-noise approximation (LNA) and the stochastic simulation algorithm (SSA). Error bars indicate 99.9% confidence intervals for the simulation-based Fano factors.

We expressly mention, without carrying out the somewhat tedious calculation, that the probability distribution (35) and the moments (36) can be written in terms of Bessel's functions via (33).

In Fig. 4, we compare the values of the Fano factor obtained in the parametric regime (11) by the quasi-steady state (QSS) approximation, the linear-noise approximation (LNA), and by the application of stochastic simulation algorithm (SSA). We observe that the QSS model predicts the SSA results more faithfully than the LNA. For $q > 0$, the LNA overestimates the dip in the Fano factor (Fig. 4, left panel). For $q = 0$, the LNA predicts a monotonous decrease of the Fano factor with the interaction strength, whereas the QSS and SSA results both show an eventual slow increase (Fig. 4, right panel).

The LNA values were calculated by the method of Sect. 3, whereby the original parameters k and δ were recovered from the values of κ and α through relations (11). Each SSA value was computed in StochPy [30] from 25 independent sample paths each consisting of 10^5 iterations of Gillespie's direct method. The QSS values were obtained from (36)–(37).

In Fig. 5, we compare simulation-based estimates of the X copy-number distribution to the QSS approximation (35) and a Poissonian benchmark. The agreement between the simulation and the QSS results improves as the value of ε is decreased: compare the left panels with $\varepsilon = 0.1$ to the right panels with $\varepsilon = 0.01$. Consistently with our previous reports of the model's sub-Poissonian behaviour, the species X copy number distributions are narrower than the Poissonian benchmark.

The values of the interaction α were selected in Fig. 5 so as to minimise the Fano factor for the given values of κ and q. Each histogram is based on 10^5 independent sample paths of the chemical system (1) generated with Gillespie's direct method. The Poisson distribution was fit to the simulation data by maximum likelihood estimation.

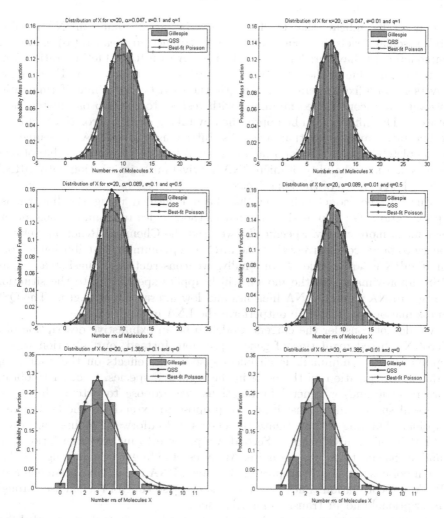

Fig. 5. Marginal distributions of X copy number obtained by Gillespie's algorithm, quasi-steady-state (QSS) approximation, and by maximum-likelihood fitting of a Poisson distribution.

6 Discussion

In this paper we considered a stochastic model for a feedforward loop driven by the interaction between mRNA and microRNA species. Using a combination of mathematical and computational methods, we investigated the effects of microRNA based regulation on the mRNA noise levels. The model behaviour depends on several parameters, namely: the gene-expression rate, the interaction strength, the microRNA to mRNA lifetime ratio, and the probability of microRNA surviving its interaction with mRNA.

Our results indicate that feedforward regulation can buffer mRNA noise to sub-Poissonian levels. The Fano factor (the variance to mean ratio) exhibits a nonmonotonic behaviour: for a fixed microRNA to mRNA lifetime ratio, there is an optimal value of the interaction strength that minimises the Fano factor; conversely, for a fixed interaction strength, there exists an optimal lifetime ratio. However, an unconstrained minimum with respect to the two parameters does not exist. The infimum can be approached by taking small microRNA to mRNA lifetime ratios and interaction strengths. Intriguingly, if mRNA is assumed to be completely stable, the Fano factor is equal to the Poissonian value of one. Decreasing the probability of microRNA survival in its interaction with mRNA leads to lower values of the Fano factor.

Much of the model behaviour has been identified using the linear-noise approximation. We also used additional methodologies to examine some of the phenomena more closely. Specifically, we used the Chemical Reaction Network Theory to prove conclusively that the mRNA copy-number distribution is Poisson if mRNAs are stable. Additionally, we constructed a quasi-steady-state (QSS) approximation of the model, which applies specifically to the situation of large mRNA to microRNA lifetimes and low interaction strengths. The QSS approximation was shown to outperform the LNA in this regime.

We do not consider the current model to be an exhaustive description of a microRNA based regulation of gene expression. Instead, our intention was to examine, using a minimalistic chemical system, the effects on the underlying feedforward regulation on the noise in the regulated species. In order to obtain more realistic and/or general formulations, we propose to extend the model in several specific directions. First, we propose to extend the model by transcriptional bursting to investigate the effects of feedforward regulation on the super-Poissonian mRNA noise. Second, we propose to include translation, and examine the effects on protein noise. Analyses of different systems suggest that protein noise is not simply proportional to the mRNA noise, but also depends on mRNA autocorrelation times [39]; proteins can also control their noise through transcriptional and/or translational feedbacks.

In summary, we studied a stochastic feedforward loop featuring a coupled production and antagonistic interaction. Using a combination of different methodologies, we examined the consequences of the interaction on gene-expression noise. We expect that analogous approaches will be helpful to understand more complex versions of the model as well as other examples of gene-regulatory motifs operating at low copy numbers.

Acknowledgements. PB acknowledges support from the Slovak Research and Development Agency under the contract No. APVV-14-0378, the VEGA grant 1/0347/18, and the EraCoSysMed project 4D-Healing. AS is supported by the National Science Foundation grant ECCS-1711548.

References

1. Abou-Jaoudé, W., Thieffry, D., Feret, J.: Formal derivation of qualitative dynamical models from biochemical networks. Biosystems **149**, 70–112 (2016)
2. Abramowitz, M., Stegun, I.: Handbook of Mathematical Functions with Formulas, Graphs, and Mathematical Tables. National Bureau of Standards, Washington, D.C. (1972)
3. Alon, U.: Network motifs: theory and experimental approaches. Nat. Rev. Genet. **8**, 450–461 (2007)
4. Anderson, D.F., Cotter, S.L.: Product-form stationary distributions for deficiency zero networks with non-mass action kinetics. Bull. Math. Biol. **78**, 2390–2407 (2016)
5. Anderson, D.F., Craciun, G., Kurtz, T.G.: Product-form stationary distributions for deficiency zero chemical reaction networks. Bull. Math. Biol. **72**, 1947–1970 (2010)
6. Bleris, L., Xie, Z., Glass, D., Adadey, A., Sontag, E., Benenson, Y.: Synthetic incoherent feedforward circuits show adaptation to the amount of their genetic template. Mol. Syst. Biol. **7**, 519 (2011)
7. Bokes, P., King, J., Wood, A., Loose, M.: Multiscale stochastic modelling of gene expression. J. Math. Biol. **65**, 493–520 (2012)
8. Bokes, P., Lin, Y., Singh, A.: High cooperativity in negative feedback can amplify noisy gene expression. Bull. Math. Biol. (2018). https://doi.org/10.1007/s11538-018-0438-y
9. Bokes, P., King, J.R., Wood, A.T., Loose, M.: Exact and approximate distributions of protein and mRNA levels in the low-copy regime of gene expression. J. Math. Biol. **64**, 829–854 (2012)
10. Bokes, P., Singh, A.: Gene expression noise is affected differentially by feedback in burst frequency and burst size. J. Math. Biol. **74**, 1483–1509 (2017)
11. Bosia, C., Osella, M., Baroudi, M.E., Cora, D., Caselle, M.: Gene autoregulation via intronic microRNAs and its functions. BMC Syst. Biol. **6**, 131 (2012)
12. Bronstein, L., Koeppl, H.: A variational approach to moment-closure approximations for the kinetics of biomolecular reaction networks. J. Chem. Phys. **148**, 014105 (2018)
13. Cardelli, L., Kwiatkowska, M., Laurenti, L.: Stochastic analysis of chemical reaction networks using linear noise approximation. Biosystems **149**, 26–33 (2016)
14. Cinquemani, E.: On observability and reconstruction of promoter activity statistics from reporter protein mean and variance profiles. In: Cinquemani, E., Donzé, A. (eds.) HSB 2016. LNCS, vol. 9957, pp. 147–163. Springer, Cham (2016). https://doi.org/10.1007/978-3-319-47151-8_10
15. Feinberg, M.: Lectures on chemical reaction networks. Notes of lectures given at the Mathematics Research Center of the University of Wisconsin (1979)
16. Ghusinga, K.R., Vargas-Garcia, C.A., Lamperski, A., Singh, A.: Exact lower and upper bounds on stationary moments in stochastic biochemical systems. Phys. Biol. **14**, 04LT01 (2017)
17. Gillespie, D.: A general method for numerically simulating stochastic time evolution of coupled chemical reactions. J. Comput. Phys. **22**, 403–434 (1976)
18. Herath, N., Del Vecchio, D.: Reduced linear noise approximation for biochemical reaction networks with time-scale separation: the stochastic tQSSA+. J. Chem. Phys. **148**, 094108 (2018)

19. Innocentini, G.C., Forger, M., Radulescu, O., Antoneli, F.: Protein synthesis driven by dynamical stochastic transcription. Bull. Math. Biol. **78**, 110–131 (2016)
20. Innocentini, G.C., Guiziou, S., Bonnet, J., Radulescu, O.: Analytic framework for a stochastic binary biological switch. Phys. Rev. E **94**, 062413 (2016)
21. Johnson, N., Kotz, S., Kemp, A.: Univariate Discrete Distributions, 3rd edn. Wiley, Hoboken (2005)
22. van Kampen, N.: Stochastic Processes in Physics and Chemistry. Elsevier, New York (2006)
23. Kan, X., Lee, C.H., Othmer, H.G.: A multi-time-scale analysis of chemical reaction networks: II. Stochastic systems. J. Math. Biol. **73**, 1081–1129 (2016)
24. Kelly, F.P.: Reversibility and Stochastic Networks. Cambridge University Press, Cambridge (2011)
25. Kim, J.K., Josić, K., Bennett, M.R.: The validity of quasi-steady-state approximations in discrete stochastic simulations. Biophys. J. **107**, 783–793 (2014)
26. Kumar, N., Jia, T., Zarringhalam, K., Kulkarni, R.V.: Frequency modulation of stochastic gene expression bursts by strongly interacting small RNAs. Phys. Rev. E **94**, 042419 (2016)
27. Kurtz, T.G.: The relationship between stochastic and deterministic models for chemical reactions. J. Chem. Phys. **57**, 2976–2978 (1972)
28. Lestas, I., Paulsson, J., Ross, N., Vinnicombe, G.: Noise in gene regulatory networks. IEEE Trans. Circuits-I **53**, 189–200 (2008)
29. Li, X., Cassidy, J.J., Reinke, C.A., Fischboeck, S., Carthew, R.W.: A microRNA imparts robustness against environmental fluctuation during development. Cell **137**, 273–282 (2009)
30. Maarleveld, T.R., Olivier, B.G., Bruggeman, F.J.: StochPy: a comprehensive, user-friendly tool for simulating stochastic biological processes. PLoS One **8**, e79345 (2013)
31. Mastny, E., Haseltine, E., Rawlings, J.: Two classes of quasi-steady-state model reductions for stochastic kinetics. J. Chem. Phys. **127**, 094106 (2007)
32. Nevozhay, D., Adams, R.M., Murphy, K.F., Josic, K., Balazsi, G.: Negative autoregulation linearizes the dose response and suppresses the heterogeneity of gene expression. Proc. Natl. Acad. Sci. U.S.A. **106**, 5123–5128 (2009)
33. Osella, M., Bosia, C., Corá, D., Caselle, M.: The role of incoherent microRNA-mediated feedforward loops in noise buffering. PLoS Comput. Biol. **7**, e1001101 (2011)
34. Paulsson, J.: Models of stochastic gene expression. Phys. Life Rev. **2**, 157–175 (2005)
35. Platini, T., Jia, T., Kulkarni, R.V.: Regulation by small RNAs via coupled degradation: Mean-field and variational approaches. Phys. Rev. E **84**, 021928 (2011)
36. Popovic, N., Marr, C., Swain, P.S.: A geometric analysis of fast-slow models for stochastic gene expression. J. Math. Biol. **72**, 87–122 (2016)
37. Schmiedel, J.M., et al.: MicroRNA control of protein expression noise. Science **348**, 128–132 (2015)
38. Singh, A.: Negative feedback through mRNA provides the best control of gene-expression noise. IEEE Trans. NanoBiosci. **10**, 194–200 (2011)
39. Singh, A., Bokes, P.: Consequences of mRNA transport on stochastic variability in protein levels. Biophys. J. **103**, 1087–1096 (2012)
40. Singh, A., Vargas-Garcia, C.A., Karmakar, R.: Stochastic analysis and inference of a two-state genetic promoter model. In: Proceedings of the American Control Conference, pp. 4563–4568 (2013)

41. Singh, A., Grima, R.: The linear-noise approximation and moment-closure approximations for stochastic chemical kinetics. arXiv preprint arXiv:1711.07383 (2017)
42. Singh, A., Hespanha, J.P.: Optimal feedback strength for noise suppression in autoregulatory gene networks. Biophys. J. **96**, 4013–4023 (2009)
43. Soltani, M., Platini, T., Singh, A.: Stochastic analysis of an incoherent feedforward genetic motif. In: American Control Conference (ACC), pp. 406–411 (2016)
44. Srivastava, R., Haseltine, E.L., Mastny, E., Rawlings, J.B.: The stochastic quasi-steady-state assumption: reducing the model but not the noise. J. Chem. Phys. **134**, 154109 (2011)
45. Stewart, A.J., Seymour, R.M., Pomiankowski, A., Reuter, M.: Under-dominance constrains the evolution of negative autoregulation in diploids. PLoS Comput. Biol. **9**, e1002992 (2013)
46. Strovas, T.J., Rosenberg, A.B., Kuypers, B.E., Muscat, R.A., Seelig, G.: MicroRNA-based single-gene circuits buffer protein synthesis rates against perturbations. ACS Synth. Biol. **3**, 324–331 (2014)
47. Veerman, F., Marr, C., Popović, N.: Time-dependent propagators for stochastic models of gene expression: an analytical method. J. Math. Biol. (2018). https://doi.org/10.1007/s00285-017-1196-4
48. Voliotis, M., Bowsher, C.G.: The magnitude and colour of noise in genetic negative feedback systems. Nucleic Acids Res. **40**, 7084–7095 (2012)
49. Yang, X., Wu, Y., Yuan, Z.: Characteristics of mRNA dynamics in a multi-on model of stochastic transcription with regulation. Chin. J. Phys. **55**, 508–518 (2017)

Stochastic Rate Parameter Inference Using the Cross-Entropy Method

Jeremy Revell and Paolo Zuliani[(✉)]

School of Computing, Newcastle University, Newcastle upon Tyne, UK
{j.d.revell1,paolo.zuliani}@ncl.ac.uk

Abstract. We present a new, efficient algorithm for inferring, from time-series data or high-throughput data (*e.g.*, flow cytometry), stochastic rate parameters for chemical reaction network models. Our algorithm combines the Gillespie stochastic simulation algorithm (including approximate variants such as tau-leaping) with the cross-entropy method. Also, it can work with incomplete datasets missing some model species, and with multiple datasets originating from experiment repetitions. We evaluate our algorithm on a number of challenging case studies, including bistable systems (Schlögl's and toggle switch) and experimental data.

1 Introduction

In this paper we are concerned with the inference of biochemical reaction stochastic rate parameters from data. Reactions are discrete events that can occur randomly at any time with a rate dependent on the chemical kinetics [40]. It has recently become clear that stochasticity can produce dynamics profoundly different from the corresponding deterministic models. This is the case, *e.g.*, in genetic systems where key species are present in small numbers or where key reactions occur at a low rate [23], resulting in transient, stochastic bursts of activity [4,24]. The standard model for such systems is the Markov jump process popularised by Gillespie [13,14]. Given a collection of reactions modelling a biological system and time-course data, the *stochastic parameter inference problem* is to find parameter values for which the Gillespie model's temporal behaviour is most consistent with the data. This is a very difficult problem, much harder, both theoretically and computationally, than the corresponding problem for deterministic kinetics—see, *e.g.*, [41, Sect. 1.3]. One simple reason is because stochastic models can behave widely differently from the same initial conditions. (The related issue of parameter non-identifiability is outside the scope of this paper, but the interested reader can find more in, *e.g.*, [37,38] and references therein.) Additionally, experimental data is usually sparse and most often involves only a limited subset of a model's species; and the system under study might exhibit multimodal behaviour. Also, data might not directly relate to a species, it might be measured in arbitrary units (*e.g.*, fluorescence measurements), thus requiring the estimation of scaling factors, or it might be described by frequency distributions (*e.g.*, high-throughput data such as flow cytometry).

M. Češka and D. Šafránek (Eds.): CMSB 2018, LNBI 11095, pp. 146–164, 2018.
https://doi.org/10.1007/978-3-319-99429-1_9

Stochastic parameter inference is thus a fundamental and challenging problem in systems biology, and it is crucial for obtaining validated and predictive models.

In this paper we propose an approach for the parameter inference problem that combines Gillespie's Stochastic Simulation Algorithm (SSA) with the cross-entropy (CE) method [27]. The CE method has been successfully used in optimisation, rare–event probability estimation, and other domains [29]. For parameter inference, Daigle *et al.* [8] combined a stochastic Expectation–Maximisation (EM) algorithm with a modified cross-entropy method. We instead develop the cross-entropy method in its own right, discarding the costly EM algorithm steps. We also show that our approach can utilise approximate, faster SSA variants such as tau-leaping [15]. Summarising, the main contributions of this paper are:

- we present a new, cross entropy-based algorithm for the stochastic parameter inference problem that outperforms previous, state–of–the–art approaches;
- our algorithm can work with multiple, incomplete, and distribution datasets;
- we show that tau-leaping can be used within our technique;
- we provide a thorough evaluation of our algorithm on a number of challenging case studies, including bistable systems (Schlögl model and toggle switch) and experimental data.

2 Background

Notation. Given a system with n chemical species, the state of the system at time t is represented by the vector $\boldsymbol{x}(t) = (x_1(t), \dots, x_n(t))$, where x_i represents the number of molecules of the ith species, S_i, for $i \in \{1, \dots, n\}$. A well-mixed system within a fixed volume at a constant temperature can be modelled by a continuous-time Markov chain (CTMC) [13,14]. The CTMC state changes are triggered by the (probabilistic) occurrences of chemical reactions. Given m chemical reactions, let $\mathcal{R}_j$ denote the jth reaction of type:

$$\mathcal{R}_j \quad : \quad \nu_{j,1}^- S_1 + \dots + \nu_{j,n}^- S_n \xrightarrow{\theta_j} \nu_{j,1}^+ S_1 + \dots + \nu_{j,n}^+ S_n,$$

where the vectors $\boldsymbol{\nu}_j^-$ and $\boldsymbol{\nu}_j^+$ represent the stoichiometries of the underlying chemical kinetics for the reactants and products, respectively. Let $\boldsymbol{\nu}_j \in \mathbb{Z}^n$ denote the overall (non-zero) state-change vector for the jth reaction type, specifically $\boldsymbol{\nu}_j = \boldsymbol{\nu}_j^+ - \boldsymbol{\nu}_j^-$, for $j \in \{1, \dots, m\}$. Assuming mass action kinetics (and omitting time dependency for $\boldsymbol{x}(t)$), the reaction $\mathcal{R}_j$ leads to the propensity [41]:

$$h_j(\boldsymbol{x}, \boldsymbol{\theta}) = \theta_j \alpha_j(\boldsymbol{x}) = \theta_j \prod_{i=1}^{n} \binom{x_i}{\nu_{j,i}^-}, \tag{1}$$

where $\boldsymbol{\theta} = (\theta_1, \dots, \theta_m)^{\mathsf{T}}$ is the vector of rate constants. In general, $\boldsymbol{\theta}$ is unknown and must be estimated from experimental data—that is the aim of our work. Our algorithm can work with propensity functions factorisable as in (1), but it is not restricted to mass action kinetics (*i.e.*, the functions α_j's can be arbitrary).

Cross-Entropy Method for Optimisation. The Kullback-Leibler divergence [20] or cross-entropy (CE) between two probability densities g and h is:

$$\mathcal{D}(g,h) = \mathbb{E}_g\left[\ln\frac{g(\boldsymbol{X})}{h(\boldsymbol{X})}\right] = \int g(\boldsymbol{x})\ln\frac{g(\boldsymbol{x})}{h(\boldsymbol{x})}d\boldsymbol{x}$$

where $\boldsymbol{X}$ is a random variable with density g, and $\mathbb{E}_g$ is expectation w.r.t. g. Note that $\mathcal{D}(g,h) \geq 0$ with equality iff $g = h$ (almost everywhere). (However, $\mathcal{D}(g,h) \neq \mathcal{D}(h,g)$.) The CE has been successfully adopted for a wide range of hard problems, including rare event simulation for biological systems [7], discrete, and continuous optimisation [28,29]. Consider the minimisation of an *objective* function J over a space χ (assuming such minimum exists), $\gamma^* = \min_{x\in\chi} J(x)$. The CE method performs a Monte Carlo search over a parametric family of densities $\{f(\cdot;\boldsymbol{v}), \boldsymbol{v} \in \mathcal{V}\}$ on χ that contains as a limit the (degenerate) Dirac density that puts its entire mass on a value $x^* \in \chi$ such that $J(x^*) = \gamma^*$—the so called *optimal* density. The key idea is to use the CE to measure how far a candidate density is from the optimal density. In particular, the method solves a sequence of optimisation problems of the type below for different values of γ by minimising the CE between a putative optimal density $g^*(\boldsymbol{x}) \propto I_{\{J(\boldsymbol{x})\leq\gamma\}}f(\boldsymbol{x},\boldsymbol{v}^*)$ for some $\boldsymbol{v}^* \in \mathcal{V}$, and the density family $\{f(\cdot;\boldsymbol{v}), \boldsymbol{v} \in \mathcal{V}\}$

$$\min_{\boldsymbol{v}\in\mathcal{V}}\mathcal{D}(g^*, f(\cdot;\boldsymbol{v})) = \max_{\boldsymbol{v}\in\mathcal{V}}\mathbb{E}_u\left[I_{\{J(\boldsymbol{X})\leq\gamma\}}\ln f(\boldsymbol{X};\boldsymbol{v})\right] \tag{2}$$

where I is the indicator function and $\boldsymbol{X}$ has density $f(\cdot;\boldsymbol{u})$ for $\boldsymbol{u} \in \mathcal{V}$. The definition of density g^* above essentially means that, for a given γ, we only consider densities that are positive only for arguments $\boldsymbol{x}$ for which $J(\boldsymbol{x}) \leqslant \gamma$. The generic CE method involves a 2-step procedure which alternates solving (2) for a candidate g^* with adaptively updating γ. In practice, problem (2) is solved approximately via a Monte Carlo adaptation, *i.e.*, by taking sample averages as estimators for $\mathbb{E}_u$. The output of the CE method is a sequence of putative optimal densities identified by their parameters $\hat{\boldsymbol{v}}_0, \hat{\boldsymbol{v}}_1, \ldots, \hat{\boldsymbol{v}}^*$, and performance scores $\hat{\gamma}_0, \hat{\gamma}_1, \ldots, \hat{\gamma}^*$, which improve with probability 1. For our problem, a key benefit of the CE method is that an analytic solution for (2) can be found when $\{f(\cdot;\boldsymbol{v}), \boldsymbol{v} \in \mathcal{V}\}$ is the exponential family of distributions. (More details in [29].)

Cross-Entropy Method for the SSA. We denote by r_j the number of firings of the jth reaction channel, τ_i the time between the ith and $(i-1)$th reaction, and τ_{r+1} the final time interval at the end of the simulation in which no reaction occurs. It can be shown that an exact SSA trajectory $\boldsymbol{z} = (\boldsymbol{x}_0, \ldots, \boldsymbol{x}_r)$, where r is the total number of reaction events $r = \sum_{j=1}^m r_j$, belongs to the exponential family of distributions [41]—whose optimal CE parameter can be found analytically. Daigle *et al.* [8] showed that the solution of (2) for the SSA likelihood yields the following Monte Carlo estimate of the optimal CE parameter v_j^*,

$$\hat{\theta}_j = \hat{v}_j^* = \frac{\sum_{k=1}^K r_{jk}I_{\{J(z_k)\leq\gamma\}}}{\sum_{k=1}^K I_{\{J(z_k)\leq\gamma\}}\left(\sum_{i=1}^{r_k+1}\alpha_j(\boldsymbol{x}_{i-1,k})\tau_{ik}\right)} \tag{3}$$

where K is the number of SSA trajectories of the Monte Carlo approximation of (2), z_k is the kth trajectory, r_{jk} and τ_{ik} are as before but w.r.t. the kth trajectory, $x_{i,k}$ denotes the state after the $(i-1)$th reaction in the kth trajectory, and the fraction is defined only when the denominator is nonzero (*i.e.*, there is at least one trajectory z_k for which $J(z_k) \leq \gamma$—so-called *elite* samples). Note for $\gamma = 0$, the CE estimator (3) coincides with the maximum likelihood estimator (MLE) for θ_j over the same trajectory. Following [7] and [26, Sect. 5.3.4], it is easy to show that a Monte Carlo estimator of the covariance matrix of the optimal parameter estimators (3) is given (written in operator style) by the matrix:

$$\hat{\Sigma}^{-1} = \left[-\frac{1}{K_E} \sum_{k \in E} \frac{\partial^2}{\partial \theta^2} - \frac{1}{K_E} \sum_{k \in E} \frac{\partial}{\partial \theta} \cdot \frac{\partial}{\partial \theta}^{\mathrm{T}} \right.$$
$$\left. + \frac{1}{K_E^2} \left(\sum_{k \in E} \frac{\partial}{\partial \theta} \right) \cdot \left(\sum_{k \in E} \frac{\partial}{\partial \theta} \right)^{\mathrm{T}} \right] (\log f(\theta | x, z_k)) \quad (4)$$

where E is the set of elite samples, $K_E = |E|$, the operator $\frac{\partial^2}{\partial \theta^2}$ returns a $m \times m$ matrix, $\frac{\partial}{\partial \theta}$ returns an m-dimensional vector ($m \times 1$ matrix), and $\frac{\partial}{\partial \theta}^{\mathrm{T}}$ denotes matrix transpose. From Eq. (4) parameter variance estimates can be readily derived. However, a more numerically stable option is to approximate the variance of the jth parameter estimator using the sample variance

$$\hat{\sigma}_j^2 = \frac{1}{K_E} \sum_{k \in E} \left(\frac{r_{jk}}{\sum_{i=1}^{r_k+1} \alpha_j(x_{i-1,k}) \tau_{ik}} - \hat{\theta}_j \right)^2 . \quad (5)$$

3   Methods

In this section, we present our *stochastic rate parameter inference with cross-entropy* (SPICE) algorithm.

Overview. To efficiently sample the parameter space, we treat each stochastic rate parameter as being log-normally distributed, *i.e.*, $\theta_j \sim$ Lognormal$(\omega_j, \mathrm{var}(\omega_j))$, where $\omega_j = \log(\theta_j)$ is the log-transformed parameter calculated analagously to (3) and (4), respectively. For the initial iteration, we sample the parameter vector θ from the (log-transformed) desired parameter search space $[\theta_{\mathrm{MIN}}^{(0)}, \theta_{\mathrm{MAX}}^{(0)}]$ using a Sobol low-discrepancy sequence [33] to ensure adequate coverage. Subsequent iterations then generate a sequence of distribution parameters $\{(\gamma_n, \theta_n, \Sigma_n)\}$ which aim to converge to the optimal parameters as follows:

1. **Updating of γ_n:** Generate K sample trajectories using the SSA, $z_1, \ldots, z_K$, from the model $f(\cdot; \theta^{(n-1)})$ with $\theta^{(n-1)}$ sampled from the lognormal distribution, and sort them in order of their performances $J_{1'} \leq \cdots \leq J_{K'}$ (see Eqs. (7) and (6) for the actual definition of the performance, or score, function we adopt). For a fixed small ρ, say $\rho = 10^{-2}$, let $\hat{\gamma}_n$ be defined as the ρth quantile of $J(z)$, *i.e.*, $\hat{\gamma}_n = J_{(\lceil \rho K \rceil)}$.

2. **Updating of θ_n:** Using the estimated level $\hat{\gamma}_n$, use the same K sample trajectories $z_1, \ldots, z_K$ to derive $\hat{\theta}_n$ and $\hat{\sigma}_n^2$ from the solution of Eqs. (3) and (4). In case of numerical issues (or undersampling) in our implementation we switch to (5) for updating the variance.

The SPICE algorithm's pseudocode is shown in Algorithm 1. This 2-step approach provides a simple iterative scheme which converges asymptotically to the optimal density. A reasonable termination criteria to take would be to stop if $\hat{\gamma}_n \not\leq \hat{\gamma}_{n-1} \not\leq \ldots$ for a fixed number of iterations. In general, more samples are required as the mean and variance of the estimates approach their optima.

Adaptive Sampling. We adaptively update the number of samples K_n taken at each iteration. The reasoning is to ensure the parameter estimates improve with statistical significance at each step. Thus, our method allows the algorithm to make faster evaluations early on in the iterative process, and concentrate simulation time on later iterations, where it becomes increasingly hard to distinguish significant improvements of the estimated parameters. We update our parameters based on a fixed number of elite samples, K_E, satisfying $J(z) \leq \gamma$. The performance of the 'best' elite sample is denoted J_n^*, while the performance of the 'worst' elite sample—previously given by the ρth quantile of $J(z)$—is $\hat{\gamma}_n$. The quantile parameter ρ is adaptively updated each iteration as $\rho_n = K_E/K_n$, where K_E is typically taken to be 1–10% of the base number of samples K_0. At each iteration, a check is made for improvement in either of the best or worst performing elite samples, i.e., if, $J_n^* < J_{n-1}^*$ or $\hat{\gamma}_n < \hat{\gamma}_{n-1}$, then we can update our parameters and proceed to the next iteration. If no improvement in either values are found, the number of samples K_n in the current iteration is increased in increments, up to a maximum $K_{\max}$. If we hit the maximum number of samples $K_{\max}$ for c iterations (e.g., $c = 3$), then this suggests no further significant improvement can be made given the restriction on the number of samples.

Objective Function. The SPICE algorithm has been developed to handle an arbitrary number of datasets. Given N time series datasets, SPICE associates N objective function scores with each simulated trajectory. Each objective value corresponds to the standard sum of L^2 distances of the trajectory across all time points in the respective dataset:

$$J_n(z) = \sum_{t=1}^{T} (y_{n,t} - x_t)^2 \qquad 1 \leq n \leq N \qquad (6)$$

where $x_t = x(t)$ and $y_{n,t}$ is the datapoint at time t in the nth dataset. To ensure adequate coverage of the data, we choose our elite samples to be the best performing quantile of trajectories for each individual dataset (with scores J_n).

In the absence of temporal correlation within the data (e.g., when measurements between time points are independent or individual cells cannot be tracked as in flow cytometry data), we instead construct an empirical Gaussian mixture model for each time point within the data. Each mixture model at time t is comprised of N multivariate normal distributions, each with a vector of mean values $\boldsymbol{y}_{n,t}$ corresponding to the *observed* species in the nth dataset, and diagonal covariance matrix $\boldsymbol{\sigma}_n^2$ corresponding to an error estimate or variance of the measurements on the species. In our experiments we used a 10% standard deviation, as we did not have any information about measurement noise. We then take the objective score function to be proportional to the negative log-likelihood of the simulated trajectory w.r.t. the data:

$$J_n(\boldsymbol{z}) = - \sum_{t=1}^{T} \ln \left(\sum_{n=1}^{N} \exp \left[-\frac{1}{2}(\boldsymbol{y}_{n,t} - \boldsymbol{x}_t)^\mathsf{T} \boldsymbol{\sigma}_n^{-2}(\boldsymbol{y}_{n,t} - \boldsymbol{x}_t) \right] \right). \qquad (7)$$

Smoothed Updates. We implement the parameter smoothing update formula

$$\hat{\boldsymbol{\theta}}^{(n)} = \lambda \tilde{\boldsymbol{\theta}}^{(n)} + (1 - \lambda)\hat{\boldsymbol{\theta}}^{(n-1)}, \quad \hat{\boldsymbol{\sigma}}^{(n)} = \beta_n \tilde{\boldsymbol{\sigma}}^{(n)} + (1 - \beta_n)\hat{\boldsymbol{\sigma}}^{(n-1)}$$

where $\beta_n = \beta - \beta \left(1 - \frac{1}{n}\right)^q$, $\lambda \in (0, 1]$, $q \in \mathbb{N}^+$ and $\beta \in (0, 1)$ are smoothing constants, and $\tilde{\boldsymbol{\theta}}, \tilde{\boldsymbol{\sigma}}$ are outputs from the solution of the cross-entropy in Eq. (2), approximated by (3) and (4), respectively. Parameter smoothing between iterations has three important benefits: (i) the parameter estimates converge to a more stable value, (ii) it reduces the probability of a parameter value tending towards zero within the first few iterations, and (iii) it prevents the sampling distribution from converging too quickly to a degenerate point probability mass at a local minima. Furthermore, [6] provide a proof that the CE method converges to an optimal solution with probability 1 in the case of smoothed updates.

Multiple Shooting and Particle Splitting. SPICE can optionally utilise these two techniques for trajectory simulation between time intervals. For multiple shooting we construct a sample trajectory comprised of T intervals matching the time stamps within the data $\boldsymbol{y}$. Originally [42], each segment from $\boldsymbol{x}_{t-1}$ to $\boldsymbol{x}_t$ was simulated using an ODE model with the initial conditions set to the previous time point of the dataset, *i.e.*, $\boldsymbol{x}_{t-1} = \boldsymbol{y}_{t-1}$. We instead treat the data as being mixture-normally distributed, thus we sample our initial conditions $\boldsymbol{x}_{t-1} \sim \mathcal{N}(\boldsymbol{y}_{n,t-1}, \boldsymbol{\sigma}_{n,t-1}^2)$, where the index of the time series n is first uniformly sampled. Using the SSA, each piecewise section of a trajectory belonging to sample k is then simulated with the same parameter vector $\boldsymbol{\theta}$. For particle splitting we adopt a multilevel splitting approach as in [8], and the objective function is calculated after the simulation of each segment from $\boldsymbol{x}_{t-1}$ to $\boldsymbol{x}_t$. The trajectories $\boldsymbol{z}_k$ satisfying $J(\boldsymbol{z}_k) \leq \hat{\gamma}$ are then re-sampled with replacement K_n times before simulation continues (recall K_n is the number of samples in the nth iteration). This process aims at discarding poorly performing trajectories in favour of those 'closest' to the data. This will in turn create an enriched sample, at the cost of introducing an aspect of bias propagation.

Hyperparameters. SPICE allows for the inclusion of hyperparameters $\boldsymbol{\phi}$ (*e.g.*, scaling constants, and non kinetic-rate parameters), which are sampled (logarithmically) alongside $\boldsymbol{\theta}$. These hyperparameters are updated at each iteration via the standard CE method.

Tau-Leaping. With inexact, faster methods such as tau-leaping [15] a degree of accuracy is traded off in favour of computational performance. Thus, we are interested in replacing the SSA with tau-leaping in our SPICE algorithm. The next Proposition shows that with a tau-leaping trajectory we get the same form for the optimal CE estimator as in (3).

Proposition 1. *The CE solution for the optimal rate parameter over a tau-leaping trajectory is the same as that for a standard SSA trajectory.*

Proof. We shall use the same notation of Sect. 2 and further assume a trajectory in which state changes occur at times t_l, for $l \in \{0, 1, \dots, L\}$. For each given time interval of size τ_l of the tau-leaping algorithm, $k_{jl} \in \mathbb{Z}^+$ firings of each reaction channel $\mathcal{R}_j$ are sampled from a Poisson process with mean $\lambda_{jl} = \theta_j \alpha_j(x_{t_l})\tau_l$. Thus, the probability of firing k_{jl} reactions, in the interval $[t_l, t_l + \tau_l)$, given the initial state x_{t_l} is $P(k_{jl}|x_{t_l}, \lambda_{jl}) = \exp\{-\lambda_{jl}\}(\lambda_{jl})^{k_{jl}}/k_{jl}!$, where $P(0|x_{t_l}, 0) = 1$. Therefore, the combined probability across all reaction channels is:

$$\prod_{j=1}^{m} P(k_{jl}|x_{t_l}, \lambda_{jl}) = \prod_{j=1}^{m} \frac{\exp\{-\lambda_{jl}\}(\lambda_{jl})^{k_{jl}}}{k_{jl}!}.$$

Extending for the entire trajectory, the complete likelihood is given by:

$$\mathcal{L} = \prod_{l=0}^{L} \prod_{j=1}^{m} P(k_{jl}|x_{t_l}, \lambda_{jl}) = \prod_{l=0}^{L} \prod_{j=1}^{m} \frac{\exp\{-\lambda_{jl}\}(\lambda_{jl})^{k_{jl}}}{k_{jl}!}.$$

We can conveniently factorise the likelihood into component likelihoods associated with each reaction channel as $\mathcal{L} = \prod_{j=1}^{m} \mathcal{L}_j$, where each component $\mathcal{L}_j$ is given by $\mathcal{L}_j = \prod_{l=0}^{L} \frac{\exp\{-\lambda_{jl}\}(\lambda_{jl})^{k_{jl}}}{k_{jl}!}$. Expanding λ_{jl}:

$$\mathcal{L}_j = \prod_{l=0}^{L} \frac{\exp\{-\theta_j \alpha_j(x_{t_l})\tau_l\}(\theta_j \alpha_j(x_{t_l})\tau_l)^{k_{jl}}}{k_{jl}!}$$

$$= \theta_j^{r_j} \exp\left\{-\theta_j \sum_{l=0}^{L} \alpha_j(x_{t_l})\tau_l\right\} \prod_{l=0}^{L} \frac{(\alpha_j(x_{t_l})\tau_l)^{k_{jl}}}{k_{jl}!},$$

where $r_j = \sum_{l=0}^{L} k_{jl}$, *i.e.*, the total number of firings of reaction channel $\mathcal{R}_j$. From [29], the solution to (2) can be found by solving:

$$\mathbb{E}_u\left[I_{\{J(X)\geq\gamma\}} \nabla \ln \mathcal{L}_j\right] = 0,$$

given that the differentiation and expectation operators can be interchanged. Expanding $\ln \mathcal{L}_j$ and simplifying, we get:

$$\mathbb{E}_u\left[I_{\{J(X)\geq\gamma\}} \nabla \left(\ln \theta_j^{r_j} - \theta_j \sum_{l=0}^{L} \alpha_j(x_{t_l})\tau_l + \ln\left\{\prod_{l=0}^{L} \frac{(\alpha_j(x_{t_l})\tau_l)^{k_{jl}}}{k_{jl}!}\right\}\right)\right] = 0.$$

Algorithm 1. SPICE — Stochastic Rate Parameter Inference with Cross-Entropy

---

    input : Datasets represented by mixture models Φ_i at times t_i for $0 \leqslant i \leqslant L$, initial parameter bounds (log-transformed) $\left[\boldsymbol{\theta}_{\mathrm{MIN}}^{(0)}, \boldsymbol{\theta}_{\mathrm{MAX}}^{(0)}\right]$, quantile ρ.

    output: Estimate of parameters $\hat{\boldsymbol{\theta}}^{(n)}$, and their variances $\hat{\boldsymbol{\Sigma}}^{(n)}$.

1 Iteration $n \leftarrow 1$

2 Generate Sobol sequence $S \leftarrow \left[\hat{\boldsymbol{\theta}}_{\mathrm{MIN}}^{(0)}, \hat{\boldsymbol{\theta}}_{\mathrm{MAX}}^{(0)}\right]$ hypercube

3 Initial sample size $K_1 \leftarrow K_{\min}$

4 Initialise $\gamma_1 \leftarrow \infty$

5 **repeat**

6   **for** $i \leftarrow 1$ **to** L **do**

7   Set initial time point $t_0 \leftarrow t_{i-1}$

8   **for** $k \leftarrow 1$ **to** K_n **do**

9   **if** $i = 1$ **then**

10   Set initial state $\boldsymbol{x} \leftarrow \boldsymbol{y}_0$

11   **if** $n = 1$ **then**

12   Sample parameters from Sobol sequence $\boldsymbol{\theta}_k \leftarrow S(k)$

13   **else**

14   Sample parameters from the parameter distribution $\boldsymbol{\theta}_k \sim \mathrm{Lognormal}\left(\hat{\boldsymbol{\theta}}^{(n-1)}, \hat{\boldsymbol{\Sigma}}^{(n-1)}\right)$

15   **else**

16   **if** *Method = Multiple Shooting* **then**

17   Sample the starting state from the distribution of the data $\boldsymbol{x} \sim \Phi_i$

18   **else**

19   Continue from the end state of the current simulation $\boldsymbol{x} \leftarrow \boldsymbol{z}_{i-1,k}$

20   Forward simulate $\boldsymbol{z}_{i,k} \leftarrow \mathrm{SSA}(\boldsymbol{x}, t_0, t_i, \boldsymbol{\theta}_k)$

21   **if** *(Method = Splitting) or $(i = L)$* **then**

                  `// depending on type of data, use (6) or (7)`

22   Calculate the cost function $d_k \leftarrow J(\boldsymbol{z}_k)$

23   **if** *Method = Splitting* **then**

24   Sample with replacement weighted trajectories satisfying $d_k < \gamma_n$

25 $\gamma_n \leftarrow \rho$th quantile of $(d_1, \ldots, d_{K_n})$

26   Compute $\hat{\boldsymbol{\theta}}^{(n)}$ and $\hat{\boldsymbol{\sigma}}^{(n)}$ by means of Eqs. (3) and (4) (or (5)), using the elite trajectories satisfying $d_k < \gamma_n$ (and taking log appropriately in (3) and (4))

27   Increment $n \leftarrow n + 1$

28   Adaptively update K_n

29 **until** *convergence detected in* $\{\gamma_1, \ldots, \gamma_{n-1}\}$

We can then take the derivative, ∇, with respect to θ_j,

$$\mathbb{E}_u\left[I_{\{J(X)\geq\gamma\}}\left(\frac{r_j}{\theta_j} - \sum_{l=0}^{L}\alpha_j(x_{t_l})\tau_l\right)\right] = 0.$$

It is simple to see that the previous entity holds when $r_j/\theta_j = \sum_{l=0}^{L}\alpha_j(x_{t_l})\tau_l$, yielding the Monte Carlo estimate,

$$\hat{\theta}_j = \frac{\sum_{k=1}^{K}I_{\{J(z_k)\leq\gamma\}}r_{jk}}{\sum_{k=1}^{K}I_{\{J(z_k)\leq\gamma\}}\sum_{l=0}^{L}\alpha_j(x_{t_l,k})\tau_{l,k}}.$$

$\square$

4 Experiments

We utilise our SPICE algorithm on four commonly investigated systems: (i) the Lotka-Volterra predator–prey model, (ii) a Yeast Polarization model, (iii) the bistable Schlögl system, and (iv) the Genetic Toggle Switch. We present results for each system obtained using both the standard SSA and optimised tau-leaping (with an error control parameter of $\varepsilon = 0.1$) to drive our simulations.

For each run of the algorithm we set the sample parameters $K_E = 10$, $K_{\min} = 1,000$, $K_{\max} = 20,000$, and set an upper limit on the number of iterations to 250. The smoothing parameters (λ, β, q) were set to $(0.7, 0.8, 5)$ respectively. For our analysis, we define the mean relative error (MRE) between a parameter estimate $\hat{\theta}$ and the truth θ^* as $\mathrm{MRE}(\%_{\mathrm{ERR}}) = M^{-1}\sum_{j}^{M}|\hat{\theta}_j - \theta_j^*|/\theta_j^* \times 100$. All our experiments were performed on a Intel Xeon 2.9GHz Linux system *without* using multiple cores—all reported CPU times are single-core. SPICE has been implemented in Julia and is open source (https://github.com/pzuliani/SPICE).

For models (i)–(iii), we use synthetic data where the true solution is known, and compare the results of SPICE against some commonly used parameter estimation techniques implemented in COPASI 4.16 [17]. Specifically, we check the performance of SPICE against the genetic algorithm (GA), evolution strategy (ES), evolutionary programming (EP), and particle swarm (PS) implementations. For the ES and EP algorithms we allow 250 generations with a population of 1,000 particles. For the GA, we run 500 generations with 2,000 particles. For the PS, we allow 1,000 iterations with 1,000 particles[1]. For model (iv), the Genetic Toggle Switch, we show results for SPICE using *real* experimental data.

All statistics presented are based on 100 runs of each algorithm using fixed datasets. For each approach we also compared the performance of using the standard SSA versus tau-leaping, alongside multiple-shooting and particle splitting approaches. However, for the models tested, neither multiple shooting nor particle splitting helped in reducing CPU times or improving the estimates accuracy.

---

[1] NB: we also tested the COPASI implementations using greater populations and more iterations (not shown), but found little improvement for the significant increase in computational cost.

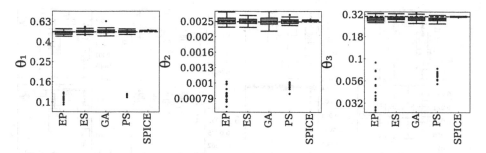

Fig. 1. Lotka-Volterra model: box plots showing the summary statistics across 100 runs of COPASI and SPICE for each of the 3 parameter estimates. We note SPICE consistently has the least variance.

Lotka-Volterra Predator–Prey Model. We implement the standard Lotka-Volterra model below with real parameters $(\theta_1, \theta_2, \theta_3) = (0.5, 0.0025, 0.3)$, and initial population $(X_1, X_2) = (50, 50)$

$$X_1 \xrightarrow{\theta_1} X_1 + X_1 \qquad X_1 + X_2 \xrightarrow{\theta_2} X_2 + X_2 \qquad X_2 \xrightarrow{\theta_3} \emptyset$$

We artificially generated 5 datasets each consisting of 40 timepoints using Gillespie's SSA, and performed parameter estimation based on these datasets. For the initial iteration, we placed bounds on the Sobol sequence parameter search space of $\theta_j \in [1e-6, 10]$, for $j = 1, 2, 3$. The minimum, maximum, and average MRE between the true parameters and their estimates across all 100 runs of each algorithm (using the standard SSA) are summarised in Table 1, together with corresponding CPU run times. Box plots summarising the obtained parameter estimates across all runs of each method are displayed in Fig. 1.

In the previous Lotka-Volterra predator–prey example, SPICE was provided with the complete data for both species X_1, X_2. However, we are also concerned with cases where the data is not fully observed, *i.e.*, when we have latent species. To compare the effects of latent species on the quality of parameter estimates, we ran SPICE again (averaging across 100 runs), this time supplying information about species X_1 alone. The results are presented in Table 1.

Yeast Polarization Model. We implement the Yeast Polarization model (see below) with real parameters $(\theta_1, \ldots, \theta_8) = (0.38, 0.04, 0.082, 0.12, 0.021, 0.1, 0.005, 13.21)$, and initial population $(R, L, RL, G, G_a, G_{bg}, G_d) = (500, 4, 110, 300, 2, 20, 90)$. The reactions of the model are [8]:

$$\emptyset \xrightarrow{\theta_1} R \qquad\qquad RL + G \xrightarrow{\theta_5} G_a + G_{bg}$$

$$R \xrightarrow{\theta_2} \emptyset \qquad\qquad G_a \xrightarrow{\theta_6} G_d$$

$$L + R \xrightarrow{\theta_3} RL + L \qquad\qquad G_d + G_{bg} \xrightarrow{\theta_7} G$$

$$RL \xrightarrow{\theta_4} R \qquad\qquad \emptyset \xrightarrow{\theta_8} RL$$

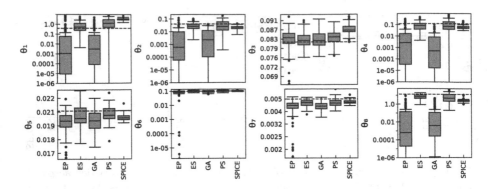

Fig. 2. Yeast Polarization parameter estimates: box plots showing the summary statistics of all 8 parameter estimates across 100 runs of COPASI's methods and SPICE. We note once again SPICE produces the least variation of obtained estimates.

We artificially generated 5 datasets each consisting of 17 timepoints using Gillespie's SSA, and performed parameter estimation based on these datasets. For the initial iteration, we placed bounds on the parameter search space of $\theta_j \in [1e-6, 10]$ for $1 \leqslant j \leqslant 7$, and $\theta_8 \in [1e-6, 100]$. The average relative errors between the estimated and the real parameters across 100 runs of the algorithm are summarised in Table 1, along with the corresponding CPU run times. The variability of the estimates obtained using SPICE (and other methods) are shown in Fig. 2.

Schlögl System. We use the Schlögl model [30] with parameters $(\theta_1, \theta_2, \theta_3, \theta_4) = (3e-7, 1e-4, 1e-3, 3.5)$, and initial population $(X, A, B) = (250, 1e5, 2e5)$. This model is well known to produce bistable dynamics (see Fig. 4).

$$2X + A \xrightarrow{\theta_1} 3X \qquad\qquad B \xrightarrow{\theta_3} X$$
$$3X \xrightarrow{\theta_2} 2X + A \qquad\qquad X \xrightarrow{\theta_4} B$$

We artificially generated 10 datasets (in order to partially capture a degree of the bistable dynamics) each consisting of 100 timepoints, and performed parameter estimation based on these datasets (also see Fig. 4). For the initial iteration, we placed bounds on the parameter search space of $\theta_1 \in [1e-9, 1e-5]$, $\theta_2 \in [1e-6, 0.01]$, $\theta_3 \in [1e-5, 10]$, $\theta_4 \in [0.01, 100]$. Unlike the previous models, we explicitly ran the Schlögl System using tau-leaping for *all* algorithms, due to the computation time being largely infeasible under the same conditions (4.5 h in SPICE, 48+ h in COPASI). The MRE of all the estimated parameters, together with CPU times for each algorithm are summarised in Table 1. Box plots of the SPICE algorithm's performance are presented in Fig. 3. Note that the Schlögl system is sensitive to the initial conditions, so even slight perturbations of its parameters can cause the system to fail in producing bimodality.

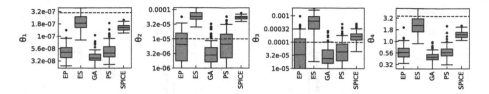

Fig. 3. Schlögl system parameter estimates: box plots comparing the parameter estimates across 100 runs of COPASI's methods and SPICE (all simulated using tau-leaping, $\varepsilon = 0.1$). Again, SPICE shows the smallest variance, with mean estimates quite close to the real values of θ_1 and θ_3. For θ_2 and θ_4, all the best mean estimates have variance much larger than SPICE estimates.

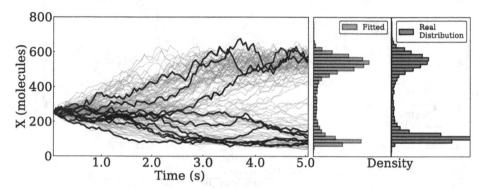

Fig. 4. Schlögl: from the left: solid black lines: the 10 datasets generated using the SSA direct method and the real parameters, and used as input for SPICE. Blue lines: 100 model runs with estimated parameters sampled by the final parameter distributions obtained by SPICE with the direct method (means = $(2.14e{-}7, 7.63e{-}5, 4.54e{-}4, 2.18)$); variances = $(7.81e{-}16, 2.81e{-}10, 4.05e{-}8, 0.13)$). Fitted: empirical distribution of 1,000 model simulations with sampled parameters from SPICE output. Real distribution: empirical distribution of 1,000 model simulations with the real parameters. (Color figure online)

Toggle Switch Model. The genetic toggle switch is a well studied bistable system, with particular importance toward synthetic biology. The toggle switch is comprised of two repressors, and two promoters, often mediated in practice through IPTG[2] and aTc[3] induction. We perform parameter inference based on real high-throughput data (see Fig. 5), implemented upon a simple model (see below) based on [12]. For our model, we define the following reaction propensities:

$$h_1 = \theta_1 \times \text{GFP} \qquad\qquad h_3 = \theta_3 \times \text{MCHERRY}$$

$$h_2 = \frac{\theta_2 \times \phi_1}{1 + \phi_1 + \phi_2 \times \text{MCHERRY}^2} \qquad h_4 = \frac{\theta_4 \times \phi_3}{1 + \phi_3 + \phi_4 \times \text{GFP}^2}$$

---

[2] Isopropyl β-D-1-thiogalactopyranoside.
[3] anhydrotetracycline.

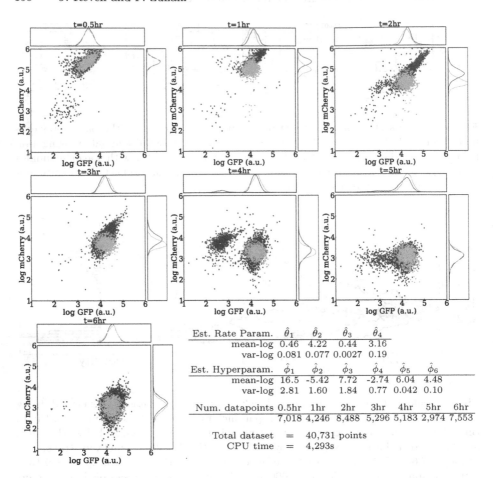

Fig. 5. Toggle switch model: blue circles: the experimental data with the $\log_{10}(\text{GFP})$ fluorescence plotted against the $\log_{10}(\text{mCherry})$ fluorescence, across all timepoints up to 6h. Orange circles: 1,000 model simulations using the direct method, with parameters sampled from the final distribution obtained by SPICE using tau-leaping ($\varepsilon = 0.1$). (Color figure online)

where GFP and mCherry are the two model species (reporter molecules), and the stochastic rate parameters are $(\theta_1, \ldots, \theta_4)$. The data used for parameter inference was obtained through fluorescent flow cytometry in [21], via the GFP and mCherry reporters, and consists 40,731 measurements across 7 timepoints over 6 h. We look specifically at the case where the switch starts in the low-GFP (high mCherry) state, and switches to the high-GFP (low-mCherry) state over the time course after aTc induction to the cells. The inclusion of real, noisy data requires a degree of additional care as the data needs to be rescaled from arbitrary units (a.u.) to discrete molecular counts. We assume a linear (multiplicative) scale, *e.g.*, such that GFP (a.u.) $= \phi_5 \times$ GFP *molecules*. Furthermore, we can no longer assume all the cells begin at the same state, and we must assume the initial state belongs to a distribution. This introduces extra so-called 'hyperparameters', specifically the GFP molecule count to fluorescent (a.u.) scale factor ϕ_5, and the respective mCherry scale factor ϕ_6. In addition, the model now contains 4 additional parameters, $\phi_1, \ldots, \phi_4$, which in turn are required to be estimated. Each hyperparameter is initially sampled as before using the low-discrepancy Sobol sequence, and updated using the means and variances of the generated elite samples as per the CE method.

The placed bounds on the initial kinetic parameter search space, based upon reported half-lives for the variants of GFP [2] and mCherry [31], were $\theta_{1,3} \in [1e-3, 1]$, and $\theta_{2,4} \in [1, 50]$. The respective bounds on the search space for the hyperparameters were $\phi_{1,2,3,4} \in [1e-3, 10]$, and $\phi_{5,6} \in [50, 500]$. To generate the parameter estimates, we used SPICE with tau-leaping ($\varepsilon = 0.1$, CPU time = 4,293 s). The estimated parameters and the resulting fit against the data for the model can be seen in Fig. 5.

Table 1. The relative errors for each stochastic rate parameter averaged across 100 runs using COPASI's Evolutionary Programming (EP), Evolution Strategy (ES), Genetic Algorithm (GA), and Particle Swarm (PS) algorithms, and our SPICE algorithm are shown. The minimum, maximum, and average mean relative error (MRE) for all parameter estimates across all runs are also given alongside the averaged CPU time.

Lotka-Volterra Model							
Alg.	θ_1	θ_2	θ_3	Min. MRE	Av. MRE	Max. MRE	Av. CPU
		(%ERR)		(%ERR)	(%ERR)	(%ERR)	(s)
EP	38.4	3.5	29.6	0.4	23.8	156.5	1200
ES	3.8	0.6	4.4	0.3	3.0	9.0	5763
GA	5.2	0.8	5.7	0.8	3.9	15.2	3640
PS	25.6	2.2	18.6	**0.1**	15.5	126.6	2689
SPICE	**3.6**	**0.4**	**0.4**	1.0	**1.5**	**2.1**	**1025**

Lotka-Volterra Latent-Species Model							
SPICE	9.4	0.41	6.4	4.1	5.4	6.8	1589

Yeast Polarization Model											
Alg.	θ_1	θ_2	θ_3	θ_4	θ_5	θ_6	θ_7	θ_8	Min. MRE	Av. MRE	Max. MRE	Av. CPU
				(%ERR)					(%ERR)	(%ERR)	(%ERR)	(s)
EP	662.9	138.4	1.7	235.4	1.7	25.3	3.4	357.0	56.2	178.2	316.9	**405**
ES	**109.8**	**18.5**	**1.2**	35.4	1.3	3.3	1.5	**27.9**	**3.6**	**24.9**	62.8	1650
GA	564.0	120.2	1.3	275.3	1.6	6.5	2.6	312.4	38.8	160.5	299.4	2696
PS	156.4	29.0	1.4	52.6	**0.9**	3.7	1.6	48.6	7.3	36.8	173.6	1755
SPICE	221.2	21.7	2.5	**34.9**	**0.9**	**1.7**	**1.1**	62.7	27.6	43.3	**54.4**	1116

Schlögl System Model								
Alg.	θ_1	θ_2	θ_3	θ_4	Min. MRE	Av. MRE	Max. MRE	Av. CPU
		(%ERR)			(%ERR)	(%ERR)	(%ERR)	(s)
EP	12.2	9.7	15.1	142.9	24.4	45.0	60.5	**307**
ES	**3.3**	15.5	19.0	**40.3**	**11.5**	**19.3**	31.7	1505
GA	13.7	11.0	14.0	159.7	32.2	49.6	66.3	987
PS	12.0	**8.5**	11.4	141.4	18.7	43.3	60.0	1095
SPICE	4.6	14.6	**6.3**	73.0	18.5	24.6	**30.9**	1054

5   Discussion

We can see from the presented results that our SPICE algorithm performs well
on the models studied. For the Lotka-Volterra model the quality of the estimates
is always good—there is no relative error larger than 2.1% in Table 1 for SPICE.
The CPU times are reasonable in absolute terms (about 20 min, single core),
and much smaller than those of the methods implemented in COPASI, and with
smaller errors. Also, having one unobserved species (X_2) in the data does not
seem to impact the results very much. In particular, from Table 1 we see that the
latent model indeed has higher error than the fully observable model. However,
the error is always smaller than 10%, which is acceptable.

The Yeast Polarization model is a more difficult system: we can indeed see
from Table 1 that a number of parameter estimates have large relative errors.
These are the same 'hard' parameters estimated by MCEM^2 [8] with similar
errors. However, in CPU time terms, our SPICE algorithm does much better
than MCEM^2: SPICE can return a quite good estimate (in line with MCEM^2's)
on average in about 18 min using the direct method, while MCEM^2 would need
about 30 days [8]—a speed-up of 2,400 times. Furthermore, for this model one
could use tau-leaping instead of the direct method, gaining a 3x speedup in
performance while giving up little on accuracy (the Min., Av., and Max. MRE
%$_{ERR}$ were 31.2, 41.5, and 56.3, respectively; Av. CPU time was 303 s).

The Schlögl system is another challenging case study, as clearly showed by
results of Table 1, which were obtained by utilising tau-leaping (as a matter of
fact, for the Schlögl model the average accuracy of SPICE increases with the use
of tau-leaping). Our choice was motivated by the large CPU time of the direct
method due to the fact that the upper steady state for X in the model has a
large molecule number (about 600), which negatively impacts the running time
of the direct method samples. The results of Table 1 show that there is no clear
winner: the Evolutionary Programming method in COPASI has the smallest
runtime, but twice the error achieved by SPICE, which has the best accuracy.
As noted before, running the COPASI implementations with larger populations
and more iterations did not significantly improve accuracy for the increased cost.

Lastly, the genetic Toggle Switch presents an interesting real-world case study
with high-throughput data. The model now comprises four hyperparameters,
each of which must be estimated alongside the four kinetic rate constants. In
addition, the non-discrete (and noisy) data is no longer known to be generated
from a convenient mathematical model. In other terms, there is no guarantee
that the model reflects the true underlying biochemical reaction network. Despite
these challenges, our SPICE algorithm does a very good job (in *little more than
an hour* of CPU time) in computing parameter estimates for which the model
quite closely matches the experimental data—we see in fact from Fig. 5 that
the model simulations fall inside the data, with very few exceptions, and the
empirical and simulated distributions closely match.

Related Work. Techniques for stochastic rate parameter estimation fall into
four categories. Early efforts included methods based on MLE: simulated max-

imum likelihood utilises Monte Carlo simulation and a genetic algorithm to maximise an approximated likelihood [34]. Efforts have been made to incorporate the Expectation-Maximisation (EM) algorithm with the SSA [18]. The stochastic gradient descent explores a Markov Chain Monte Carlo sampler with a Metropolis-Hastings update step [39]. In [25] a hidden Markov model is used for the system state, which is then solved by (approximate) likelihood maximisation. Lastly, a recent work [8] has combined an ascent-based EM algorithm with a modified cross-entropy method. Another category of methodologies include Bayesian inference. In particular, approximate Bayesian computation (ABC) gains an advantage by becoming 'likelihood free', and recent advances in sequential Monte Carlo (SMC) samplers have further improved these methods [32, 35]. We note the similarities between ABC(-SMC) approaches and SPICE. Both methods can utilize 'elite' samples to produce better parameter estimates. A key difference is that ABC(-SMC) uses accepted simulation parameters to construct a posterior distribution, while SPICE utilizes complete trajectory information to compute optimal updates of an underlying parameter distribution. The Bayesian approach presented in [5] can handle partially observed systems, including notions of experimental error. Linear noise approximation techniques have been used alongside Bayesian analysis [19]. A very recent work [36] combines Bayesian analysis with statistical emulation in an attempt at reducing the cost due to the SSA simulations. A third class of methodologies center around the numerical solution of the chemical master equation (CME), which is often intractable for all but the simplest of systems. One approach is to use dynamic state space truncation [3] or finite state projection methods [9] that truncate the CME state space by ignoring the smallest probability states. Another variation is to use a method of moments approximation [10, 16] to construct ordinary differential equations (ODEs) describing the time evolution for the mean, variance, *etc.*, of the underlying distribution. Other CME approximations are system size expansion using van Kampen's expansion [11], and solutions of the Fokker-Planck equation [22] using a form of linear noise approximation. Finally, another method [42] treats intervals between time measurements piecewise, and within each interval an ODE approximation is used for the objective function. This method has been recently extended using linear noise approximation [43]. A recent work [1], tailored for high-throughput data, proposes a stochastic parameter inference approach based on the comparison of distributions.

6   Conclusions

In this paper we have introduced the SPICE algorithm for rate parameter inference in stochastic reaction networks. Our algorithm is based on the cross-entropy method and Gillespie's algorithm, with a number of significant improvements. Key strengths of our algorithm are its ability to use multiple, possibly incomplete datasets (including distribution data), and its (theoretically justified) use of tau-leaping methods for model simulation. We have shown that SPICE works well in practice, in terms of both computational cost and estimate accuracy (which was often the best in the models tested), even on challenging case studies involving bistable systems and real high-throughput data. On a non-trivial case study, SPICE can be orders of magnitude faster than other approaches, while offering comparable accuracy in the estimates.

Acknowledgements. This work has been supported by a BBSRC DTP PhD studentship and the EPSRC Portabolomics project (EP/N031962/1).

References

1. Aguilera, L.U., Zimmer, C., Kummer, U.: A new efficient approach to fit stochastic models on the basis of high-throughput experimental data using a model of IRF7 gene expression as case study. BMC Syst. Biol. **11**(1), 26 (2017)
2. Andersen, J.B., Sternberg, C., Poulsen, L.K., Bjørn, S.P., Givskov, M., Molin, S.: New unstable variants of green fluorescent protein for studies of transient gene expression in bacteria. Appl. Environ. Microbiol. **64**(6), 2240–2246 (1998)
3. Andreychenko, A., Mikeev, L., Spieler, D., Wolf, V.: Approximate maximum likelihood estimation for stochastic chemical kinetics. EURASIP J. Bioinform. Syst. Biol. **2012**(1), 9 (2012)
4. Blake, W.J., KAErn, M., Cantor, C.R., Collins, J.J.: Noise in eukaryotic gene expression. Nature **422**(6932), 633–637 (2003)
5. Boys, R., Wilkinson, D., Kirkwood, T.: Bayesian inference for a discretely observed stochastic kinetic model. Stat. Comput. **18**, 125–135 (2008)
6. Costa, A., Jones, O.D., Kroese, D.: Convergence properties of the cross-entropy method for discrete optimization. Oper. Res. Lett. **35**(5), 573–580 (2007)
7. Daigle, B.J., Roh, M.K., Gillespie, D.T., Petzold, L.R.: Automated estimation of rare event probabilities in biochemical systems. J. Chem. Phys. **134**(4), 044110 (2011)
8. Daigle, B.J., Roh, M.K., Petzold, L.R., Niemi, J.: Accelerated maximum likelihood parameter estimation for stochastic biochemical systems. BMC Bioinform. **13**(1), 68 (2012)
9. Dandach, S.H., Khammash, M.: Analysis of stochastic strategies in bacterial competence: a master equation approach. PLoS Comput. Biol. **6**(11), 1–11 (2010)
10. Engblom, S.: Computing the moments of high dimensional solutions of the master equation. Appl. Math. Comput. **180**(2), 498–515 (2006)
11. Fröhlich, F., Thomas, P., Kazeroonian, A., Theis, F.J., Grima, R., Hasenauer, J.: Inference for stochastic chemical kinetics using moment equations and system size expansion. PLoS Comput. Biol. **12**(7), 1–28 (2016)
12. Gardner, T.S., Cantor, C.R., Collins, J.J.: Construction of a genetic toggle switch in *Escherichia coli*. Nature **403**(6767), 339–342 (2000)
13. Gillespie, D.T.: A general method for numerically simulating the stochastic time evolution of coupled chemical reactions. J. Comput. Phys. **22**(4), 403–434 (1976)
14. Gillespie, D.T.: Exact stochastic simulation of coupled chemical reactions. J. Phys. Chem. **81**(25), 2340–2361 (1977)
15. Gillespie, D.T.: Approximate accelerated stochastic simulation of chemically reacting systems. J. Chem. Phys. **115**(4), 1716–1733 (2001)
16. Hasenauer, J., Wolf, V., Kazeroonian, A., Theis, F.J.: Method of conditional moments (MCM) for the chemical master equation. J. Math. Biol. **69**(3), 687–735 (2014)

17. Hoops, S., et al.: COPASI - a complex pathway simulator. Bioinformatics **22**(24), 3067–3074 (2006)
18. Horváth, A., Martini, D.: Parameter estimation of kinetic rates in stochastic reaction networks by the EM method. In: BMEI, pp. 713–717. IEEE (2008)
19. Komorowski, M., Finkenstädt, B., Harper, C.V., Rand, D.A.: Bayesian inference of biochemical kinetic parameters using the linear noise approximation. BMC Bioinform. **10**(1), 343 (2009)
20. Kullback, S., Leibler, R.A.: On information and sufficiency. Ann. Math. Stat. **22**(1), 79–86 (1951)
21. Leon, M.: Computational design and characterisation of synthetic genetic switches. Ph.D. thesis, University College London, UK (2017). http://discovery.ucl.ac.uk/1546318/1/Leon_Miriam_thesis_final.pdf
22. Liao, S., Vejchodský, T., Erban, R.: Tensor methods for parameter estimation and bifurcation analysis of stochastic reaction networks. J. Roy. Soc. Interface **12**(108), 20150233 (2015)
23. McAdams, H.H., Arkin, A.: Stochastic mechanisms in gene expression. PNAS **94**(3), 814–819 (1997)
24. Pirone, J.R., Elston, T.C.: Fluctuations in transcription factor binding can explain the graded and binary responses observed in inducible gene expression. J. Theoret. Biol. **226**(1), 111–112 (2004)
25. Reinker, S., Altman, R.M., Timmer, J.: Parameter estimation in stochastic biochemical reactions. IEE Proc. - Syst. Biol. **153**(4), 168–178 (2006)
26. Robert, C., Casella, G.: Monte Carlo Statistical Methods. Springer, Heidelberg (2004). https://doi.org/10.1007/978-1-4757-4145-2
27. Rubinstein, R.Y.: Optimization of computer simulation models with rare events. Eur. J. Oper. Res. **99**(1), 89–112 (1997)
28. Rubinstein, R.Y.: The cross-entropy method for combinatorial and continuous optimization. Methodol. Comput. Appl. Prob. **1**(2), 127–190 (1999)
29. Rubinstein, R.Y., Kroese, D.P.: The Cross-Entropy Method. Springer, Heidelberg (2004)
30. Schlögl, F.: Chemical reaction models for non-equilibrium phase transitions. Zeitschrift für physik **253**(2), 147–161 (1972)
31. Shaner, N.C., Campbell, R.E., Steinbach, P.A., Giepmans, B.N.G., Palmer, A.E., Tsien, R.Y.: Improved monomeric red, orange and yellow fluorescent proteins derived from *Discosoma* sp. red fluorescent protein. Nat. Biotechnol. **22**, 1567–1572 (2004)
32. Sisson, S.A., Fan, Y., Tanaka, M.M.: Sequential Monte Carlo without likelihoods. PNAS **104**(6), 1760–5 (2007)
33. Sobol', I.M.: On the distribution of points in a cube and the approximate evaluation of integrals. USSR Comput. Math. Math. Phys. **7**(4), 86–112 (1967)
34. Tian, T., Xu, S., Gao, J., Burrage, K.: Simulated maximum likelihood method for estimating kinetic rates in gene expression. Bioinformatics **23**(1), 84–91 (2007)
35. Toni, T., Welch, D., Strelkowa, N., Ipsen, A., Stumpf, M.P.H.: Approximate Bayesian computation scheme for parameter inference and model selection in dynamical systems. J. Roy. Soc. Interface **6**(31), 187–202 (2009)
36. Vernon, I., Liu, J., Goldstein, M., Rowe, J., Topping, J., Lindsey, K.: Bayesian uncertainty analysis for complex systems biology models: emulation, global parameter searches and evaluation of gene functions. BMC Syst. Biol. **12**(1), 1 (2018)
37. Villaverde, A.F., Banga, J.R.: Reverse engineering and identification in systems biology: strategies, perspectives and challenges. J. Roy. Soc. Interface **11**(91), 20130505 (2013)

38. Voit, E.O.: The best models of metabolism. Wiley Interdisc. Rev.: Syst. Biol. Med. **9**(6), e1391 (2017)
39. Wang, Y., Christley, S., Mjolsness, E., Xie, X.: Parameter inference for discretely observed stochastic kinetic models using stochastic gradient descent. BMC Syst. Biol. **4**(1), 99 (2010)
40. Wilkinson, D.J.: Stochastic modelling for quantitative description of heterogeneous biological systems. Nat. Rev. Genet. **10**(2), 122–133 (2009)
41. Wilkinson, D.J.: Stochastic Modelling for Systems Biology. CRC Press, Boca Raton (2012)
42. Zimmer, C., Sahle, S.: Parameter estimation for stochastic models of biochemical reactions. J. Comput. Sci. Syst. Biol. **6**(1), 11–21 (2012)
43. Zimmer, C., Sahle, S.: Deterministic inference for stochastic systems using multiple shooting and a linear noise approximation for the transition probabilities. IET Syst. Biol. **9**, 181–192 (2015)

Experimental Biological Protocols
with Formal Semantics

Alessandro Abate[2], Luca Cardelli[1,2], Marta Kwiatkowska[2], Luca Laurenti[2(✉)],
and Boyan Yordanov[1]

[1] Microsoft Research Cambridge, Cambridge, UK
[2] Department of Computer Science, University of Oxford, Oxford, UK
luca.laurenti@cs.ox.ac.uk

Abstract. Both experimental and computational biology is becoming increasingly automated. Laboratory experiments are now performed automatically on high-throughput machinery, while computational models are synthesized or inferred automatically from data. However, integration between automated tasks in the process of biological discovery is still lacking, largely due to incompatible or missing formal representations. While theories are expressed formally as computational models, existing languages for encoding and automating experimental protocols often lack formal semantics. This makes it challenging to extract novel understanding by identifying when theory and experimental evidence disagree due to errors in the models or the protocols used to validate them. To address this, we formalize the syntax of a core protocol language, which provides a unified description for the models of biochemical systems being experimented on, together with the discrete events representing the liquid-handling steps of biological protocols. We present both a deterministic and a stochastic semantics to this language, both defined in terms of hybrid processes. In particular, the stochastic semantics captures uncertainties in equipment tolerances, making it a suitable tool for both experimental and computational biologists. We illustrate how the proposed protocol language can be used for automated verification and synthesis of laboratory experiments on case studies from the fields of chemistry and molecular programming.

1 Introduction

The classical cycle of observation, hypothesis formulation, experimentation, and falsification, which has driven scientific and technical progress since the scientific revolution, is lately becoming automated in all its separate components. Data gathering is conducted by high-throughput machinery. Models are automatically synthesized, at least in part, from data [4,7,11]. Experiments are selected to maximize knowledge acquisition. Laboratory protocols are run under reproducible and auditable software control. However, integration between these automated components is lacking. Theories are not placed in the same formal context as the (coded) protocols that are supposed to test them. Theories talk about changes

© The Author(s) 2018
M. Češka and D. Šafránek (Eds.): CMSB 2018, LNBI 11095, pp. 165–182, 2018.
https://doi.org/10.1007/978-3-319-99429-1_10

in physical quantities, while protocols talk about steps carried out by machines: neither knows about the other, although they both try to describe the same process. The consequence is that often it is hard to tell what happened when experiments and models do not match: was it an error in the model, or an error in the protocol? Often both the model and the protocol have unknown parameters: do we use the experimental data to fit the model or to fit the protocol? When most activities are automated, we need a way to answer those questions that is equally automated.

In this paper we present a novel language to model experimental bio-chemical protocols that gives an integrated description of the protocol and of the underlying molecular process. From this integrated representation both the model of a phenomenon (for possibly automated mathematical analysis), and the steps carried out to test it (for automated execution by lab equipment) can be separately extracted. This is essential to perform automated model synthesis and falsification by taking also into account uncertainties in both the model structure and equipment tolerances. Our goal in this paper is to define a simple core language for modelling biological protocols and formalize its semantics. We then show how our language can easily be extended to collect observations of the process and to model complicate protocols. An example of an experimental biological protocol is shown in Example 1.

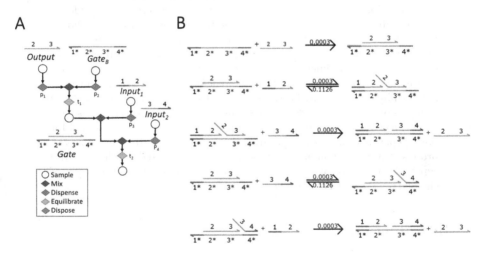

Fig. 1. (A) Graphical representation of the protocol. Dispose operations discard a sample and are implicitly considered inside Dispense operations (See Sect. 7 for details) (B) Graphical representation of the Chemical Reaction Network (CRN) between the different DNA strands in the considered solution. The CRN is written according to the language for modelling composable DNA system presented in [23]. For example, in the second reaction, strand $\{1^*\}[2\,3]\{4^*\}$ reacts with $\langle 1^*\,2 \rangle$ at a rate 0.0003, and there exists an inverse reaction with rate 0.1126.

Example 1. We consider an experimental protocol for DNA strand displacement. DNA strand displacement (DSD) is a design paradigm for DNA nano-devices [10]. In such a paradigm, single-stranded DNA acts as signals and double-stranded (or more complex) DNA structures act as gates. The interactions between signals and gates allow one to generate computational mechanisms that can operate autonomously at the molecular level [26]. The DSD programming language has been developed as a means of formally programming and analyzing such devices [10,19]. In Fig. 1, we consider an AND circuit implemented in DSD, which can be represented with the Chemical Reaction Network (CRN) in Fig. 1b. Strands $Input_1 = \langle 1^* \ 2 \rangle$ and $Input_2 = \langle 3 \ 4^* \rangle$ represent the two inputs, while strand $Output = \langle 2 \ 3 \rangle$ is the output. Strand $Gate = \{1^*\}[2 \ 3]\{4^*\}$ is an auxiliary strand. The $Output$ strand is released only if both the inputs react with the $Gate$ gate. The protocol in Fig. 1a proceeds as follow: $Output$ and $Gate_B$ strands are dispensed from the original samples. Then, they are let evolve for t_1 seconds to create $Gate$ strands. Then, the two inputs are dispensed from their samples. The resulting samples are mixed and the resulting solution evolves for t_2 seconds. Finally, we collect the final sample and observe the results.

We present two semantics for the introduced language: a deterministic semantics and a stochastic semantics. In both cases, the resulting mathematical model is an hybrid system, where the discrete dynamics are used to map the discrete operations of a lab protocol, while the continuous dynamics model the evolution of the physical variables. In the deterministic semantics, physical variables are modeled in terms of *ordinary differential equations (ODEs)* given by the *rate equations* [15], while discrete operations are mapped into discrete events that are triggered by some deterministic guards. The stochastic semantics extends the deterministic semantics: it allows one to model uncertainties that are intrinsic in the discrete operations of the protocol, such as those due to lab equipment and whose error ranges have also been standardized

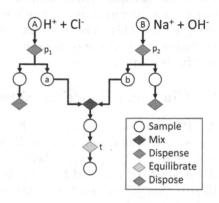

Fig. 2. Graphical representation of an acid-base titration protocol. The protocol is initialized with samples A (containing H^+ and Cl^-) and B (containing Na^+ and OH^-). Some fraction of each sample (p_1 and p_2) is mixed together and the resulting sample is let to equilibrate for t seconds.

(standards ISO 17025 and 8655). Thus, in the resulting stochastic model, the time at which a discrete event happens, may be a random variable with exponential distribution. We show that the resulting stochastic semantics is a *Piecewise Deterministic Markov Process* (PDMP). That is, a class of Markov stochastic hybrid processes where the continuous variables evolve according to ODEs and the discrete variables evolve by means of random jumps [12].

On examples from chemistry and molecular programming, we demonstrate how our integrated representation allows one to perform analysis and synthesis of both the discrete steps of the protocol and of the underlying biological system.

Related Work. Several factors contribute to the growing need for a formalization of experimental protocols in biology. First, better record-keeping of experimental operations is recognized as a step towards tackling the 'reproducibility crisis' in biology [16]. Second, the emergence of 'cloud labs' [17] creates a need for precise, machine-readable descriptions of the experimental steps to be executed. To address these needs, frameworks allowing protocols to be recorded, shared, and reproduced locally or in a remote lab have been proposed. These frameworks introduce different programming languages for experimental protocols including BioCoder [3], Autoprotocol, and Antha [24]. These languages provide expressive, high-level protocol descriptions but consider each experimental sample as a labelled 'black-box'. This makes it challenging to study a protocol together with the biochemical systems it manipulates in a common framework.

In contrast, we consider a simpler set of protocol operations but capture the details of experimental samples, enabling us to track properties of chemical species (e.g. amounts, concentrations, etc.) as they react during the execution of a protocol. This allows us to formalize and verify requirements for the correct execution of a protocol or to optimize various protocol or system parameters to satisfy these specifications.

2   Chemical Reaction Networks

A CRN $\mathcal{C} = (\mathcal{A}, \mathcal{R})$ is a pair of finite sets, where $\mathcal{A}$ denotes a set of *chemical species*, $|\mathcal{A}|$ is its cardinality, and $\mathcal{R}$ denotes a set of reactions. A *reaction* $\tau \in \mathcal{R}$ is a triple $\tau = (r_\tau, p_\tau, k_\tau)$, where $r_\tau \in \mathbb{N}^{|A|}$ is the *source complex*, $p_\tau \in \mathbb{N}^{|A|}$ is the *product complex* and $k_\tau \in \mathbb{R}_{)0}$ is the coefficient associated with the rate of the reaction. The quantities r_τ and p_τ represent the stoichiometry of reactants and products. Given a reaction $\tau_1 = ([1,0,1],[0,2,0],k_1)$, we often refer to it visually as $\tau_1 : \lambda_1 + \lambda_3 \rightarrow^{k_1} 2\lambda_2$. The *net change* associated to τ is defined by $v_\tau = p_\tau - r_\tau$.

Many models have been introduced to study CRNs [6,8,9,15]. Here we consider the *rate equations* [15], which describe the time evolution of the concentration of the species in $\mathcal{C}$, in a sample of temperature T and volume V, as follows:

$$\frac{d\Phi(t)}{dt} = F(t) = \sum_{\tau \in \mathcal{R}} v_\tau \cdot \gamma_S(\Phi(t), k_\tau, V, T), \tag{1}$$

where $\gamma_S(\Phi(t), k_\tau, V, T))$ is the propensity rate, and in case of mass action kinetics we have

$$\gamma_S(\Phi(t), k_\tau, V, T)) = k_\tau(T) \prod_{S \in A} \Phi_S^{r_{S,\tau}}(t),$$

where Φ_S and $r_{S,\tau}$ are the components of vectors Φ and r_τ relative to species S, and where in $k_\tau(T)$ we make explicit the dependence from temperature T.

Definition 1 *(Chemical Reaction System). A chemical reaction system (CRS) $C = (\mathcal{A}, \mathcal{R}, x_0)$ is defined as a tuple, where $(\mathcal{A}, \mathcal{R})$ is a CRN and $x_0 \in \mathbb{N}^{|\mathcal{A}|}$ represents its initial condition.*

Example 2. Consider the CRS $C = (\mathcal{A}, \mathcal{R}, x_0)$, evolving in a volume V and at temperature T, where $\mathcal{A} = \{H_2O, Na^+, OH^-, Cl^-, H^+\}$ and $\mathcal{R}$ is composed of the following reactions:

$$Na^+ + OH^- + H^+ + Cl^- \rightarrow^k H_2O + Na^+ + Cl^-$$

where $k = 2.81e^{-10}$ is the rate constant at temperature $T = 298$ Kelvin. Then, according to Eq. (1), we have that the state of H^+ is given by the solution of the following ordinary differential equation:

$$\frac{dH^+(t)}{dt} = -kNa^+(t)OH^-(t)H^+(t)Cl^-(t), \tag{2}$$

with $H^+(0) = \frac{x_{0,H^+}}{V}$, where x_{0,H^+} is the component of x_0 corresponding to H^+. Note that Eq. (2) is given in terms of concentrations of species instead of molecular numbers.

In order to introduce a formal semantics for experimental protocols, we first need to define a formal semantics for a CRS, which has been only introduced informally in the previous section. Let $S = (\mathbb{R}^{|\mathcal{A}|} \times \mathbb{R}_{\geq 0} \times \mathbb{R}_{\geq 0})$ be a sample. Then, we define the semantics for a CRS as follows.

Definition 2 *(CRS Semantics). Let $\mathcal{C} = (\mathcal{A}, \mathcal{R})$ be a CRN, $x_0 \in \mathbb{R}^{|\mathcal{A}|}_{\geq 0}$, $V, T \in \mathbb{R}_{\geq 0}$ be the initial concentration (moles per litre), volume (liters) and temperature (degrees Kelvin). Call $F(V,T) : \mathbb{R}^{|\mathcal{A}|} \rightarrow \mathbb{R}^{|\mathcal{A}|}$ the drift at volume V and temperature T for $\mathcal{C}$. Then, the semantics of the CRS $(\mathcal{A}, \mathcal{R}, x_0)$ at volume V, temperature T and time t, for a time horizon $H \in \mathbb{R}_{\geq 0} \cup \{\infty\}$,*

$$[\![\cdot]\!] : ((((CRS \times \mathbb{R}_{\geq 0} \times \mathbb{R}_{\geq 0}) \rightarrow \mathbb{R}_{\geq 0} \cup \{\infty\}) \rightarrow \mathbb{R}_{\geq 0}) \rightarrow S)$$

is defined as

$$[\![((\mathcal{A}, \mathcal{R}, x_0), V, T)]\!](H)(t) = (G(t), V, T)$$

where $G : [0...H] \rightarrow \mathbb{R}^{|\mathcal{A}|}$ is the solution of $G(t') = x_0 + \int_0^{t'} F(V,T)(G(s))ds$.

If for such an H, G is not unique, then we say that $[\![((\mathcal{A}, \mathcal{R}, x_0), V, T)]\!] (H)(t)$ is ill posed.

In Definition 2 we have explicitly introduced a dependence on the time horizon H, because it may happen that the solution of the rate equations is defined only for a finite time horizon [15]: For instance, the CRN given by the reaction $A + A \rightarrow A + A + A$, with initial concentration for A of 0.1 mol/l, is ill-posed for $H = \infty$ since it does not admit a unique solution over an infinite time horizon.

3   A Language for Experimental Biological Protocols

We introduce the syntax of a new language for modelling experimental protocols. A formal semantics of the language, based on denotational semantics [25], is then discussed. The physical process underlying a biological experimental protocol is modeled as a Chemical Reaction Systems (CRS) (Definition 1).

Definition 3 *(Syntax of a Protocol). Given a set of variables Var, the syntax of a protocol P for a given fixed CRN $C = (\mathcal{A}, \mathcal{R})$ is*

$$
\begin{aligned}
P = \qquad\qquad &x &&(sample\ variable)\\
&(x_0, V, T) &&(initial\ condition)\\
&Mix(P_1, P_2) &&(mix\ samples)\\
&let\ x = P_1\ in\ P_2 &&(define\ variable)\\
let\ x, y = &Dispense(P_1, p)\ in\ P_2 &&(dispense\ samples)\\
&Equilibrate(P, t) &&(equilibrate\ for\ t\ seconds)\\
&Dispose(P) &&(discard\ P)
\end{aligned}
$$

where $T, V, t \in \mathbb{R}_{\geq 0}, x_0 \in \mathbb{R}^{|\mathcal{A}|}, x, y \in Var$, and $p \in \mathbb{R}_{(0,1)}$ is a unit-less fraction. Moreover, let-bound variables must occur (as free variables) exactly once in P_2.

A protocol P yields a sample, which is the result of operations of Equilibrate, Mix, Dispose and Dispense, over a CRS. This syntax allows one to create and manipulate new samples using Mix (put together different samples), Dispense (separate samples) and Dispose (discard samples) operations. Note that the CRN is common for all samples. However, different samples may have different initial conditions. The single-occurrence (linearity) restriction implies that a sample cannot be duplicated or eliminated from the pool.

Remark 1. In the syntax presented in Definition 3, we are implicitly assuming that all the samples evolve at constant temperature, volume and pressure, and we are not considering the effect of having samples with different heat capacities. This is due to the fact that in this work we mainly target dilute aqueous solutions, and for these solutions the heat capacities are very similar to those of water. Thus, assuming the heat capacity is constant for all samples is a reasonable approximation. However, for more general protocols the heat capacity will need to be taken into account explicitly. This can be easily done by storing the heat capacity of each sample in the protocol and then computing the heat capacity of the resulting sample after a mixing operation [27].

Example 3. We use $let\ x, \_ = Dispense(P_1, p)\ in\ P_2$ as a short-hand for $let\ x, y = Dispense(P_1, p)\ in\ Mix(Dispose(y), x)$. Given a CRN $\mathcal{C} = (\{H^+, Cl^-, Na^+, OH^-, H_2O\}, \mathcal{R})$, where $\mathcal{R} = \{Na^+ + OH^- + H^+ + Cl^- \to^k H_2O + Na^+ + Cl^-\}$, the protocol (call it Pro_1) represented graphically in Fig. 2 is defined formally as

$$Pro_1 = let\ A = ([(H^+, 0.1M); (Cl^-, 0.1M)], 1.0mL, 298.15K)\ in$$
$$let\ B = ([(Na^+, 0.1M); (OH^-, 0.1M)], 1.0mL, 298.15K)\ in$$
$$let\ a, \_ = Dispense(A, p_1)\ in$$
$$let\ b, \_ = Dispense(B, p_2)\ in$$
$$Equilibrate(Mix(a, b), t).$$

In the formula above, $[(H^+, 0.1M); (Cl^-, 0.1M)]$ is a short-hand for vector $[0.1, 0.1, 0, 0, 0]$ representing the initial concentration of the species in sample A for CRN $\mathcal{C}$, where we made clear that the concentration is specified in molar units (M).

The following equivalences can be shown structurally, namely based on the standard definitions of substitution ($P\{x \leftarrow P'\}$) and free-variables ($FV(P)$):

Proposition 1 *(Equivalence Relationships).*

$$let\ x = P_1\ in\ P_2\ =\ P_2\{x \leftarrow P_1\}$$
$$let\ x = P_1\ in\ P_2\ =\ let\ y = P_1\ in\ (P_2\{x \leftarrow y\})\ for\ y \notin FV(P_2),$$

where $P_2\{x \leftarrow P_1\}$ is the capture-avoiding substitution of P_2 for x in P_1, and $FV(P_2)$ are the free variables of P_2.

We stress that in order to define a semantics for the protocol language in Definition 3, we require a pair $(P, \mathcal{C})$, where P is a protocol and $\mathcal{C}$ is a CRN. In the next Section we formally introduce CRNs. However, we should also stress that many languages exist to represent CRNs. For instance, graphical languages or implicit representations, as those that we use in Example 1, where the set of reactions can be determined just from the structure of the initial DNA strands, by the rules of DNA strand displacement [23]. In this paper, we do not require a particular representation language for CRNs. We simply assume that we can always extract a representation of a CRN, which matches the definition given in the next Section.

4   Deterministic Semantics of Experimental Protocols

In an experimental protocol discrete operations are mixed with physical variables, namely concentration of the species of a CRN that evolve continuously in time. We first consider a deterministic semantics for the language presented in Definition 3. Then, in the next section, we extend such a semantics in order to take into account errors and inaccuracies within a protocol, which in practice can be quite relevant: this leads to probabilistic semantics.

The deterministic semantics of a protocol P for a CRN $\mathcal{C} = (\mathcal{A}, \mathcal{R})$, under a given environment $\rho : Var \rightarrow S$, is a function $[\![P]\!]^\rho : (Var \rightarrow S) \rightarrow S$, where S is a sample as defined in Sect. 2, defined inductively as follows.

Definition 4 *(Deterministic Semantics of a Protocol).* *Let* $S = (\mathbb{R}^{|\mathcal{A}|} \times \mathbb{R}_{\geq 0} \times \mathbb{R}_{\geq 0})$, *then the deterministic semantics of a protocol P for CRN* $\mathcal{C} = (\mathcal{A}, \mathcal{R})$, *under environment* $\rho : Var \to S$ *is defined inductively as follows*

$$[\![x]\!]^\rho = \rho(x)$$

$$[\![x_0, V, T]\!]^\rho = (x_0, V, T)$$

$$[\![Mix(P_1, P_2)]\!]^\rho = \left(\frac{x_0^1 V_1 + x_0^2 V_2}{V_1 + V_2}, V_1 + V_2, \frac{T_1 V_1 + T_2 V_2}{V_1 + V_2}\right)$$

$$where\,(x_0^1, V_1, T_1) = [\![P_1]\!]^\rho \, and \,(x_0^2, V_2, T_2) = [\![P_2]\!]^\rho$$

$$[\![let\,x = P_1\,in\,P_2]\!]^\rho = [\![P_2]\!]^{\rho_1}$$

$$where\,(x_0, V, T) = [\![P_1]\!]^\rho \, and\, \rho_1 = \rho\{x \leftarrow (x_0, V, T)\}$$

$$[\![let\,x, y = Dispense(P_1, p)\,in\,P_2]\!]^\rho = [\![P_2]\!]^{\rho_1}$$

$$where\,(x_0, V, T) = [\![P_1]\!]^\rho$$

$$and\,\rho_1 = \rho\{x \leftarrow (x_0, V \cdot p, T), y \leftarrow (x_0, V \cdot (1-p), T)\}$$

$$[\![Equilibrate(P, t)]\!]^\rho = [\![(\mathcal{A}, \mathcal{R}, x_0), V, T]\!](H)(t)$$

$$where\,(x_0, V, T) = [\![P]\!]^\rho$$

$$[\![Dispose(P)]\!]^\rho = (0^{|\mathcal{A}|}, 0, 0),$$

where $H \in \mathbb{R}_{\geq 0}$ *is such that for any* $Equilibrate(P, t)$, $[\![(\mathcal{A}, \mathcal{R}), x_0, V, T]\!](H)(t)$ *is well posed. If such an* H *does not exist, we say that* P *is ill posed.*

The above semantics identifies a protocol with the concentration of the species, the volume, and the temperature of the sample at final time. Note that in Definition 4 we are assuming that the temperature stays constant during each equilibration step. This is reasonable for many lab protocols, where temperature is carefully regulated. Alternatively, the above semantics can easily be extended by introducing an additional ODE to model the evolution of the temperature over time. Nevertheless, this would also require updating our definition of CRNs, as the reaction rates are generally influenced by the temperature.

Example 4. Consider the protocol Pro_1 introduced in Example 3. The CRN of the system comprises the reactions given in the CRN in Example 2. According to Definition 4, the state of variable H^+ at the end of Pro_1 is given by the solution of the following equation:

$$[\![Pro_1]\!]^\rho(H^+) = H^+(0) - \int_0^t k Na^+(s) OH^-(s) H^+(s) Cl^-(s)\,ds,$$

where $H^+(0) = \frac{p_1 0.1 + p_2 10^{-7.4}}{p_1 + p_2}$, ρ is any environment, and $[\![Pro_1]\!]^\rho(H^+)$ stands for the component relative to H^+ of the sample after the execution of the protocol.

5   Stochastic Semantics of Experimental Protocols

In this Section we introduce the stochastic semantics for an experimental protocol, and show that any program defined according to Definition 3 can be mapped onto a Piecewise Deterministic Markov Processes (PDMPs) [12]. PDMPs, introduced in Sect. 5.1, are a class of stochastic hybrid systems where continuous variables evolve deterministically according to a system of ordinary differential equations (ODEs), while discrete operations may be probabilistic, and introduce noise in the system.

5.1 Piecewise Deterministic Markov Process

The syntax of a PDMP is given as follows.

Definition 5. *A Piecewise Deterministic Markov Process (PDMP)* $\mathcal{H}$ *is a tuple* $\mathcal{H} = (\mathcal{Q}, d, \mathcal{G}, F, \Lambda, R)$, *where*

- $\mathcal{Q} = \{q_1, ..., q_{|\mathcal{Q}|}\}$ *is the set of* discrete modes
- $d \in \mathbb{N}$ *is the dimension of the state space of the* continuous dynamics. *The hybrid state space is defined as* $\mathcal{S} = \cup_{q \in \mathcal{Q}} \{q\} \times \mathbb{R}^d$
- $\mathcal{G} : \mathcal{Q} \times \mathbb{R}^d \to \{0, 1\}$ *is a set of guards*
- $F : \mathcal{Q} \times \mathbb{R}^d \to \mathbb{R}^d$ *is a family of vector fields*
- $\Lambda : \mathcal{S} \times \mathcal{Q} \to \mathbb{R}_{\geq 0}$ *is an* intensity function, *where for* $(q_i, x) \in \mathcal{S}, q_j \in \mathcal{Q}$, *we define* $\Lambda((q_i, x), q_j) = \lambda_{i,j}(x)$ *and* $\lambda_{q_i}(x) = \sum_{q_j \neq q_i} \lambda_{i,j}(x)$
- $R : \mathcal{B}(\mathcal{S}) \times \mathcal{S} \to [0.1]$ *is the* reset function, *which assigns to each* $(q, x) \in \mathcal{S}$ *a probability measure* $R(\cdot, q, x)$ *on* $(\mathcal{S}, \mathcal{B}(\mathcal{S}))$, *where* $\mathcal{B}(\mathcal{S})$ *is the smallest* σ−algebra *on* $\mathcal{S}$ *containing all the sets of the form* $\cup_{q \in \mathcal{Q}} \{q\} \times A_q$, *where* A_q *is a measurable subset of* $\mathbb{R}^d$.

For $t \in \mathbb{R}_{\geq 0}, q \in \mathcal{Q}, x \in \mathbb{R}^d$, we call $\Phi(q, t, x)$ the solution of the following differential equation:

$$\frac{d\Phi(q, t, x)}{dt} = F(q, \Phi(q, t, x)), \quad \Phi(q, 0, x) = x.$$

The solution of a PDMP is a stochastic process $Y = (\alpha, X)$, whose semantics is classically defined according to the notion of execution (see Definition 6 below) [13]. In order to introduce such a notion, we define the exit time $t^*(q, x, \mathcal{G})$ as

$$t^*(q, x, \mathcal{G}) = \inf\{t \in \mathbb{R}_{\geq 0} \mid \mathcal{G}(q, \Phi(q, t, x)) = 1\} \tag{3}$$

and the *survival function* $f(q, t, x) = \begin{cases} e^{-\int_0^t \lambda_q(\Phi(q, \tau, x)) d\tau} & \text{if } t \langle t^*(q, x, \mathcal{G}) \\ 0 & \text{otherwise.} \end{cases}$

Here $t^*(q, x, \mathcal{G})$ represents the first time instant, starting from state (q, x), when the guard set is reached by a solution of the process; further $f(q, t, x)$ denotes the probability that the system remains within q, starting from x, at time t [12], which depends on random arrivals induced by the intensity function Λ. The semantics of a PDMP for initial condition (q_0, x_0) is provided next.

Definition 6 *(Execution of PDMP* $\mathcal{H}$).
Set $t := 0$
Set $(\alpha(0), X(0)) := (q_0, x_0)$

While $t\langle\infty$
   Sample $\mathbb{R}_{\geq 0}$-*valued random variable* T *such that*

$$Prob(T\rangle\bar{t}) = f(\alpha(t), \bar{t}, X(t))$$

   $\forall \tau \in [t, t+T)$ **Set** $(\alpha(\tau), X(\tau)) := (\alpha(t), \Phi(\alpha(t), \tau - t, X(t)))$
   If $t + T\langle\infty$
      Sample $(\alpha(t+T), X(t+T))$ *according to*
        $R(\cdot, (\alpha(t), \Phi(\alpha(t), T, X(t))))$
   End If
   Set $t := t + T$
End While

For further details on PDMPs and on their measure theoretic properties we refer to [12].

5.2 Stochastic Semantics

Let us recall that the semantics of Definition 4 are fully deterministic. However, it is often the case that operations of *Dispense* and *Equilibrate* are stochastic in nature, due to experimental inaccuracies related to lab equipment. In what follows, we encompass these features by extending the semantics, previously defined as deterministic, with stochasticity. More precisely, we account for the following:

– in the *Equilibrate*(P, t) step, time is sampled from a distribution;
– the resulting volume after a *Dispense* step is sampled from a distribution.

The first characteristic models the fact that in real experiments the system is not equilibrated for exactly t seconds, as it may start or be stopped at different time instants, and it accounts for the fact that after a mix of samples well mixed conditions are not reached instantaneously; whereas the second feature takes into account the experimental errors associated to pipetting devices whose ranges have been standardized (standard ISO 8655). For the first feature, consider the function $\mathcal{T}(t', t) = e^{-\frac{t'}{t}}$, defined for two values $t', t \in \mathbb{R}_{\geq 0}$: this corresponds to the density function of an exponential random variable, modelling random arrivals. For the second feature, let $\mathcal{B}(\mathbb{R}_{\geq 0}^m)$ be the Borel sigma-algebra over $\mathbb{R}_{\geq 0}^m$, $m\rangle 0$. Then, we consider the following function $\mathcal{D} : \mathcal{B}(\mathbb{R}_{[0,1]}) \times \mathbb{R}_{\geq 0} \times \mathbb{R}_{[0,1]} \to [0,1]$, which assigns to $\mathcal{D}(\cdot, V, p)$ a probability measure in $\mathcal{B}(\mathbb{R}_{[0,1]})$. Function $\mathcal{D}$ is used to reset the volume randomly, after a discrete operation. (As an anticipation of upcoming results, notice that both functions $\mathcal{T}$ and $\mathcal{D}$ can be mapped to elements in the syntax of a PDMP model.)

We define the *Stochastic Semantics* of a protocol as an extension of the deterministic ones from Definition 4. For the sake of compactness, we write explicitly only the operators that differ from the earlier definition.

Definition 7 *(Stochastic Semantics of a Protocol). Let* $S = (\mathbb{R}^{|\mathcal{A}|} \times \mathbb{R}_{\geq 0} \times \mathbb{R}_{\geq 0})$, *then the semantics of a protocol* P *for CRN* $\mathcal{C} = (\mathcal{A}, \mathcal{R})$, *under environment* $\rho : Var \rightarrow S$ *and functions* $\mathcal{T}$, $\mathcal{D}$, *as defined above, is defined inductively as follows*

$$[\![let\, x, y = Dispense(P_1, p)\, in\, P_2]\!]^\rho = [\![P_2]\!]^{\rho_1}$$
$$where\, (x_0, V, T) = [\![P_1]\!]^\rho$$
$$and\, \rho_1 = \rho\{x \leftarrow (x_0, V \cdot p', T), y \leftarrow (x_0, V \cdot (1 - p'), T)\}$$
$$for\, p'\, being\, sampled\, from\, \mathcal{D}(\cdot, V, p)$$
$$[\![Equilibrate(P, t)]\!]^\rho = [\![(\mathcal{A}, \mathcal{R}, x_0), V, T]\!](H)(\mathcal{I})$$
$$where\, (x_0, V, T) = [\![P]\!]^\rho$$
$$and\, \mathcal{I}\, is\, a\, \mathbb{R}_{\geq 0} - valued\, random\, variable\, such\, that\, for\, s \in \mathbb{R}_{\geq 0}$$
$$Prob(\mathcal{I} > s) = \mathcal{T}(s, t)$$

where $H \in \mathbb{R}_{\geq 0}$ *is such that for any* $Equilibrate(P, t)$, *and any* $\mathcal{I}$ *random variable such that* $Prob(\mathcal{I}\rangle s) = \mathcal{T}(s, t)$, $[\![(\mathcal{A}, \mathcal{R}), x_0, V, T]\!](H)(\mathcal{I})$ *is well posed with probability 1. If such an* H *does not exist, we say that* $[\![P]\!]^\rho$ *is ill posed.*

$\mathcal{D}$ is a transition kernel that depends only on the current state of the system. $\mathcal{T}$ is the cumulative probability distribution of a random variable with exponential distribution. As a consequence, according to Definition 6, $[\![P]\!]^\rho$ induces semantics that are solution of a PDMP. $\mathcal{T}$ determines the probability of changing discrete state and $\mathcal{D}$ acts as a probabilistic reset, there are no guards, and the continuous dynamics evolve according to the ODE in Definition 2. More formally, given a protocol P and an environment ρ, $[\![P]\!]^\rho$ induces semantics that correspond to the solution of a PDMP $\mathcal{H} = (\mathcal{Q}, d, \mathcal{G}, F, \Lambda, R)$ as per Definitions 5 and 6. In $\mathcal{H}$, $\mathcal{Q}$ represents the set of discrete operations, $d = |\mathcal{A}| + 1$ denotes the continuous dimension (the number of continuous variables plus one ODE for modeling the time evolution). The vector field F is given by Definition 2, with an additional clock variable *time* representing time as $\frac{dtime}{dt} = 1$. For each $Equilibrate(P, t)$ step, t is sampled from $\mathcal{T}$. $\mathcal{D}$ is a reset associated to Dispense operations. It is also worth stressing that in Definition 7 all dispense operations are sampled from the same distribution. However, it would be a trivial extension to have different distributions for different dispense steps. This would for instance model a scenario where different instruments are used.

We can now leverage results from the analysis of PDMP models and export them over the protocol language. The following assumptions guarantee that $[\![P]\!]^\rho$ exists, is a strong Markov process, and allow us to exclude pathological Zeno behaviours [12,18].

Assumption 1 – *Let* $A_0, A_1 \subset \mathcal{B}(\mathbb{R}_{[0,1]})$ *be the smallest sets in* $\mathcal{B}(\mathbb{R}_{[0,1]})$ *containing respectively 0 and 1. Then,* $\mathcal{D}(A_0, V, p) = \mathcal{D}(A_1, V, p) = 0$ *for any* $p \in (0, 1), V \neq 0$. *That is, the Volume of a sample after a dispense is zero with probability zero.*
– *Let* F *be the drift term of the rate equations (Definition 2). Then,* F *is a globally Lipschitz function.*

- *For any Equilibrate($\cdot, t$) we have that $t \rangle 0$.*

Let us interpret these assumptions over the protocol languages. The first assumption guarantees that the volume of a non-empty sample is almost-surely not equal to 0. The second assumption guarantees that the solution of Definition 2 exists and does not hit infinity in finite time. This excludes non-physical reactions like $X + X \rightarrow X + X + X$. The third assumption guarantees that we have a finite number of jumps over a finite time, thus excluding Zeno behaviours [12,13].

Example 5. Consider the protocol introduced in Example 3. For $\sigma_1 \rangle 0, A \subset \mathbb{R}_{[0.1, 0.8]}$. Assume that $\mathcal{D}(A, p, \bar{V}) = \dfrac{\int_A e^{-\frac{x-p}{2\sigma_1^2}} dx}{\int_{0.1}^{0.8} e^{-\frac{x-p}{2\sigma_1^2}} dx}$. That is, $\mathcal{D}(\cdot, p, V)$ is a truncated Gaussian measure centered at p and independent of the volume. Then, according to Definition 7, we have the following final value for H^+:

$$H^+(\mathcal{I}) = H^+(0) - \int_0^{\mathcal{I}} k Na^+(s) OH^-(s) H^+(s) Cl^-(s) ds,$$

with $HCl(0) = \frac{V_1 0.1 + V_2 10^{-7.4}}{V_1 + V_2}$. Here $\mathcal{I}$ is a random variable with an exponential distribution with rate $\frac{1}{T}$, V_1 is a random variable sampled from $\mathcal{D}(\cdot, p_1, 1)$, and V_2 is a random variable sampled from $\mathcal{D}(\cdot, p_2, 1)$.

Remark 2. The deterministic semantics (Definition 4) can also be mapped into the framework of PDMPs. More specifically, Definition 4 induces a PDMP $\mathcal{H} = (\mathcal{Q}, d, \mathcal{G}, F, \Lambda, R)$, where $\mathcal{Q}$ is the set of discrete operations, $\Lambda((q_i, x), q_j) = 0$ for any $q_i, q_j \in \mathcal{Q}, x \in \mathbb{R}^d$, $\mathcal{G}$ is a set of guards hitting the changes in the discrete locations when the variable modelling the time reaches a threshold, F is given by Definition 2, and the reset R is a Dirac delta function.

6   Extending the Protocol Language with Observations

The language introduced in Sect. 3 can be extended in a number of directions, according to specific scenarios envisioned for the protocols. For instance, a common laboratory task is to take observations of the state of the samples handled by a protocol. That is, often it is useful to store the state of the system at different times or when a particular event happens. As some of the events may be stochastic, in general it is not possible to know before the simulation starts when a particular event happens. Consequently, observations need to be included in the language.

Definition 8 *(Extended Syntax). Given a set of variables Var, the syntax of a protocol for a given fixed CRN $C = (\mathcal{A}, \mathcal{R})$ and $idn \in \mathbb{N}$ is*

$$
\begin{aligned}
P = \quad & x && (\textit{sample variable}) \\
& (x_0, V, T) && (\textit{initial condition}) \\
& Mix(P_1, P_2) && (\textit{mix samples}) \\
& let \; x = P_1 \; in \; P_2 && (\textit{define variable}) \\
& let \; x, y = Dispense(P_1, p) \; in \; P_2 && (\textit{dispense samples}) \\
& Equilibrate(P, t) && (\textit{let time pass}) \\
& Dispose(P) && (\textit{discard } P) \\
& Observe(P, idn) && (\textit{observe sample})
\end{aligned}
$$

where $T, V, t \in \mathbb{R}_{\geq 0}, x, y \in Var, p \in \mathbb{R}_{[0,1]}$. Moreover, let-bound variables must occur exactly once (that is, be free) in P_2.

$Observe(P, idn)$ makes an observation of protocol P after its execution, and identifies such an observation with identifier idn. In order to include observations we extend the semantics as detailed next, where we consider in detail just the deterministic semantics, focusing on a few key operators. Extensions to the other operators and to Stochastic Semantics follow intuitively.

Definition 9 *(Extended Deterministic Semantics). For CRN $C = (\mathcal{A}, \mathcal{R})$ let $S = \mathbb{R}^{|\mathcal{A}|} \times \mathbb{R}_{\geq 0} \times \mathbb{R}_{\geq 0}$, $Obs = \mathbb{R}^{|\mathcal{A}|} \times \mathbb{N} \times \mathbb{R}_{\geq 0}$, Obs^*, an eventually empty set of Obs and $\mathcal{M} = S \times Obs^* \times \mathbb{R}_{\geq 0}$. The semantics of a protocol P, under environment $\rho : Var \to \mathcal{M}$, is a function $[\![P]\!] : (Var \to \mathcal{M}) \times \mathbb{R}_{\geq 0} \to \mathcal{M}$ defined inductively as follows*

$$
[\![Mix(P_1, P_2)]\!]_t^\rho =
$$
$$
\left(\left(\frac{x_0^1 V_1 + x_0^2 V_2}{V_1 + V_2}, V_1 + V_2, \frac{T_1 V_1 + T_2 V_2}{V_1 + V_2} \right), Obs_1 :: Obs_2, \max(t_1, t_2) \right)
$$
$$
\text{where } ((x_0^1, V_1, T_1), Obs_1, t_1) = [\![P_1]\!]_t^\rho \text{ and } ((x_0^2, V_2, T_2), Obs_2, t_2) = [\![P_2]\!]_t^\rho
$$
$$
[\![Observe(P, idn)]\!]_t^\rho = ((x_0, V, T), Obs \cup O, t_1)
$$
$$
\text{where } ((x_0, V, T), Obs, t_1) = [\![P]\!]_t^\rho \text{ and } O = (x_0, idn, t_1)
$$
$$
[\![Equilibrate(P, t)]\!]_{t'}^\rho = ([\![(\mathcal{A}, \mathcal{R}), x_0, V, T]\!](H)(t), Obs, t_1 + t)
$$
$$
\text{where } ((x_0, V, T), Obs, t_1) = [\![P]\!]_{t'}^\rho,
$$

where $H \in \mathbb{R}_{\geq 0}$ is such that for any $Equilibrate(P, t)$, $[\![(\mathcal{A}, \mathcal{R}), x_0, V, T]\!](H)(t)$ is well posed. If such H does not exist, we say that P is ill posed.

Observations are stored as a list of strings, each of which memorizing the concentration of the species at the observation, the identificator of the observation, and the observation time. Note that the above syntax does not prevent the programmer to assign the same identifier to two distinct observations. We further stress that often observations of the state of an experiment are not exact, but corrupted by sensing noise. For instance, this is what happens with noisy fluorescence measurements. This noise can be easily taken into account at a semantical level by sampling an observation from a distribution with additive noise, where the noise level depends on the particular measure technique or instrumentation. Finally, we can also extend the sample semantics to take into account noise in Dispense operations.

7   Case Study

As a case study we consider the experimental protocol for DNA strand displacement presented in Fig. 1. The protocol in Fig. 1a can be written formally as follows. We use $let\,x,\_ = Dispense(P1,p)\,in\,P2$ as a short-hand for $let\,x,y = Dispense(P1,p)\,in\,Mix(Dispose(y),x)$

$$P_1 = let\,In1 = ((Input1, 100.0nM), 0.1mL, 298.15K)\,in$$
$$let\,In2 = ((Input2, 100.0nM), 0.1mL, 298.15K)\,in$$
$$let\,GA = ((Output, 100.0nM), 0.1mL, 298.15K)\,in$$
$$let\,GB = ((Gate_B, 100.0nM), 0.1mL, 298.15K)\,in$$
$$let\,sGA, \_ = Dispense(GA, p_1)\,in$$
$$let\,sGB, \_ = Dispense(GB, p_2)\,in$$
$$let\,sIn1, \_ = Dispense(In1, p_3)\,in$$
$$let\,sIn2, \_ = Dispense(In1, p_4)\,in$$
$$Observe(Equilibrate(Mix(Mix(Equilibrate(Mix(sGA, sGB), t_1),$$
$$sIn1), sIn2), t_2), idn),$$

where Input1, Input2, Output, $Gate_B$ are species of the CRN represented graphically in Fig. 1, $t_1 = 3000$, and $t_2 = 5 \cdot 10^6$. According to the standard ISO 8655 for a volume of $1\,mL$, the maximum standard deviation of a particular pipetting device is $0.3\,\mu L$ per single operation. In order to incorporate such an error in our model, we make use of the stochastic semantics. Thus, the concentration of the *Output* strand at the end of the protocol is a random variable. It is also common that the reaction rates of the physical system are not known exactly and they may be affected by extrinsic noise [22]. This leads to another source of uncertainty in the output of the protocol, which can be easily incorporated in our semantics. We assume that the rate of each reaction has a normal distribution with variance equal to half of its mean (sub-Poisson noise). In Fig. 3a we plot 4500 executions resulting from the protocol. From the figure it is easy to realize how the two difference sources of noise may have a distinctive effect on the final outcome of the experiments.

In many experimental protocols, one of the key challenges is to synthesize the optimal discrete parameters to maximize the probability of obtaining desired behaviours. From now on, we assume perfect knowledge of the reaction rates of the physical system, while the discrete operations of the protocol and the times in each equilibration operation are still noisy. We assume $p_1 = p_2 = 0.4$, and our goal is to see how the concentration of the *Output* changes while varying $(p_3, p_4) \in [0.45, 0.65] \times [0.45, 0.65]$. We are interested in the following property

$$P_{Safe}([3.0 \cdot 10^{-4}, 3.5 \cdot 10^{-4}]) = Prob(Output(t') \in [3.0 \cdot 10^{-4}, 3.5 \cdot 10^{-4}] | t' = t_{final}),$$

where t_{final} is the final time of the protocol. The following probability is estimated using Statistical Model Checking [21] in Fig. 3b, which in this context reduces to Monte-Carlo sampling. From Fig. 3b it is easy to infer that the optimal value for such property is not unique (it is attained at values over the yellow band) and obtained, for instance, at $(p_3, p_4) = (0.5, 0.54)$.

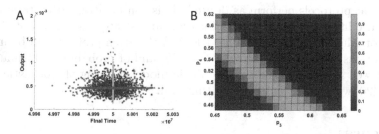

Fig. 3. (A): (red) 1500 execution of the protocol assuming the physical model is fully known, and the only source of noise is in the discrete parameters of the protocols (p_1, p_2, p_3, p_4). (yellow) 1500 executions of the protocol when the rates of the physical system are sampled from a sub-Poisson distribution, and discrete operations are not affected by noise. (blue) 1500 simulations of the protocol when both sources of noise are active.(B): $P_{Safe}([3.0 \cdot 10^{-4}, 3.5 \cdot 10^{-4}])$ as a function of p_3 and p_4. Each cell is estimated from 20000 executions of the protocol. (Color figure online)

8 Discussion

We have presented a language to formalize experimental biological protocols, and provided both a deterministic and a stochastic semantics to this language. Our language provides a unified description of the system being experimented on, together with the discrete events representing parts of biological protocols dealing with the handling of samples. Moreover, we allow the modeller to take into account uncertainties in both the model structure and the equipment tolerances. This makes our language a suitable tool for both experimental and computational biologists. Our objective has been that of providing a basic language with an integrated representation of an experimental biological protocol. To this end, we have kept the language as simple as possible, showing how different extensions can be easily integrated. For instance, in our denotational semantics, the dynamics of a physical process is given by a set of ODEs. This is accurate when the number of involved molecules is large enough, as in the discussed example of DNA strand displacement (DSD). However, in other scenarios, such as localized computation or gene expression, this might be unsatisfactory, as stochasticity becomes relevant [5,14]: the semantics presented here can be easily extended to incorporate such stochasticity, which can be achieved by considering more general classes of stochastic hybrid processes, such as switching diffusions [1,20,28] or continuous-time Markov chains (CTMCs) [2].

One of the main advantages in providing a language with formal semantics for experimental protocols is that protocols can now be quantitatively analyzed inexpensively in-silico, and classical problems of analysis of·CRNs, such as parameter estimation [7], can be studied within the corresponding modelling framework; this can also take into account the discrete operations of the protocol, which influence the dynamics of the system. An additional target of this work is to provide automated techniques to synthesise optimal protocols, or to certify that protocols perform as desired. This can be attained by tapping into the mature literature on formal verification and strategy synthesis of PDMPs, or that of other more specialised models that the given protocol can be mapped onto. Notions of finite-state abstractions [29] and of probabilistic bisimulations [1,2], as well as algorithms for probabilistic model checking of stochastic hybrid models [20] will be relevant towards this goal.

References

1. Abate, A.: Probabilistic bisimulations of switching and resetting diffusions. In: Proceedings of the 49th IEEE Conference of Decision and Control, pp. 5918–5923 (2010)
2. Abate, A., Brim, L., Češka, M., Kwiatkowska, M.: Adaptive aggregation of Markov chains: quantitative analysis of chemical reaction networks. In: Kroening, D., Păsăreanu, C.S. (eds.) CAV 2015. LNCS, vol. 9206, pp. 195–213. Springer, Cham (2015). https://doi.org/10.1007/978-3-319-21690-4_12
3. Ananthanarayanan, V., Thies, W.: Biocoder: a programming language for standardizing and automating biology protocols. J. Biol. Eng. 4(1), 13 (2010)
4. Andreychenko, A., Mikeev, L., Spieler, D., Wolf, V.: Parameter identification for Markov models of biochemical reactions. In: Gopalakrishnan, G., Qadeer, S. (eds.) CAV 2011. LNCS, vol. 6806, pp. 83–98. Springer, Heidelberg (2011). https://doi.org/10.1007/978-3-642-22110-1_8
5. Arkin, A., Ross, J., McAdams, H.H.: Stochastic kinetic analysis of developmental pathway bifurcation in phage λ-infected *Escherichia coli* cells. Genetics **149**(4), 1633–1648 (1998)
6. Bortolussi, L., Cardelli, L., Kwiatkowska, M., Laurenti, L.: Approximation of probabilistic reachability for chemical reaction networks using the linear noise approximation. In: Agha, G., Van Houdt, B. (eds.) QEST 2016. LNCS, vol. 9826, pp. 72–88. Springer, Cham (2016). https://doi.org/10.1007/978-3-319-43425-4_5
7. Cardelli, L., et al.: Syntax-guided optimal synthesis for chemical reaction networks. In: Majumdar, R., Kunčak, V. (eds.) CAV 2017. LNCS, vol. 10427, pp. 375–395. Springer, Cham (2017). https://doi.org/10.1007/978-3-319-63390-9_20
8. Cardelli, L., Kwiatkowska, M., Laurenti, L.: A stochastic hybrid approximation for chemical kinetics based on the linear noise approximation. In: Bartocci, E., Lio, P., Paoletti, N. (eds.) CMSB 2016. LNCS, vol. 9859, pp. 147–167. Springer, Cham (2016). https://doi.org/10.1007/978-3-319-45177-0_10

9. Cardelli, L., Kwiatkowska, M., Laurenti, L.: Programming discrete distributions with chemical reaction networks. Nat. Comput. **17**(1), 131–145 (2018)
10. Chen, Y.-J., et al.: Programmable chemical controllers made from DNA. Nat. Nanotechnol. **8**(10), 755–762 (2013)
11. Dalchau, N., Murphy, N., Petersen, R., Yordanov, B.: Synthesizing and tuning chemical reaction networks with specified behaviours. In: Phillips, A., Yin, P. (eds.) DNA 2015. LNCS, vol. 9211, pp. 16–33. Springer, Cham (2015). https://doi.org/10.1007/978-3-319-21999-8_2
12. Davis, M.H.: Piecewise-deterministic Markov processes: a general class of non-diffusion stochastic models. J. Roy. Stat. Soc. Ser. B (Methodol.) 353–388 (1984)
13. Davis, M.H.: Markov Models & Optimization, vol. 49. CRC Press, Boca Raton (1993)
14. Dunn, K.E., Dannenberg, F., Ouldridge, T.E., Kwiatkowska, M., Turberfield, A.J., Bath, J.: Guiding the folding pathway of DNA origami. Nature **525**(7567), 82 (2015)
15. Ethier, S.N., Kurtz, T.G.: Markov Processes: Characterization and Convergence, vol. 282. Wiley, Hoboken (2009)
16. Freedman, L.P., Cockburn, I.M., Simcoe, T.S.: The economics of reproducibility in preclinical research. PLOS Biol. **13**(6), 1–9 (2015)
17. Hayden, E.C.: The Automated Lab. Nature **516**(7529), 131–132 (2014)
18. Kouretas, P., Koutroumpas, K., Lygeros, J., Lygerou, Z.: Stochastic hybrid modeling of biochemical processes. Stochast. Hybrid Syst. **24**(9083) (2006)
19. Lakin, M.R., Youssef, S., Polo, F., Emmott, S., Phillips, A.: Visual DSD: a design and analysis tool for DNA strand displacement systems. Bioinformatics **27**(22), 3211–3213 (2011)
20. Laurenti, L., Abate, A., Bortolussi, L., Cardelli, L., Ceska, M., Kwiatkowska, M.: Reachability computation for switching diffusions: finite abstractions with certifiable and tuneable precision. In: Proceedings of the 20th International Conference on Hybrid Systems: Computation and Control, pp. 55–64. ACM (2017)
21. Legay, A., Delahaye, B., Bensalem, S.: Statistical model checking: an overview. In: Barringer, H., et al. (eds.) RV 2010. LNCS, vol. 6418, pp. 122–135. Springer, Heidelberg (2010). https://doi.org/10.1007/978-3-642-16612-9_11
22. Paulsson, J.: Summing up the noise in gene networks. Nature **427**(6973), 415–418 (2004)
23. Phillips, A., Cardelli, L.: A programming language for composable DNA circuits. J. Roy. Soc. Interface **6**(Suppl. 4), S419–S436 (2009)
24. Sadowski, M.I., Grant, C., Fell, T.S.: Harnessing QbD, programming languages, and automation for reproducible biology. Trends Biotechnol. **34**(3), 214–227 (2017)
25. Scott, D.S., Strachey, C.: Toward a Mathematical Semantics for Computer Languages, vol. 1. Oxford University Computing Laboratory, Programming Research Group, Oxford (1971)
26. Seelig, G., Soloveichik, D., Zhang, D.Y., Winfree, E.: Enzyme-free nucleic acid logic circuits. Science **314**(5805), 1585–1588 (2006)
27. Teja, A.S.: Simple method for the calculation of heat capacities of liquid mixtures. J. Chem. Eng. Data **28**(1), 83–85 (1983)
28. Yin, G., Zhu, C.: Hybrid Switching Diffusions: Properties and Applications, vol. 63. Springer, New York (2010). https://doi.org/10.1007/978-1-4419-1105-6
29. Zamani, M., Abate, A.: Symbolic models for randomly switched stochastic systems. Syst. Control Lett. **69**, 38–46 (2014)

Robust Data-Driven Control of Artificial Pancreas Systems Using Neural Networks

Souradeep Dutta, Taisa Kushner$^{(\boxtimes)}$, and Sriram Sankaranarayanan

University of Colorado, Boulder, CO, USA
{souradeep.dutta,taisa.kushner,sriram.sankaranarayanan}@colorado.edu

Abstract. In this paper, we provide an approach to data-driven control for artificial pancreas systems by learning neural network models of human insulin-glucose physiology from available patient data and using a mixed integer optimization approach to control blood glucose levels in real-time using the inferred models. First, our approach learns neural networks to predict the future blood glucose values from given data on insulin infusion and their resulting effects on blood glucose levels. However, to provide guarantees on the resulting model, we use quantile regression to fit multiple neural networks that predict upper and lower quantiles of the future blood glucose levels, in addition to the mean.

Using the inferred set of neural networks, we formulate a model-predictive control scheme that adjusts both basal and bolus insulin delivery to ensure that the risk of harmful hypoglycemia and hyperglycemia are bounded using the quantile models while the mean prediction stays as close as possible to the desired target. We discuss how this scheme can handle disturbances from large unannounced meals as well as infeasibilities that result from situations where the uncertainties in future glucose predictions are too high. We experimentally evaluate this approach on data obtained from a set of 17 patients over a course of 40 nights per patient. Furthermore, we also test our approach using neural networks obtained from virtual patient models available through the UVA-Padova simulator for type-1 diabetes.

1 Introduction

This paper investigates the use of neural networks as data-driven models of insulin glucose regulation in the human body. Furthermore, we show how the networks can be used as part of a *robust* control design that can compute optimal insulin infusion for patients with type-1 diabetes based on the predictive models. We test our approach in instances of large disturbances such as unannounced meals consisting of over $130\,\mathrm{g}$ of carbs. Such a scheme is commonly known as *model-predictive control* (MPC) [8]. Neural networks have emerged as a versatile data-driven approach to modeling numerous processes, including biological processes. They can model nonlinear functions, and be trained from data using the well-known backpropagation algorithm [22]. This has led to numerous machine learning algorithms that use neural networks for regression and classification tasks.

© Springer Nature Switzerland AG 2018
M. Češka and D. Šafránek (Eds.): CMSB 2018, LNBI 11095, pp. 183–202, 2018.
https://doi.org/10.1007/978-3-319-99429-1_11

We first consider the process of learning neural network models from data. To do so, we used two different data sets: a *synthetic* dataset using a well-known differential equation model of human insulin regulation was simulated under randomly varying conditions [15,17]; and a clinical dataset consisting of 17 patients with 40 nights per patient [34]. Both datasets report blood glucose (BG) levels at 5 min intervals along with the insulin infusions provided. From these datasets, we infer a feedforward neural network that employs 180 min of BG and insulin values to predict the likely BG value 30 min into the future. However, a key challenge lies in capturing the uncertainty inherent in this value. To do so, we infer multiple networks including a network that is trained to predict the mean BG level, as well as networks that are trained to predict upper and lower quantiles using *quantile regression* [30]. The upper (lower) α quantile network attempts to predict a value that will lie above (below) the actual value with α probability.

Next, we use the prediction models to calculate the optimal insulin infusion for a patient given the initial history of BG values to (a) maintain a target BG level well inside the normal range of $[70, 180]$ mg/dL; (b) ensure that the risk of hypoglycemia (BG ≤ 70 mg/dL) and high BG levels (BG ≥ 210 mg/dL) are bounded. The latter objectives are achieved by enforcing the constraint that the lower and upper quantile networks must predict values that are within the range.

We evaluate our approach by exploring an optimal network structure to improve prediction accuracy. Our approach yields relatively small networks with 16 neurons in two hidden layers that can predict the BG levels with a mean prediction error of about 7 mg/dL. This error is comparable to the measurement error of the sensor used in the prediction.

Next, we evaluate the resulting control scheme on two different types of experiments. In one experiment, a neural network based predictive model is learned offline against a popular ODE-based simulation model proposed by Dalla-Man et al. [15,17]. In another set of experiments, two different neural network models are trained against the same data set using stochastic gradient descent. One of the models is used as the predictive model and the other is used as a stand-in for a real patient. In both instances, we simulate the controllers under varying initial conditions and disturbances of sensor drop out and unannounced meals. The simulation results show an average time in range of about 73.2% with hypoglycemic incidence rate of only 2% over 2580 simulations.

1.1   Related Work

There has been a recent upsurge of work on data-driven model inference and control synthesis. Data driven inference techniques are being used in applications ranging from high level demand-response strategies for cyber-physical energy systems [3], to artificial pancreas models [37]. In light of this paper, we focus on applications to the latter. Recent work by Paoletti et al. [37], used data-driven methods to learn uncertainty sets from historic meal and exercise patterns in order to eliminate the need for meal announcements by the patient. Griva et al. utilized data-driven ARX models to assist the predictive functional control algorithm used in their artificial pancreas. This model achieved good prediction for a 30 minute look ahead time within 10 mg/dL [23]. Perez et al. used past BG levels and a three-layer feed-forward neural network to predict BG values 30–45 min out, with accuracy 18–27 mg/dL. [39]. In comparison, our models achieve accuracy of about 7–10 mg/dL. Our previous work investigated a data driven modeling approach using linear ARMAX models for verifying closed-loop PID controllers [32]. Therein, we focused entirely on the verification of controllers and tuning of gains for specific patients. We employed a nondeterministic model with intervals around the prediction used to model errors. In contrast, the neural networks used here are deterministic. However, we use multiple models to predict the mean, and the quantiles.

Model Predictive Control (MPC) is a well known approach to control synthesis [8]. Numerous MPC schemes for insulin infusion control have been constructed based on ODE-based models (see survey by Bequette [4]). Multiple groups have developed nonlinear MPC strategies using neural networks. Piche et al. [40] combined a linear dynamic model with a steady-state model learned on historic data to develop MPC solutions. Other have used neural networks to replace either, or both, the controller and plant models in an MPC framework with varied success rates [7,41,47]. To our knowledge this work is the first to propose MPC using multiple neural networks to construct non-deterministic models, and compute optimal infusion schedules using integer linear optimization solvers.

Another contribution of this work is to use multiple predictive models for the mean and the quantiles of the distribution. In this regard, our work is related to that of Cameron et al., who also use multiple models [9]. However, their models are instantiations of multiple possible meal scenarios weighted by their likelihoods during the prediction horizon. In contrast, we use models for the mean and the upper/lower quantiles, which are able to handle large disturbances from unannounced meals.

Lastly, it's worth mentioning that, although neural networks are popular as data-driven models, their applications to safety-critical domains has been limited by the lack of guarantees. Recent work has sought to provide such guarantees by solving verification problems posed on neural networks [19,29,33].

2   Preliminaries

In this section, we present some preliminary notions involving the "artificial pancreas" that controls insulin delivery for patients with type-1 diabetes, and discuss various modeling approaches, including physiological models. Next, we discuss preliminary concepts about neural network and the encoding of neural networks into mixed integer optimization problems.

2.1   Type-1 Diabetes and Artificial Pancreas

Patients with type-1 diabetes depend on external insulin delivery to maintain their blood glucose (BG) levels within a *euglycemic* range of $[70, 180]$ mg/dL. BG levels below 70 mg/dL lead to hypoglycemia which is characterized by a loss of consciousness, coma or even death [11]. On the other side, levels above 180 mg/dL constitute hyperglycemia which leads to longer term damage to the eyes, kidneys, peripheral nerves and the heart. Levels above 300 mg/dL are associated with a condition called ketoacidosis, where, due to insufficient insulin, the body breaks down fat for energy resulting in a buildup of ketones. To treat type-1 diabetes, patients receive artificial insulin externally either through multiple daily injections, or insulin pumps [11]. The latter allows a continuous basal infusion at a pre-programmed rate through the day along with large insulin boluses delivered to counteract the effect of meals. For the most part, the management of BG levels is performed manually by the patient. This requires careful counting of meal carbohydrates, and almost constant vigilance on the part of the patient, as relatively minor errors result in poor BG control at best, and life threatening hypoglycemia or ketacidosis, in the worst cases.

Artificial pancreas (AP) systems look to ease the burden of BG control by automating the delivery of insulin through an insulin pump, using continuous glucose sensors to periodically measure the BG levels and employing a control algorithm to decide on how much insulin to deliver [14,28,44]. A variety of strategies have been proposed for the artificial pancreas, ranging from relatively simple pump shutoff controllers [35], PID-based algorithms [42,43,48], rule-based systems [2,36], predictive controllers that use a model to forecast future BG trends in the patient against planned future insulin infusions [4,9,26]. Furthermore, the control is classified as either fully closed loop, wherein the user is (in theory) not needed to announce impending meals or exercise versus hybrid closed loops which continue to rely on users to bolus for meals or announce impending exercise [31]. Numerous control algorithms are currently in various stages of clinical trials. The Medtronic 670G device, based on a PID control algorithm, was recently approved by the US FDA and is available as a commercial product [21,25].

A key consideration for many AP devices involves adapting the insulin delivery to the personal characteristics of the patient. Patients display a wide range of variability in their response to insulin [37]. This variability is crudely summarized into numbers such as the daily insulin requirement and the insulin to carbohydrate ratio, in order to calculate insulin requirements for basal insulin and meal boluses. However, in order to achieve safe and effective control, it is important to model many additional aspects of the patient's insulin-glucose response [32]. Thus, the challenge involves how to build mathematical models that capture important aspects of a specific patient's physiology.

2.2 Mathematical Models

Mathematical models of insulin-glucose response have a long history. Bergman proposed a minimal ODE-based model of insulin-glucose physiology [6,12]. His model involves three state-variables, captures the nonlinear insulin-glucose response. At the same time, it does not model aspects such as endogenous glucose production by the liver, insulin dependent vs. insulin independent uptake of glucose by various tissues in the body, and the effect of renal clearance of glucose that happens during hyperglycemia. Finally, the model assumes direct glucose and insulin inputs into the blood stream.

More detailed physiological models have been proposed, notably by Hovorka et al. [27] and Dalla Man et al. [15,18]. These models address many of the missing aspects of the original Bergman model. A recently updated version of the Dalla Man model accounts for the effect of fasting and exercise through the counter-regulatory hormone glucagon [16]. The Dalla-Man ODE model has been approved by the US FDA as a replacement for animal trials in testing control algorithms at the pre-clinical stage [38].

A key critique of ODE-based physiological models involves the estimation of model parameters corresponding to the available patient data. The Dalla-Man model involves upwards of 40 patient parameters which need to be identified to model a particular individual. Some of these parameters may in fact be time varying. Furthermore, the model involves state variables that cannot be directly measured without intrusive radiological tracer studies. For most patients, available data consists mainly of noisy measurements of BG levels coupled with insulin infusion logs.

2.3   Neural Network

Neural networks are a *connectionist*, data-driven model that represents functions from a domain of inputs to outputs. There are two types of neural networks: (a) feedforward neural networks which do not have internal memory; and (b) recurrent neural networks that have internal memory in the form of units called *long short term memory* (LSTM). In this paper, we focus exclusively on feedforward neural network models, but briefly discuss recurrent neural networks since they form a viable modeling option that was not chosen for this paper.

A feedforward network $\mathcal{N}$ consists of n inputs, m outputs, and k hidden layers with $N_1, \ldots, N_k$ neurons in each of the hidden layers. The jth neuron in the ith hidden layer is denoted $N_{i,j}$ for $1 \leq i \leq k$ and $1 \leq j \leq N_i$. The inputs are connected to the first hidden layer, each hidden layer i is connected to the subsequent hidden layer $i + 1$ for $1 \leq i \leq k - 1$, and finally, the last hidden layer is connected to the output layer. The connections have associated weights denoted by matrices W_i and biases denoted by a vector $\mathbf{b}_i$.

Definition 1 (Feedforward Neural Networks). *Formally, a feedforward network is a tuple $\langle n, m, k, (N_i)_{i=1}^k, (W_i, \mathbf{b}_i)_{i=0}^k \rangle$, modeling a function $F_N : \mathbb{R}^n \mapsto \mathbb{R}^m$ wherein the weights of the connection from input layer to the first hidden layer are given by $(W_0, \mathbf{b}_0)$ with $W_0 \in \mathbb{R}^{N_0, n}, \mathbf{b}_0 \in \mathbb{R}^{N_0}$, the connection from layer i to $i + 1$ is given by $(W_i, \mathbf{b}_i)$ for $1 \leq i \leq k - 1$, where $W_i \in \mathbb{R}^{N_{i+1}, N_i}$ and $\mathbf{b}_i \in \mathbb{R}^{N_{i+1}}$, and finally $(W_k, \mathbf{b}_k)$ are the weights connecting the last hidden layer to the output layer.*

Each neuron $N_{i,j}$ is associated with a nonlinear activation function $\sigma_{i,j} : \mathbb{R} \mapsto \mathbb{R}$. Table 1 lists the commonly used activation function, with the ReLU being the most popular, recently. For convenience, we assume that all neurons have the same activation function σ. We lift σ to vectors of variables as $\sigma(\mathbf{x}) : \begin{pmatrix} \sigma(x_1) \\ \vdots \\ \sigma(x_n) \end{pmatrix}$. A neural

Table 1. Commonly used activation functions.

Name	$\sigma(z)$
ReLU	$\max(z, 0)$
Sigmoid	$\frac{1}{1+e^{-z}}$
Tanh	$\tanh(z)$
PReLU	$\max(\alpha z, z)$
SoftPlus	$\log(1 + \exp(z))$

network N computes a function $F_N : \mathbb{R}^n \mapsto \mathbb{R}^m$. Given a feedforward neural network, the function computed is defined as a composition of two types of functions (a) hidden layer functions $F_i : \sigma(W_i \mathbf{x}_i + \mathbf{b}_i)$, and (b) the output function $F_{out} : W_k \mathbf{x}_k + \mathbf{b}_k$. The overall function computed by the network is $F_N(\mathbf{x}) : F_{out}(F_{k-1}(\cdots (F_0(\mathbf{x})) \cdots))$.

Feedforward vs. Recurrent: Feedforward networks are useful in modeling functions from input to output. They are used in problems such as learning a function from data through regression and classifying between different categories. As such, they do not have internal states. Modeling sequences involves recurrent neural networks that augment feedforward networks with "feedback connections" through a series of memory units which can remember values between successive time steps. In analogy with digital circuits, feedforward networks are analogous to combinational circuits made from logic gates such as and/or/not, whereas recurrent networks correspond to sequential circuits that involve feedback using memory elements such as flip-flops. In this paper, we focus entirely on data-driven models using feedforward neural networks. We justify our choice on the basis of three factors: (a) the networks are easier to model and we encode some of the existing knowledge in terms of the structure of the network to avoid overfitting; (b) the networks are easier to train and (c) finally, we provide simpler approaches to reason about these networks. On the other hand, using a recurrent network can simplify some of the choices made in our model.

2.4   Encoding Networks into Constraints

A core primitive in this paper is to solve control problems involving neural networks as predictive models. We recall how the function computed by a network can be systematically modeled as a set of mixed integer linear constraints [45]. This linear programming (LP) encoding is standard, and has been presented in details elsewhere [19,33].

Let $\mathcal{N}$ be a network with n inputs, $k-1$ hidden layers, and a single output. We restrict our discussion in this section to ReLU activation units. Let $\mathbf{x} \in \mathbb{R}^n$ be the input to a neural network, represented as n (LP) variables, $\{F_1, F_2, \ldots, F_{k-1}\}$ represent the outputs of the hidden layers, and $y \in \mathbb{R}$ be the (LP) variables representing the output of the network. Let, $\{W_0 \ldots W_k\}$, and $\{\mathbf{b}_0 \ldots \mathbf{b}_k\}$, be the weights and biases, as described in Definition 1.

We introduce binary (LP) variable vectors $\{\mathbf{v}_1, \ldots, \mathbf{v}_{k-1}\}$. Such that for each variable in the vector, $v_i[j] \in \{0, 1\}$. The binary variables are introduced in order to model the piecewise linear nature of ReLU units. That is, if $v_i[j]$ is 1, the ReLU is off, and on otherwise.

At a hidden layer i, the network constraints require that $F_{i+1} = ReLU(W_i F_i + b_i)$. We use the binary (LP) variables v_{i+1} to encode the piecewise linear behavior of ReLU units.

$$
\mathsf{C}_{i+1} : \begin{cases} F_{i+1} \geq W_i F_i + b_i, \\ F_{i+1} \leq W_i F_i + b_i + \mathsf{M}v_{i+1}, \\ F_{i+1} \geq \mathbf{0}, \\ F_{i+1} \leq \mathsf{M}(\mathbf{1} - v_{i+1}) \end{cases}
$$

where M is a "large enough" constant. In practice, the number M often decides the performance of the solver, and it is possible to derive tight estimates of M using interval analysis on the network [33]. Thus we can combine these constraints, to form the encoding of the entire network: $C : \mathsf{C}_0 \wedge \ldots \wedge \mathsf{C}_{k+1}$. Next, we can use additional constraints on the inputs and outputs of the neural network $\mathcal{N}$, to find feasible assignments.

3 Data Driven Models with Neural Networks

In order to predict future blood glucose values for each patient, we used the various patient datasets to train neural-network models. In this section, we describe the data sources, followed by a description of our prediction models. Next, we will describe and justify our choice of the neural network structure. Finally, we will describe the training process and the results obtained.

3.1 Data Sources

The paper examines two different sources of data: (a) synthetic data obtained from running the UVA-Padova simulation model over a randomly selected set of meals and outputting the insulin input and BG values encountered over time; and (b) clinical trial data for a pump shutoff control algorithm recorded "longitudinal data" for $n = 17$ patients with 40 nightly sessions for each patient.

Synthetic Data Collection: The synthetic data collection is based on the Dalla-Man model [15,17,18], a 10 state variable nonlinear ODE model with upwards of 40 patient parameters. We selected 6 parameter sets describing virtual adult patients to perform our simulation. Each daytime simulation run for a patient involved two randomly sized simulated meals with the first meal having up to [20, 35] grams of carbohydrates (CHO) and the second meal involving [35, 135] grams of CHO. The insulin infusion was controlled by the "multi-basal" controller reported in our previous work [13]. The output data involved the BG values reported in 5 min intervals, and the insulin infusion also reported in 5 min intervals. Overall, the collected data consists of 6 different patients with 500 simulation runs per patient.

PSO3 Clinical Trial Data: The patient data was obtained from a previously held home trial of a predictive pump shutoff algorithm [10]. This pump is not an artificial pancreas and does not adjust insulin delivery rates, however, it does shut off delivery when BG is predicted (via Kalman filter) to drop below a threshold. We use a total of 17 patients with 40 nights of data per patient [34]. The collected data reports the BG levels collected every 5 min and the insulin infusion logs reporting basal insulin and boluses delivered. The logs were converted into insulin infusions at 5 min intervals (units delivered/5 min). For each nightly session, we discarded the first three hours worth of data to mitigate the effect of an unknown meal size and insulin bolus delivered just before the start of the session. This provided roughly 8 h of glucose and insulin data for each session. Entire nightly sessions were discarded from consideration for two main reasons: (a) the patient suffered from a hypoglycemia and were treated using "rescue" carbohydrates; and (b) the sensor readings were incomplete due to dropouts or data recording issues.

3.2   Prediction Model Structure

We now describe the structure of the prediction model. There are two aspects to the prediction model. Let t be the current time in minutes. Our goal is to predict the glucose value at a future "lookahead" time $t + T$, using the history of past N glucose values with a "past" step size of h: $G(t), G(t - h), \ldots, G(t - Nh)$ and past M insulin values $u(t), u(t - h), \ldots, u(t - Mh)$. The nature of the dataset, described previously, restricts the value of $h = 5$ min.

To decide on the value of the future lookahead, we note some known physiological details of insulin action: (a) insulin infused subcutaneously has a delay of 20–30 min before it reaches the patient's blood stream and starts action; (b) insulin has an action profile with peak action time of about 75 min and persists in the blood stream up to 4 h [20]. Therefore, to model the effect of the insulin administered at time t, we set the future lookahead $T = 30$ min. Our attempts to infer models for shorter lookahead such as $T = 5$ min leads to accurate but ultimately useless models that predict $G(t + 5) = G(t)$. Next, we set $Mh = 180$ min to account for the entire past history of insulin which may affect the value at $G(t + 30)$. Finally, to simplify the structure of the prediction model, we set $N = M$.

Let $\overleftarrow{G}_N(t)$ denote the column vector consisting of the N past glucose values $(G(t), G(t - h), \ldots, G(t - Nh))$ and likewise $\overleftarrow{u}_N(t)$ denote the past N insulin values $(u(t), \ldots, u(t - Nh))$. Thus, the desired prediction model is a distribution-valued function:

$$G(t + T) \sim F(\overleftarrow{G}_N(t), \overleftarrow{u}_N(t)).$$

wherein $F(\overleftarrow{G}_N(t), \overleftarrow{u}_N(t))$ is a random variable that models the distribution of the predicted values. Unfortunately, modeling and reasoning about distributions

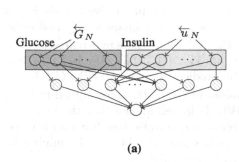

		# neurons in layer 1				
		3	6	8	20	200
# neurons in layer 2	3	1152	-	-	-	-
	4	-	623	8558	4518	-
	5	-	5.57	6.82	5.98	-
	6	-	10.34	3.37	2.15	-
	8	-	6.20	1.34	15.85	-
	10	-	3.21	7.45	5.99	-
	20	-	-	-	-	323.98

(a)   (b)

Fig. 1. (a) Neural network structure, (b) Mean training error (mg/dL) with varying combinations of neurons in each layer. Values presented are averaged over three trials for real patient ID: PSO3-001-0001.

is a hard problem. Therefore, we use a simpler approach using finitely many models that predict the mean and quantiles of the distribution.

The mean predictive model captures the expected value at time $t + T$:

$$\mathbb{E}(G(t+T)) = F_m(\overleftarrow{G}_N(t), \overleftarrow{u}_N(t)).$$

Likewise, we wish to formulate α upper quantile models $\overline{F}_\alpha(\overleftarrow{G}_N(t), \overleftarrow{u}_N(t))$ such that

$$\mathbb{P}\left\{ G(t+T) \geq \overline{F}_\alpha(\overleftarrow{G}_N(t), \overleftarrow{u}_N(t)) \right\} \leq (1-\alpha).$$

For instance, for $\alpha = 0.95$, we train an upper quantile model that predicts a level $\overline{F}_{0.95}$ which can be exceeded with probability at most 0.05. Similarly, an α lower quantile model $\underline{F}_\alpha(\overleftarrow{G}_N(t), \overleftarrow{u}_N(t))$ seeks a prediction such that

$$\mathbb{P}\left\{ G(t+T) \leq \underline{F}_\alpha(\overleftarrow{G}_N(t), \overleftarrow{u}_N(t)) \right\} \leq (1-\alpha).$$

In other words, $\underline{F}_{0.95}$ model predicts a value such that the actual value will lie below the predicted value with probability at most 0.05. The overall distribution is represented by a *mean* predictive model F_m and a set of upper quantile models $\overline{F}_\alpha$ and lower quantile models $\underline{F}_\beta$ for selected values of α and β.

Neural Network Structure: Figure 1(a) shows the overall structure of the neural network used in our model. Our neural networks involve two layers with the input layer partitioned into two parts: glucose inputs and insulin inputs. Next, based on existing ODE models [5, 15, 27] we partition the first hidden layer into two parts meant to model *insulin transport* and *glucose transport*. Finally, we add a joint hidden layer to model the insulin action on glucose. Besides this, note that our model does not have internal states. Rather, we assume that this state is well captured by the history of glucose and insulin inputs of the model. The number of neurons in the hidden layer was chosen by training a series of networks with different hidden layer sizes and choosing the best performing model, as will be described subsequently in this section.

Loss Functions: Our goal is to train neural network predictive models F_m for the mean prediction and the quantile models for lower quantiles $\underline{F}_\beta$ and upper quantiles $\overline{F}_\alpha$, respectively. These are achieved using the same regression process that chooses unknown network weights $\mathcal{W} : \{(W_0, \mathbf{b}_0), (W_1, \mathbf{b}_1), (W_2, \mathbf{b}_2)\}$ that minimize a loss function $\mathcal{L}$ over the prediction error $e(t; \mathcal{W}) : G(t + T) - F_{\mathcal{W}}(\overleftarrow{G}_N, \overleftarrow{u}_N)$ at time t, wherein the values of $G(t), u(t)$ are obtained from the training data: $\mathcal{W} :$ argmin $\sum_t \mathcal{L}(e(t; \mathcal{W}))$. It is well known that the mean of a distribution is obtained as the minimizer of a quadratic loss function over the samples $\mathcal{L}_m(e) : \|e\|_2^2$. Likewise, the quantiles are also obtained by minimizing loss functions corresponding to α upper quantiles [30].

$$\mathcal{L}_\alpha(e) = \max(-\alpha e, (1 - \alpha)e).$$

Similarly, the loss function for the β lower quantile minimizes the upper quantile loss function for $1 - \beta$: $\mathcal{L}_{1-\beta}$. It must be remarked that precise quantile regression requires large amounts of data since we are fitting a function at the tails of the distributions we wish to model, and consequently the process can be sensitive to outliers.

Training: The network weights are initialized at random and trained using off the shelf backpropagation algorithms implemented in the popular package TensorFlow [1]. The training process supports user specified loss functions as long as they are differentiable. To this end, a differentiable approximation of $\mathcal{L}_\alpha$ was obtained by replacing the max with a *softmax* operator: $\mathsf{softmax}(x, y) = \frac{e^x}{e^x + e^y}$. The training is performed by first partitioning the given data into different training and test sets. Next, gradient descent is performed over randomly selected batches of data points. Once a model is trained, its accuracy is evaluated over the test data. The training is performed by selecting loss functions for the mean model, the $\alpha = 0.95$ upper quantile and $\beta = 0.8$ lower quantile. The reason for the asymmetry is that our training results for $\beta = 0.95$ yielded a model with very poor predictive accuracy over the test data, rendering it useless. Training with $\beta = 0.8$ yielded a tighter lower bound in 94% of tests cases. The approach results in 3 neural network models F_m, $\overline{F}_{0.95}$ and $\underline{F}_{0.8}$.

3.3   Evaluation Results

In order to select number of neurons in each hidden layer, we tested networks with 4–200 neurons. Figure 1(b) shows mean error for various selected network structures, centered around those which gave lowest test error. Choosing 8 neurons per hidden layer produces the optimal result, a relatively small network with low prediction error.

Overall, we have found a two-layer network topology to be sufficient for obtaining fairly high prediction accuracies of 7.2 ± 3.0 mg/dL, which is competitive with the error bounds on current predictive models [23,24].

We were able to successfully train neural network models to fit all but two real-data patients, wherein success is defined as test error <12 mg/dL, and all synthetic data patients. Simple factors such as the amount of data or the variability in the BG levels fail to account for the two failure cases. Further in-depth analysis will be is being completed for future work. For the successful instances, the mean prediction error is 7.1867 ± 3.028 mg/dL. Training a model for each patient required 328 s on average, as measured on a MacBook Pro laptop with 3.3 GHz Intel Core i7 processor.

In Fig. 2 we present example model predictions, and in Fig. 3, we present histograms of error for the quantile networks.

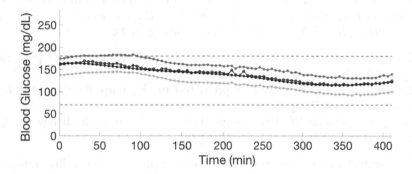

Fig. 2. Sample plot comparing our neural network model predictions to real data. Legend: filled black circles: actual data, open black circles: mean network prediction, open blue circles: upper 0.95 quantiles, open yellow circles: lower 0.8 quantiles. The dashed red lines show the normal blood glucose ranges. (Color figure online)

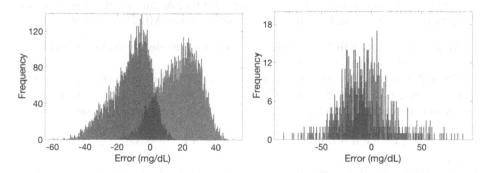

Fig. 3. Histograms of error for the lower quantile regression (orange), and the upper quantile regression (blue) models learned on synthetic data (left) and real data (right). Error is defined as $y_{data} - y_{mdl}$, positive error indicates our model is predicting below the data, and negative error indicates our model prediction is above the data. (Color figure online)

4   Model Predictive Control

In this section, we present the overall controller design that uses the neural networks inferred in the previous section as a starting point. The predictive controller takes as inputs:

1. Neural network models F_m, $\overline{F}_\alpha$ and $\underline{F}_\beta$ for the mean, upper quantile and lower quantile predictions, respectively, for time $t + T$ given historical data for times in the range $[t - Nh, t]$; and
2. Historical data $\overleftarrow{G}_N(t)$ and $\overleftarrow{u}_N(t)$ for glucose and insulin.

We wish to compute an *optimal schedule* of future insulin infusions over some time horizon Kh, where h is the time period for the controller: $u(t + h)$, $u(t + 2h)$, ..., $u(t + (K - 1)h)$, to achieve the following goals:

1. The mean predicted BG levels at time $T = Kh$ lie in some goal range $[G^*_{min}, G^*_{max}]$.
2. The lower quantile BG levels as predicted by $\underline{F}_\beta$ must lie above the hypoglycemia limit of $70\,\text{mg/dL}$.
3. The upper quantile BG levels as predicted by $\overline{F}_\alpha$ must lie below an upper hyperglycemia limit taken to be $210\,\text{mg/dL}$.

This control can be computed by solving a mixed integer linear program (MILP). The unknowns for this problem are as follows:

1. The predicted mean BG values for the future $G(t + h), \ldots, G(t + Kh)$. Thus, will assume that $G(s)$ denotes a known data point if $s \leq t$ and an unknown variable in the MILP if $s > t$.
2. The lower quantile BG values $G_l(t + h), \ldots, G_l(t + Kh)$,
3. The upper quantile BG values $G_u(t + h), \ldots, G_u(t + Kh)$.
4. The insulin infusions $u(t+h), \ldots, u(t+Kh)$. Once again $u(s)$ denotes a known data point if $s \leq t$ and an unknown variable in the MILP, otherwise.

Other unknowns will be introduced when we encode networks as MILP, as we will describe subsequently.

We will now describe the constraints that need to be enforced; fixed parameters are shown in blue. Figure 4 shows how the networks for the mean, upper and lower quantiles interact in the system of constraints shown below.

1. The final predicted mean value must be within bounds: $G^*_{min} \leq G(t+Kh) \leq G^*_{max}$.
2. The lower and upper quantiles must lie within hypo and hyper glycemia limits: $70\,\text{mg/dL} \leq G_l(t + Kh)$, and $G_u(t + Kh) \leq 210\,\text{mg/dL}$.
3. The values of the insulin infused must remain within limits: $u_l \leq u(s) \leq u_h$, $s \in \{t + h, \ldots, t + (K - 1)h\}$.
4. The total insulin delivered must be within limit: $\sum_{s=t-Nh}^{t+(K-1)h} u(s) \leq U_{tot}$.
5. The glucose level $G(s)$ must be predicted according to the network F_m: $G(t + jh) = F_m(\overleftarrow{G}(t + jh - T), \overleftarrow{u}(t + jh - T)), s \in \{t + h, \ldots, t + (K - 1)h\}$.

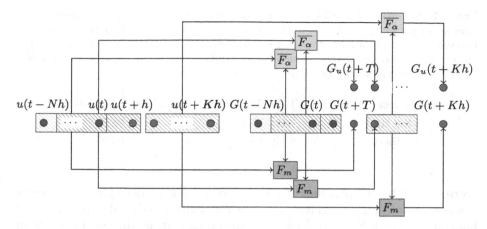

Fig. 4. Schematic illustration of the recursive unfolding setup for the model predictive control optimization problem. The red dots represent unknown variables, whereas the blue dots represent known data at optimization time. The unknown mean glucose values $G_m(t + h), \ldots, G(t + Kh)$ are constrained to be the mean predictions through the network F_m. Similarly, the unknown upper bounds $G_u(t + h), \ldots, G_u(t + Kh)$ are obtained using the network $\overline{F_\alpha}$. The lower bounds $G_l(t)$ are not shown in this figure. (Color figure online)

6. The upper limit $G_u(s)$ must be as predicted using the network $\overline{F_\alpha}$: $G_u(t + jh) = \overline{F_\alpha}(\overleftarrow{G}(t + jh - T), \overleftarrow{u}(t + jh - T)), s \in \{t + h, \ldots, t + (K - 1)h\}$.
7. The lower limit $G_l(s)$ must be predicted using the network $\underline{F_\beta}$: $G_l(t + jh) = \underline{F_\beta}(\overleftarrow{G}(t + jh - T), \overleftarrow{u}(t + jh - T)), s \in \{t + h, \ldots, t + (K - 1)h\}$.

The overall MILP formulation combines the constraints noted above with the objective that minimizes the deviation at time $t + Kh$ from a desired target value: $|G(t + Kh) - G^*|$. Note that constraints involve equalities written as $z = F(x, y)$, wherein F is the function computed by a neural network. These constraints are converted into MILP constraints through the introduction of fresh binary variables, as described in Sect. 2.4. The objective involves minimizing the absolute value of a linear expression, and can be converted to a linear objective in a standard manner [45].

Filling-in Prediction Gap: Since our approach uses history up to time t to predict the value at $t + T$, there is a potential "gap" in our predictions for BG values in the set $\{t+h, \ldots, t+T-h\}$ that require historical values past what was assumed to be available. There are two solutions: (a) Rather than use historical values spanning $[t - Nh, t]$, we will extend our history to $[t - Nh - T, t]$ and thus eliminate the gap; (b) Alternatively, we will assume unknown values for the "missing" history in the range $[t - Nh - T, t - (N + 1)h]$ that lie inside a admissible range $[0, 500]$. Furthermore, we can constrain the increase/decrease of successive BG values to a physiologically feasible range.

Table 2. Summary of results for clinical trial and synthetic datasets. Legend: #P - number of patients, # Tr - number of random trials per patient, T.H. - total time horizon for the simulations, Avg. Hyper - Average number of trials resulting in hyper glycemia, Avg. Hypo - Average number of trials resulting in hypoglycemia, TiR - Percent time in the range, BG $\in [70, 180]$ mg/dL.

Dataset	#P	# Tr	T.H.	Avg. Hyper	Avg. Hypo	Avg. TiR
Clin. trial	15	100	7.5 h	15.3	2.7	68.9
Synthetic	6	180	8.5 h	1.6	3.6	92.95

Deploying the Controller: There are two ways of deploying the control scheme: (a) compute a "single shot" insulin infusion schedule over the time horizon $[t + h, t + Kh]$ and deliver this to the patient; or (b) use some part of the computed insulin infusion schedule and update the schedule in real-time as new BG measurements are obtained. The latter option is called *receding horizon control*, and is preferred since it can act against errors between the predictive model and the actual patient. In practice, we use a receding horizon scheme that computes a control schedule for the entire time horizon $[t + h, t + Kh]$. The first insulin value of the resulting solution is delivered to the patient. At time $t + h$, a new BG measurement is obtained, and the entire computation is restarted over the new horizon $[t + 2h, t + (K + 1)h]$.

Handling Infeasibilities: The optimization problem can be infeasible when-ever the quantile models are inconsistent w.r.t the mean model prediction. This can happen for two key reasons: (a) the uncertainty in the future BG values is too large for us to guarantee that both the lower and upper bounds will be safe; (b) the networks are "extrapolating" on inputs that lie far from the training data. Situation (a) can be addressed by dropping the constraints from the problem, starting with the constraint on the upper bound network and hyperglycemia limit. If this does not resolve the infeasibility, we drop the mean value in range constraint. Failing this, we conclude that the predictions are potentially incon-sistent with each other. Thus, we revert to a safe insulin infusion, often given by the patient's basal insulin or complete shutoff of insulin.

5   Evaluation

Table 2 summarizes the overall results obtained across the two datasets. For each patient in a given dataset, we inferred predictive models, as described in Sect. 3. Additionally, for the clinical trial patients, we learned an additional neural net-work by repeating the backpropagation process. For the synthetic patient, the ODE model was used as the patient model in our simulation. The randomized nature of this process provides us with different results and accuracies for each run. For each patient, we simulated numerous trials with different randomly cho-sen initial conditions. For all trials, the value of G^* was taken to be 140 mg/dL with the tolerance limit D of 40 mg/dL.

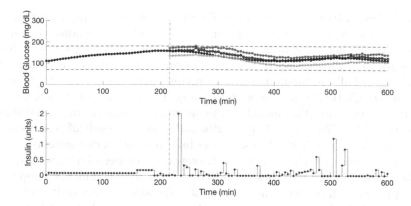

Fig. 5. Example of an instance when the controller gave a "correction bolus" of 2 units at $t = 220$ min insulin after the patient ate a meal, then reverted back to injecting only basal insulin of $< 0.5U/5$ min until glucose began to rise again near $t = 500$, where smaller "correction boluses" are given.

Clinical Trial Data: For this controller, the total insulin allowed was set to $U : 7$ IU over 3.5 h, with insulin limits $u_l = 0$ and $u_h = 12$ U/hr. The initial conditions consist of 180 min of randomly chosen insulin and BG data in the range $[60, 250]$ mg/dL. Whenever the initial condition starts above 180 mg/dL, the controller is able to bring it inside euglycemic range 30% of the time. Averaging across patients, 2.7 out of 100 trials resulted in hypoglycemia. In all instances, the controller delivered no insulin whenever BG fell below 90 mg/dL. About 6% of the optimization instances resulted in infeasibilities that required us to drop all constraints and revert to 0 insulin delivery. However, in all cases, we were able to recover feasibility and complete the simulation. The controller took a maximum of 2.87 s on our hardware platform to execute the control algorithm.

Synthetic Data: The total insulin allowed was set to $U : 5$ IU over 3.5 h, with insulin limits $u_l = 0$ and $u_h = 24$ U/hr. For this data, the Dalla Man et al. ODE model was used as the plant model in lieu of a real patient. For each virtual patient, we simulated 180 different combinations of initial BG values and randomly chosen set of two meals over 10 h. The controller itself was "blind" to these meals, and meals were made particularly large (eg. 135 g CHO) in order to maximize disturbances and BG variability, and hence stress the control algorithm. Hyperglycemic events that could likely be attributed by our controller - i.e, events where a patient has $BG < 180$ mg/dL initially, and $BG > 180$ mg/dL at the end occurred in 2.24% of the trials. Infeasibilities were limited to 2.82% on an average. The control computation step took a maximum of 4.65 s on our hardware platform. The controller action typically treats high BG by delivering a large insulin bolus and adjusts future insulin delivery down in anticipation of a hypoglycemia (see Fig. 5).

Instabilities in Predictive Models: For the clinical trial dataset, certain models exhibited high sensitivity to large jumps in the BG value, leading to apparent instabilities characterized by growing oscillations defined as repeated change in BG values exceeding $20\,\mathrm{mg/dL}$ over each $5\,\mathrm{min}$ period. This was observed in roughly 33% of all the simulation runs involving clinical trial patients. An example of such oscillations is presented in Fig. 6(a). The presence of these were correlated to data instances that included "discontinuous" jumps in the initial history, characterized by $|G(t+h) - G(t)| \geq 20\,\mathrm{mg/dL}$, likely a result of sensor calibration events. Across all initial histories that involved such a discontinuity, 73% of cases led to such unstable oscillations. However, a number of instances involved instabilities although the initial conditions did not have significant jumps.

Despite instabilities in some models, all simulations remained stable for at least for $t = 135\,\mathrm{min}$, and on average models remained stable $t = 221\,\mathrm{min}$, after which growing oscillations began. At the same time, not all cases of discontinuities in the initial history led to such oscillations, as shown in Fig. 6(b). Hence, if models are updated with new initial history data every 90–120 min, these instabilities should not pose significant issue.

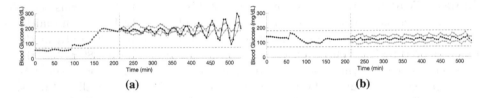

(a)   **(b)**

Fig. 6. (a) Instance where a calibration error occurring at $t = 95\,\mathrm{min}$, can over time cause large ($>20\,\mathrm{mg/dL}$) oscillations in the simulation. (b) the calibration error occurring at $t = 55\,\mathrm{min}$, did not cause large ($>20\,\mathrm{mg/dL}$) oscillations in the simulation. For each figure, the vertical black line separates the initial data from the simulated data.

Instabilities in the Synthetic Data: We note that these instabilities did not occur in the models learned from the synthetic data. This is consistent with our theory since no discontinuous jumps are possible in the ODE output. This is notable as the synthetic data included large jumps in BG due to meal disturbances, yet no instabilities arose in model prediction.

6   Limitations and Future Work

The first limitation lies in the way the data was collected. Although the PSO3 trial was a home trial of a device, the conditions of the trial could arguably result in extra scrutiny and presumably stricter adherence to clinical norms. This naturally biases the BG values to a tighter range and limits the number of hypoglycemia incidents seen (since it is often a stopping criterion for a participant in a trial). It is also well known that neural networks often perform poorly when tested on data that is "far" away from training data. This is one of the biggest limitations of our approach. To mitigate this, our results must be reproduced on longitudinal data obtained over a longer time periods, so that we may include the vast majority of clinically relevant situations.

Another limitation lies in the use of a neural network as the plant model. Although the network was not identical to the predictive model, it is nevertheless trained from the same dataset. This can arguably lead to better results than what is possible in reality. However, the inverse is also likely: by using a learned model as the plant, additional noise is introduced, and this could have contributed to instabilities observed in model prediction for the clinical data set. Hence, future work includes improvement to plant model construction.

The Dalla Man model is considered state-of-the-art. Nevertheless, it is hard to fit real patient data to it, due to the numerous parameters involved in the model [46].

Future work will consider alternative network topologies, or varied the history length needed. The use of recurrent neural network is yet another direction for future work. One of the limitations of our approach is the lack of support for meal detection or meal announcements for clinical data, though our controllers were able to bolus for unannounced meals in the case of synthetic data. We will attempt to integrate our work with related and complementary approaches for this problem proposed by Paoletti et al. [37], as well as obtain a patient data set which includes daytime data.

7 Conclusions

Thus, we have presented a data-driven approach that infers neural networks for the mean, upper and lower quantiles for predicting future BG levels from data. We formulate a robust control scheme for calculating safe and optimal infusions on this data in the presence of unannounced meals, and sensor errors. Finally, we have evaluated our performance over a variety of datasets, initial histories, patient models, and meal sizes. Our approach shows promising results. However, we also noted instabilities in the model that must be addressed as part of future work.

Acknowledgments. This work was supported by the US National Science Foundation (NSF) through awards 1446900, 1646556, and 1815983. All opinions expressed are those of the authors and not necessarily of the NSF.

References

1. Abadi, M., et al.: Tensorflow: large-scale machine learning on heterogeneous distributed systems. CoRR abs/1603.04467 (2016). http://arxiv.org/abs/1603.04467
2. Atlas, E., Nimri, R., Miller, S., Grunberg, E.A., Phillip, M.: MD-logic artificial pancreas system: a pilot study in adults with type 1 diabetes. Diab. Care **33**(5), 1072–1076 (2010)

3. Behl, M., Jain, A., Mangharam, R.: Data-driven modeling, control and tools for cyber-physical energy systems. In: Proceedings of the 7th International Conference on Cyber-Physical Systems, ICCPS 2016, pp. 35:1–35:10. IEEE Press, Piscataway (2016)

4. Bequette, B.W.: Algorithms for a closed-loop artificial pancreas: the case for model predictive control. J. Diab. Sci. Technol. **7**, 1632–1643 (2013)

5. Bergman, R.N., Urquhart, J.: The pilot gland approach to the study of insulin secretory dynamics. Recent Progress Hormon. Res. **27**, 583–605 (1971)

6. Bergman, R.N.: Minimal model: perspective from 2005. Hormon. Res. **64**(suppl 3), 8–15 (2005)

7. Bhat, N., McAvoy, T.J.: Use of neural nets for dynamic modeling and control of chemical process systems. Comput. Chem. Eng. **14**(4–5), 573–582 (1990)

8. Camacho, E., Bordons, C., Alba, C.: Model Predictive Control. Advanced Textbooks in Control and Signal Processing. Springer, London (2004). https://doi.org/10.1007/978-0-85729-398-5

9. Cameron, F., Niemeyer, G., Bequette, B.W.: Extended multiple model prediction with application to blood glucose regulation. J. Process Control **22**(8), 1422–1432 (2012)

10. Cameron, F., et al.: Inpatient studies of a Kalman-filter-based predictive pump shutoff algorithm. J. Diab. Sci. Technol. **6**(5), 1142–1147 (2012)

11. Chase, H.P., Maahs, D.: Understanding Diabetes (Pink Panther Book), 12 edn. Children's Diabetes Foundation, Denver (2011). Available online through CU Denver Barbara Davis Center for Diabetes

12. Chee, F., Fernando, T.: Closed-Loop Control of Blood Glucose. Springer, Heidelberg (2007). https://doi.org/10.1007/978-3-540-74031-5

13. Chen, X., Dutta, S., Sankaranarayanan, S.: Formal verification of a multi-basal insulin infusion control model. In: Workshop on Applied Verification of Hybrid Systems (ARCH), p. 16. Easychair (2017)

14. Cobelli, C., Dalla Man, C., Sparacino, G., Magni, L., Nicolao, G.D., Kovatchev, B.P.: Diabetes: models, signals and control (methodological review). IEEE Rev. Biomed. Eng. **2**, 54–95 (2009)

15. Dalla Man, C., Camilleri, M., Cobelli, C.: A system model of oral glucose absorption: validation on gold standard data. IEEE Trans. Biomed. Eng. **53**(12), 2472–2478 (2006)

16. Dalla Man, C., Micheletto, F., Lv, D., Breton, M., Kovatchev, B., Cobelli, C.: The UVa/Padova type I diabetes simulator: new features. J. Diab. Sci. Technol. **8**(1), 26–34 (2014)

17. Dalla Man, C., Raimondo, D.M., Rizza, R.A., Cobelli, C.: Gim, simulation software of meal glucose-insulin model (2007)

18. Dalla Man, C., Rizza, R.A., Cobelli, C.: Meal simulation model of the glucose-insulin system. IEEE Trans. Biomed. Eng. **1**(10), 1740–1749 (2006)

19. Dutta, S., Jha, S., Sankaranarayanan, S., Tiwari, A.: Output range analysis for deep feedforward neural networks. In: Dutle, A., Muñoz, C., Narkawicz, A. (eds.) NFM 2018. LNCS, vol. 10811, pp. 121–138. Springer, Cham (2018). https://doi.org/10.1007/978-3-319-77935-5_9

20. Freeman, J.S.: Insulin analog therapy: improving the match with physiologic insulin secretion. J. Am. Osteopath. Assoc. **109**(1), 26–36 (2009)

21. Garg, S.K., et al.: Glucose outcomes with the in-home use of a hybrid closed-loop insulin delivery system in adolescents and adults with type 1 diabetes. Diab. Technol. Ther. **19**(3), 1–9 (2017)

22. Goodfellow, I., Bengio, Y., Courville, A.: Deep Learning. MIT Press, Cambridge (2016)
23. Griva, L., Breton, M., Chernavvsky, D., Basualdo, M.: Commissioning procedure for predictive control based on arx models of type 1 diabetes mellitus patients. IFAC-PapersOnLine 50(1), 11023–11028 (2017)
24. van Heusden, K., Dassau, E., Zisser, H.C., Seborg, D.E., Doyle III, F.J.: Control-relevant models for glucose control using a priori patient characteristics. IEEE Trans. Biomed. Eng. 59(7), 1839–1849 (2012)
25. Hakami, H.: FDA approves MINIMED 670G system - world's first hybrid closed loop system (2016)
26. Hovorka, R., et al.: Nonlinear model predictive control of glucose concentration in subjects with type 1 diabetes. Physiol. Measur. 25, 905–920 (2004)
27. Hovorka, R., et al.: Partitioning glucose distribution/transport, disposal and endogenous production during IVGTT. Am. J. Physiol. Endocrinol. Metab. 282, 992–1007 (2002)
28. Hovorka, R.: Continuous glucose monitoring and closed-loop systems. Diab. Med. 23(1), 1–12 (2005)
29. Katz, G., Barrett, C., Dill, D.L., Julian, K., Kochenderfer, M.J.: Reluplex: an efficient SMT solver for verifying deep neural networks. In: Majumdar, R., Kunčak, V. (eds.) CAV 2017. LNCS, vol. 10426, pp. 97–117. Springer, Cham (2017). https://doi.org/10.1007/978-3-319-63387-9_5
30. Koenker, R.: Quantile Regression. Econometric Society Monographs, no. 38, p. 342 (2005)
31. Kowalski, A.: Pathway to artificial pancreas revisited: moving downstream. Diab. Care 38, 1036–1043 (2015)
32. Kushner, T., Bortz, D., Maahs, D., Sankaranarayanan, S.: A data-driven approach to artificial pancreas verification and synthesis. In: International Conference on Cyber-Physical Systems (ICCPS 2018). IEEE Press (2018)
33. Lomuscio, A., Maganti, L.: An approach to reachability analysis for feed-forward relu neural networks. CoRR abs/1706.07351 (2017). http://arxiv.org/abs/1706.07351
34. Maahs, D.M., et al.: A randomized trial of a home system to reduce nocturnal hypoglycemia in type 1 diabetes. Diab. Care 37(7), 1885–1891 (2014)
35. Medtronic Inc.: "paradigm" insulin pump with low glucose suspend system (2012). cf. http://www.medtronicdiabetes.ca/en/paradigm_veo_glucose.html
36. Nimri, R., et al.: Night glucose control with md-logic artificial pancreas in home setting: a single blind, randomized crossover trial-interim analysis. Pediatric Diab. 15(2), 91–100 (2014)
37. Paoletti, N., Liu, K.S., Smolka, S.A., Lin, S.: Data-driven robust control for type 1 diabetes under meal and exercise uncertainties. In: Feret, J., Koeppl, H. (eds.) CMSB 2017. LNCS, vol. 10545, pp. 214–232. Springer, Cham (2017). https://doi.org/10.1007/978-3-319-67471-1_13
38. Patek, S., et al.: In silico preclinical trials: methodology and engineering guide to closed-loop control in type 1 diabetes mellitus. J. Diab. Sci. Technol. 3(2), 269–82 (2009)
39. Pérez-Gandía, C., et al.: Artificial neural network algorithm for online glucose prediction from continuous glucose monitoring. Diab. Technol. Ther. 12(1), 81–88 (2010)
40. Piche, S., Sayyar-Rodsari, B., Johnson, D., Gerules, M.: Nonlinear model predictive control using neural networks. IEEE Control Syst. 20(3), 53–62 (2000)

41. Psichogios, D.C., Ungar, L.H.: Direct and indirect model based control using artificial neural networks. Indus. Eng. Chem. Res. **30**(12), 2564–2573 (1991)
42. Ruiz, J.L., et al.: Effect of insulin feedback on closed-loop glucose control: a crossover study. J. Diab. Sci. Technol. **6**(5), 1123–1130 (2012)
43. Steil, G.M., Rebrin, K., Darwin, C., Hariri, F., Saad, M.F.: Feasibility of automating insulin delivery for the treatment of type 1 diabetes. Diabetes **55**, 3344–3350 (2006)
44. Teixeira, R.E., Malin, S.: The next generation of artificial pancreas control algorithms. J. Diabetes Sci. Tech. **2**, 105–112 (2008)
45. Vanderbei, R.J.: Linear Programming: Foundations & Extensions, Second Edn. Springer, Heidelberg (2001). https://doi.org/10.1007/978-1-4614-7630-6, cf. http://www.princeton.edu/~rvdb/LPbook/
46. Visentin, R., Dalla Man, C., Cobelli, C.: One-day Bayesian cloning of type 1 diabetes subjects: toward a single-day UVa/Padova type 1 diabetes simulator. IEEE Trans. Biomed. Eng. **63**(11), 2416–2424 (2016)
47. Wang, T., Gao, H., Qiu, J.: A combined adaptive neural network and nonlinear model predictive control for multirate networked industrial process control. IEEE Trans. Neural Netw. Learn. Syst. **27**(2), 416–425 (2016)
48. Weinzimer, S., Steil, G., Swan, K., Dziura, J., Kurtz, N., Tamborlane, W.: Fully automated closed-loop insulin delivery versus semiautomated hybrid control in pediatric patients with type 1 diabetes using an artificial pancreas. Diab. Care **31**, 934–939 (2008)

Programming Substrate-Independent Kinetic Barriers with Thermodynamic Binding Networks

Keenan Breik[1](✉), Cameron Chalk[1], David Doty[2], David Haley[2],
and David Soloveichik[1](✉)

[1] University of Texas at Austin, Austin, USA
ksb@cs.utexas.edu, david.soloveichik@utexas.edu
[2] University of California, Davis, USA

Abstract. Engineering molecular systems that exhibit complex behavior requires the design of kinetic barriers. For example, an effective catalytic pathway must have a large barrier when the catalyst is absent. While programming such energy barriers seems to require knowledge of the specific molecular substrate, we develop a novel substrate-independent approach. We extend the recently-developed model known as thermodynamic binding networks, demonstrating programmable kinetic barriers that arise solely from the thermodynamic driving forces of bond formation and the configurational entropy of forming separate complexes. Our kinetic model makes relatively weak assumptions, which implies that energy barriers predicted by our model would exist in a wide variety of systems and conditions. We demonstrate that our model is robust by showing that several variations in its definition result in equivalent energy barriers. We apply this model to design catalytic systems with an arbitrarily large energy barrier to uncatalyzed reactions. Our results yield robust amplifiers using DNA strand displacement, a popular technology for engineering synthetic reaction pathways, and suggest design strategies for preventing undesired kinetic behavior in a variety of molecular systems.

1 Introduction

Abstract mathematical models of molecular systems, such as chemical reaction networks, have long been useful in *natural* science to study the properties of natural molecules. With recent experimental advances in synthetic biology and DNA nanotechnology [1,3,8,9], such models have come to be viewed also as *programming languages* for describing the desired behavior of synthetic molecules.

We can describe a chemical program with abstract chemical reactions such as

$$A + C \rightarrow B + C \tag{1}$$

$$A \rightarrow B. \tag{2}$$

© Springer Nature Switzerland AG 2018
M. Češka and D. Šafránek (Eds.): CMSB 2018, LNBI 11095, pp. 203–219, 2018.
https://doi.org/10.1007/978-3-319-99429-1_12

In particular, a program may require (1) and forbid (2). But what remains hidden at this level of abstraction is a well-known chemical constraint: if (1) is possible, then (2) must also be, no matter the exact substances. Knowing this, we might try to slow (2) by ensuring B has high free energy. But then $B + C$ must also have high free energy, so (1) slows in tandem. The only option to slow (2) but not (1) is to use a *kinetic barrier*: designing A so that, although it is possible for A to reconfigure into B, the system must traverse a higher energy (less favorable) intermediate in the absence of C.

To develop a substrate-independent approach to engineering kinetic barriers we need to rely on a universal thermodynamic property that would be relevant in a wide variety of chemical systems. We focus on the entropic penalty of association (decreasing the number of separate complexes). Intuitively, the entropic penalty is due to decreasing the number of microstates corresponding to the independent three-dimensional positions of each complex (configurational entropy). This thermodynamic penalty can be made dominant compared with other factors by decreasing the concentration.

To formalize this entropic penalty, we use the *thermodynamic binding networks* (TBN) model [5]. TBNs represent molecules as abstract *monomers* with binding sites that allow them to bind to other monomers. For a configuration γ, the TBN model defines $H(\gamma)$ as the number of bonds formed, and $S(\gamma)$ as the number of free complexes,[1] and the *energy* $E(\gamma) = -wH(\gamma) - S(\gamma)$ as a (negative) weighted sum of the two.[2] To be applicable to a wide variety of chemical systems, the TBN model does not impose geometric constraints on bonding (monomers are simply multisets of binding sets). Implementation of TBNs requires choosing a concrete physical substrate and geometric arrangement that permits the desired configurations to form.

---

[1] The quantities $H(\gamma)$ and $S(\gamma)$ are meant to evoke the thermodynamic quantities of enthalpy and entropy, although the mapping is not exact. Indeed, there are other contributions to physical entropy besides the number of separate complexes, and the free energy contribution of forming additional bonds typically contains substantial enthalpic and entropic parts.

[2] In typical DNA nanotechnology applications, the Gibbs free energy $\Delta G(\gamma)$ of a configuration γ can be estimated as follows. Bonds correspond to domains of length l bases, and forming each base pair is favorable by ΔG°_{bp}. Thus, the contribution of $H(\gamma)$ to $\Delta G(\gamma)$ is $(\Delta G^\circ_{bp} \cdot l)H(\gamma)$. At 1 M, the free energy penalty due to decreasing the number of separate complexes by 1 is ΔG°_{assoc}. At effective concentration C M, this penalty increases to $\Delta G^\circ_{assoc} + RT \ln(1/C)$. As the point of zero free energy, we take the configuration with no bonds, and all monomers separate. Thus, the contribution of $S(\gamma)$ to $\Delta G(\gamma)$ is $(\Delta G^\circ_{assoc} + RT \ln(1/C))(|\gamma| - S(\gamma))$, where $|\gamma|$ is the total number of monomers. To summarize,

$$\Delta G(\gamma) = (\Delta G^\circ_{bp} \cdot l)H(\gamma) + (\Delta G^\circ_{assoc} + RT \ln(1/C))(|\gamma| - S(\gamma)).$$

Note that, as expected, this is a linear combination of $H(\gamma)$ and $S(\gamma)$, and that increasing the length of domains l weighs $H(\gamma)$ more heavily, while decreasing the concentration C weighs $S(\gamma)$ more heavily. Typically $G^\circ_{bp} \approx -1.5 \, \text{kcal/mol}$, and $G^\circ_{assoc} \approx 1.96 \, \text{kcal/mol}$ [7].

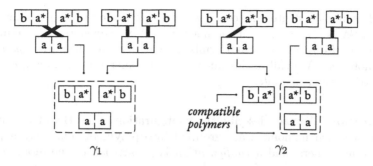

γ_1 γ_2

Fig. 1. Two configurations γ_1 and γ_2 of the TBN $\mathcal{T} = \{\{a, a\}, \{a^*, b\}, \{a^*, b\}\}$. Note that $\mathcal{T}$ has 3 monomers but 2 monomer types and 6 sites but 3 site types. A dashed box indicates monomers that are part of the same polymer. A single configuration (bottom) can correspond to multiple ways of binding complementary sites (top), which are not distinguished in our model. In γ_2 the polymer on the left has exposed sites $\{b, a^*\}$ and the polymer on the right $\{a, b\}$; they are thus compatible since the exposed site a^* of the left is complementary to exposed site a of the right. Since γ_2 has compatible polymers it is not saturated, but γ_1 is.

We augment the TBN model with a notion of kinetic paths (changes in configuration) due to merging of different complexes and splitting them up (and in the full version of this paper, making, breaking, or exchanging bonds). This gives rise to a notion of *paths* of configurations, with different energies. Define the *height* of a path starting at γ as the maximum value of $E(\delta) - E(\gamma)$ over all configurations δ on the path. Then the kinetic energy *barrier* separating configuration δ from configuration γ is the height of the minimum-height path from γ to δ.

In Sect. 2 we introduce our main kinetic model. We further show that when $w \geq 1$, it is sufficient to consider only fully bonded configurations in the energy barrier analysis. In Sects. 3.1 and 3.2 we develop two constructions for catalytic systems. Both constructions yield families of TBNs parametrized by a complexity parameter n such that the uncatalyzed energy barrier scales linearly with n. The catalyzed energy barrier is always 1. We show a direct DNA strand displacement implementation of one of the constructions. Finally we show an autocatalytic TBN, with an arbitrarily large energy barrier to undesired triggering, that exponentially amplifies its input signal (Sect. 3.2.1).

2 Kinetic Model

Our kinetic models build on thermodynamic binding networks (TBN) [5]. Intuitively, we model a chemical system as a collection of molecules, each of which has a collection of binding sites, which can bind if they are complementary. Although the TBN model is more general, DNA domains can be thought of as the prototypical example of binding sites. No geometry is enforced, which allows the model to handle topologically complex structures, such as pseudoknots.

TBN. (Figure 1 illustrates the concepts of the following two subsections.) Formally, a *TBN* is a multiset of monomer types. A *monomer type* is a multiset of site types. A *site type* is a formal symbol, such as a, and has a *complementary type*, denoted a^*. We call an instance of a monomer type a *monomer* and an instance of a site type a *site*.

Configuration. We may describe the configuration of a TBN at any moment in terms of which monomers are grouped into polymers. This way a *polymer* is a set of monomers, and a *configuration* is a partition of the monomers into polymers.[3]

The *exposed sites* of a polymer is the multiset of site types that would remain if one were to remove as many complementary pairs of sites as possible. Each such pair is counted as a *bond*. Note that bonds are not specified as part of a configuration, and intuitively we think of polymers as being maximally bonded. Two polymers are *compatible* if they have some complementary exposed sites. A configuration is *saturated* if no two polymers are compatible. This is equivalent to having the maximum possible number of bonds.

Notice that a polymer may have two incompatible halves. This represents spontaneous co-localization and comes with an energy penalty, as discussed later.

Path. (Figure 2 illustrates the concepts of the following three subsections.) One configuration can change into another by a sequence of small steps. If γ can become δ by replacing two polymers with their union, then γ *merges* to δ and δ *splits* to γ, and we write $\gamma <^1 \delta$. We denote by $\leq^1$, $<$, $\leq$ the reflexive, transitive, and reflexive transitive closures of $<^1$. A *path* is a nonempty sequence of configurations where each merges or splits to the next. Note that there is a path between any two configurations.[4]

We could imagine smaller steps that manipulate individual bonds. But surprisingly, a bond-aware model leads to essentially equivalent kinetic barriers, which we prove in the full version of this paper. Thus keeping track of bonds is an unnecessary complication.

Energy. For a configuration γ, denote by $H(\gamma)$ the number of bonds summed over all polymers. Denote by $S(\gamma)$ the number of polymers. Note that a saturated configuration has maximum $H(\gamma)$. The *energy* of γ is

$$E(\gamma) = -wH(\gamma) - S(\gamma),$$

---

[3]Note that swapping two monomers of the same type between different polymers produces a different configuration. Distinguishing different monomers of the same type allows us to equate the space of configurations with the lattice of partitions, which is a key tool in the full version of this paper.

[4]For instance, although this path is likely energetically unfavorable, we can merge all initial polymers into one, and then split into the desired end polymers.

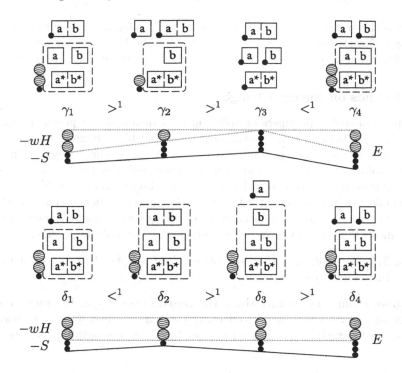

Fig. 2. A path p consisting of the configurations γ_1, γ_2, γ_3, γ_4 and a path p' consisting of the configurations δ_1, δ_2, δ_3, δ_4 of the TBN $\mathcal{T} = \{\{a\}, \{b\}, \{a, b\}, \{a^*, b^*\}\}$. The energy of each configuration is shown below it. A large wavy disc indicates energy due to a bond. A small solid disc indicates energy due to a polymer. Here bond strength $w = 2$, so a wavy disc is twice as tall as a solid disc. The height of p is $h(p) = E(\gamma_3) - E(\gamma_1) = (-4) - (-2w - 2) = 2$. The height of p' is $h(p') = E(\delta_2) - E(\delta_1) = 1$.

where the bond strength w represents the benefit from gaining a bond relative to gaining a polymer. Note that $H(\gamma) \geq 0$ and $S(\gamma) > 0$, so $E(\gamma) < 0$, and that lower energy, which results from more bonds or more polymers, is more favorable. The choice to make favorability correspond to lower energy (more negative) is motivated by consistency with the standard physical chemistry notion of free energy. We call a minimum energy configuration *stable*.

Merging incompatible polymers forms no additional bonds and so is always unfavorable, since $S(\gamma)$ drops without $H(\gamma)$ rising. In contrast, when bond strength $w > 1$, merging compatible polymers is always favorable. So every stable (that is, minimum energy) configuration is saturated. This regime is typical of many real systems, and in particular, we can engineer DNA strand displacement systems [10] to have large bond strength w by increasing the length of domains (see also footnote 2).

Barrier. With notions of paths and energy, we can establish the difficulty of passing from a configuration γ to another δ. The *height* $h(p)$ of a path p starting from γ is the greatest energy difference $E(\delta) - E(\gamma)$ from γ to any configuration δ along p. Notice that $h(p) \geq E(\gamma) - E(\gamma) = 0$. The *barrier* $b(\gamma, \delta)$ from γ to δ is the least height of any path from γ to δ. Notice that $b(\gamma, \delta) \geq 0$ as well.

2.1 Bounds on Energy Change

Merging compatible polymers or splitting into incompatible polymers changes energy in a predictable way. Splitting into incompatible polymers keeps all bonds and results in one more polymer, so overall it drops energy by 1. Similarly, merging compatible polymers results in one fewer polymer but at least one more bond. So overall it drops energy by $w - 1$ when bond strength $w \geq 1$.

To make this precise, we introduce two other partial orders on configurations. Let $\gamma \trianglelefteq \delta$ mean that γ can become δ by merges *of compatible polymers*. Let $\gamma \preceq \delta$ mean that γ can become δ by merges of *in*compatible polymers.

Claim 2.1. If $\gamma \preceq \delta$, then $E(\gamma) = E(\delta) - (S(\gamma) - S(\delta))$. If $\gamma \trianglelefteq \delta$, then $E(\delta) \leq E(\gamma) - (w - 1)(S(\gamma) - S(\delta))$.

The above claim says nothing about the general case $\gamma \leq \delta$. In order to apply the bounds, it will prove useful to decompose $\leq$ into $\preceq$ and $\trianglelefteq$. Any sequence of merges can be modified so that all merges between compatible polymers come first:

Claim 2.2. If $\gamma \leq \delta$, then some α has $\gamma \trianglelefteq \alpha \preceq \delta$.

Proof (sketch). Intuitively we want to reorder merges. In the full version of the paper, we show how to treat a merge as an object that can be applied in a context other than its original configuration, where the original polymers involved may not exist. With this machinery, the overall argument shows that if $\gamma \leq \delta$, then we can form α by starting with γ and doing as many merges as possible while preserving $\gamma \trianglelefteq \alpha \leq \delta$. No additional merge forms a bond, so $\alpha \preceq \delta$.   □

2.2 Saturated Paths

We prefer to reason about saturated configurations because there are substantially fewer of them and they have special properties, which simplifies proofs. In this section we show that the barrier remains essentially the same even if we consider paths that traverse only saturated configurations. This may be surprising since breaking some bonds might seem to allow a path to bypass an otherwise large barrier.

We see an example of the special properties of saturated configurations in the following claim, which is used in later sections to show a large energy barrier in our constructions.

Claim 2.3. If γ and δ are saturated, then $b(\gamma, \delta) \geq S(\gamma) - S(\delta)$.

Proof. Consider a path p from γ to δ. Since γ and δ are saturated, $H(\gamma) = H(\delta)$, so $h(p) \geq E(\delta) - E(\gamma) = S(\gamma) - S(\delta)$. So $b(\gamma, \delta) \geq S(\gamma) - S(\delta)$.   □

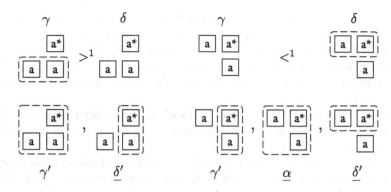

Fig. 3. An example of the two cases of the proof of Claim 2.5. An underline indicates a quantity constructed in the proof.

Now we turn to the main result of this section. A *saturated path* is a path along which every configuration is saturated. For example, the bottom path p' in Fig. 2 is saturated. If γ and δ are saturated, then let $b_{\mathrm{sat}}(\gamma, \delta)$ denote the barrier from γ to δ when allowing only saturated paths. Since a saturated path is a path, $b_{\mathrm{sat}}(\gamma, \delta) \geq b(\gamma, \delta)$. It turns out that if bond strength $w \geq 2$, then the reverse inequality also holds, so $b_{\mathrm{sat}}(\gamma, \delta) = b(\gamma, \delta)$. And if $w \geq 1$, then the reverse inequality "almost" holds.

We first need a technical result proven in the full version of this paper:

Claim 2.4. If $\gamma <^1 \delta$ and $\gamma \leq \gamma'$, then some α has $\delta \leq \alpha$ and $\gamma' <^1 \alpha$.

As in the proof of Claim 2.2, the proof of the above claim relies on treating merges as objects that can be applied in different contexts. In particular, the same merge that changes γ to δ is applied to get from γ' to α.

To connect $b_{\mathrm{sat}}(\gamma, \delta)$ to $b(\gamma, \delta)$, we first focus on a single step along a path and then extend to the full path. Let $[\gamma]$ denote the set of saturated γ' with $\gamma \trianglelefteq \gamma'$.

Claim 2.5. Let bond strength $w \geq 1$. If γ merges or splits to δ and $\gamma' \in [\gamma]$, then some $\delta' \in [\delta]$ and saturated path p' from γ' to δ' has

$$E_{\max}(p') \leq \max\{E(\gamma), E(\delta)\} + \max\{0, 2 - w\},$$

where $E_{\max}(p')$ is the maximum energy of any configuration along p'.

Proof. Let bond strength $w \geq 1$, and suppose $\gamma' \in [\gamma]$.

First consider the case where γ splits to δ (see Fig. 3, left). Then $\delta <^1 \gamma$. So $\delta \leq \gamma$. By assumption $\gamma' \in [\gamma]$, so $\gamma \leq \gamma'$, so transitively $\delta \leq \gamma'$. By Claim 2.2, there is δ' with $\delta \trianglelefteq \delta' \preceq \gamma'$. By assumption γ' is saturated, and now $\delta' \preceq \gamma'$, so δ' is saturated, and so $\delta' \in [\delta]$. So let p' be the path from γ' to δ' by splits into incompatible polymers guaranteed to exist by $\delta' \preceq \gamma'$. Each such split drops energy, so the claim holds.

Next consider the case where γ merges to δ (see Fig. 3, right). Then $\gamma <^1 \delta$. If $\gamma = \gamma'$, then let $\delta' = \delta$, and let p' be γ followed by δ. Otherwise $\gamma < \gamma'$. In that case let α be the configuration guaranteed by Claim 2.4. Then $\delta \leq \alpha$. So by Claim 2.2, there is δ' with $\delta \trianglelefteq \delta' \preceq \alpha$. Now by assumption γ' is saturated, and by construction $\gamma' \leq \alpha$, so α is saturated. Next $\delta' \preceq \alpha$ means δ' is saturated, and so $\delta' \in [\delta]$.

Let p' be the concatenation of two paths p'_1 and p'_2 defined as follows. Let p'_1 be γ' followed by α. This merge can at worst result in one less polymer and no additional bonds, so $E(\alpha) \leq E(\gamma') + 1$.

Then let p'_2 be the path from α to δ' by splits into incompatible polymers guaranteed to exists by $\delta' \preceq \alpha$. Each such split drops energy, so α is the highest energy intermediate configuration. But by assumption $\gamma \trianglelefteq \gamma'$ and $\gamma \neq \gamma'$, so $S(\gamma) - S(\gamma') \geq 1$. So by Claim 2.1, and assuming bond strength $w \geq 1$, we get $E(\gamma') \leq E(\gamma) - (w - 1)$. So

$$E(\alpha) \leq E(\gamma') + 1$$
$$\leq E(\gamma) + 2 - w,$$

which implies the claim.   □

To extend the result to a full path, we apply it to each configuration along the path.

Claim 2.6. For bond strength $w \geq 1$ and saturated γ and δ, we have

$$b_{\mathrm{sat}}(\gamma, \delta) \leq b(\gamma, \delta) + \max\{0, 2 - w\}.$$

Proof. Suppose bond strength $w \geq 1$ and γ and δ are saturated, and consider a path p from γ to δ. By assumption γ is saturated, so $\gamma \in [\gamma]$. So we can apply Claim 2.5 to each configuration of p in turn to get a saturated path p'.

Let E' denote the maximum energy along p' and E denote the maximum energy along p. Then Claim 2.5 ensures $E' \leq E + \max\{0, 2 - w\}$. And both p' and p start in the same configuration with the same energy, so $h(p') \leq h(p) + \max\{0, 2 - w\}$. So $b_{\mathrm{sat}}(\gamma, \delta) \leq b(\gamma, \delta) + \max\{0, 2 - w\}$.   □

Since $b(\gamma, \delta) \leq b_{\mathrm{sat}}(\gamma, \delta)$, we have the following corollary of Claim 2.6.

Corollary 2.7. For bond strength $w \geq 2$ and saturated γ and δ, we have $b_{\mathrm{sat}}(\gamma, \delta) = b(\gamma, \delta)$.

For Claim 2.6 and Corollary 2.7, bond strength $w \geq 1$ is necessary. If $w < 1$, then $b_{\mathrm{sat}}(\gamma, \delta)$ can be larger than $b(\gamma, \delta)$ by an arbitrary amount.

3   TBNs with Programmable Energy Barriers

We present two constructions for TBNs with equal energy (stable) "initial" and "triggered" configurations, such that the energy barrier to get from one to the

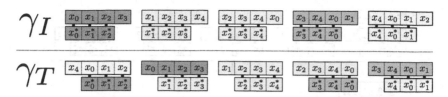

Fig. 4. The two stable configurations of a translator cycle with complex length $z = 3$ and number of complex types $c = 5$.

other can be made arbitrarily large: both constructions are parameterized by n, with the energy barrier scaling linearly with n. Further, both constructions admit catalysts that reduce the energy barrier to 1.

The first construction (translator cycle), discussed Sect. 3.1, is based on a DNA strand displacement catalyst, and the progress from the initial to triggered configurations with the catalyst can be physically implemented as a strand displacement cascade. Although this system has been previously proposed in [11], for the first time we rigorously prove an energy barrier.

The second construction (grid gate), discussed in Sect. 3.2, does not have an evident physical implementation (e.g., as a strand displacement system), but surpasses the translator cycle system in the following ways: (1) a proof of *copy tolerance*[5]: the energy barrier is proven in more general contexts where multiple copies of monomers are present, in any ratio, (2) *autocatalysis*: the grid gate can be modified so that the catalyst transforms the gate into a polymer that has the same excess domains as the catalyst, which can itself catalyze the transformation of additional gates (leading to exponential amplification).

Throughout both sections, we make the assumption that $w \geq 2$, so that by Corollary 2.7, it is sufficient to describe energy barriers by pathways in the saturated model; if we weaken this assumption to $w \geq 1$, then by Claim 2.6 the barrier proved is within 1 of the barrier in the unrestricted pathway model (allowing unsaturated configurations).

The constructions demonstrate that catalysts and autocatalysts with arbitrarily high energy barriers can be engineered solely by reference to the general thermodynamic driving forces of binding and formation of separate complexes, which are captured in the TBN model.

3.1 Translator Cycle

Consider the TBN illustrated in Fig. 4. There are two particular configurations that interest us, an initial configuration γ_I and a triggered configuration γ_T. The two configurations are stable. In the presence of a catalyst monomer $\{x_4, x_0, x_1, x_2\}$ (or an extra copy of any *top monomer*—any of the monomers with unstarred domains), a height 1 pathway exists to reach γ_T, illustrated by

---

[5]Our proof technique for the barrier of the translator cycle does not seem sufficient to prove copy tolerance, although we conjecture the translator cycle is copy tolerant.

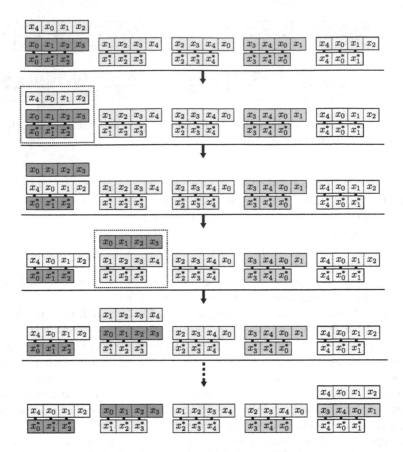

Fig. 5. A segment of the height 1 path which is possible because an extra copy of a top monomer, $\{x_4, x_0, x_1, x_2\}$, is present to act as a catalyst. If no catalyst is initially present, if one complex splits—thus reducing the number of bonds by z—the top monomer from the split complex can be used as a catalyst, resulting in a path of height $z + 1$ similar to the path shown in the figure.

Fig. 5. Further, this catalytic system is realizable as a DNA strand displacement cascade; more information about this connection can be found in the full version of this paper, and in [11]. In the case of many copies of each complex, since the catalyst is in fact any of the top monomers, the system may be used as an amplifier: at the end of the pathway shown in Fig. 5, another monomer with domains $\{x_4, x_0, x_1, x_2\}$ becomes free which can catalyze another set of complexes which are in the initial configuration. Note, however, that proving the energy barrier for the translator cycle in the multi-copy setting remains open.

To program a large energy barrier, we give a formal definition for generalizing the translator cascade, parameterized by *complex length z* and *number of complex types c*. Given $z \leq c$, a (z, c)-*translator cycle* is a TBN with monomers $\mathcal{M} = \{m_i, m_i^* \mid i \in \{0, \cdots, c-1\}$, where $m_i = \{x_i, x_{i+1 \text{ (mod } c)}, \cdots, x_{i+z \text{ (mod } c)}\}$

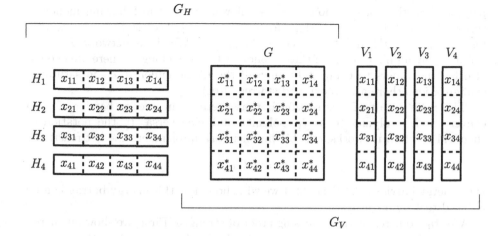

Fig. 6. The monomer types in the grid gate TBN for the case $n = 4$.

and $m_i^* = \{x_i^*, x_{i+1 \ (\text{mod } c)}^*, \cdots, x_{i+z-1 \ (\text{mod } c)}^*\}$. The *initial configuration* is $\gamma_I = \{\{m_i, m_i^*\} \mid i \in \{0, \cdots, c-1\}\}$ and the *triggered configuration* is $\gamma_T = \{\{m_{i-1 \ (\text{mod } c)}, m_i^*\} \mid i \in \{0, \cdots, c-1\}\}$.

What we want is that when the catalyst is absent, there is a large energy barrier to move from γ_I to γ_T. This barrier depends on the complex length z and the number of complex types c, and can be made arbitrarily large.

We can begin a path to γ_T by splitting any complex apart, thus reducing the number of bonds by z, and then use the top monomer as a catalyst in the same way as the with-catalyst pathway shown in Fig. 5. So, if z is not large, then there is a small barrier. We can also bring all complexes together while reducing the number of polymers by $c - 1$, and then split into the triggered complexes, so if c is not large, there is also a small barrier. Somewhat surprisingly, it is not sufficient to set $z = c = n$ to attain a barrier of $\Omega(n)$; a complicated path exists which has height $\Theta(\frac{c}{z})$ which is illustrated in the full version of the paper.

Note that if the cascade has $c = z^2$, then the uncatalyzed paths described above have height $\Omega(z)$. Are there other paths with smaller heights (that is, is the energy barrier $\Omega(z)$ in that case)? We prove that indeed the energy barrier is $\Omega(z)$:

Theorem 3.1. *If $z = n$ and $c = n^2$, then $b(\gamma_I, \gamma_T) = \Omega(n)$.*

The proof appears in the full version of this paper.

3.2 Grid Gate

Consider the TBN illustrated in Fig. 6. We focus on two polymer types G_H and G_V depicted in the figure, and show that there is a barrier to convert G_H to G_V and vice versa. To generalize to arbitrarily high energy barriers, we parameterize the construction by n as follows. Define the following monomer types: "horizontal" $H_i := \{x_{ij}\}_{j=1}^n$ for $i \in \{1, \ldots, n\}$, "vertical" $V_j := \{x_{ij}\}_{i=1}^n$ for $j \in \{1, \ldots, n\}$, and "gate" $G := \{x_{ij}^*\}_{i,j=1}^n$. We fix a network $\mathcal{T}$ which contains any number of any of these monomer types, so long as there are enough of other monomers to completely bind all the G monomers (i.e., in saturated configurations there are no exposed starred sites).

The main result of this section is that there is an energy barrier of n to convert a G_H polymer to a G_V polymer and vice versa. In the notation of chemical reaction networks, this binding network implements the net reaction

$$G_H \rightleftharpoons G_V$$

with energy barrier n. In Sect. 3.2.1 we will show how this energy barrier can be reduced to 1 by a catalyst.

We state our results in increasing order of strength. First, we show an energy barrier to changing a single G_H to a single G_V (and vice versa), in the presence of exactly one of each V_j monomer (i.e., just enough to create a G_V). Let γ_H denote the configuration of the TBN depicted in Fig. 6 in which G and each H_i are grouped into a single polymer G_H, and each V_j is in its own polymer. Let γ_V denote the symmetric configuration in which G and each V_j are grouped into a single polymer G_V, and each H_i is in its own polymer.

Claim 3.2. $b(\gamma_H, \gamma_V) = b(\gamma_V, \gamma_H) = n$.

Proof. Consider a saturated path p from γ_H to γ_V. Let δ denote the first configuration in the path with some H_i separate from G. Since δ is saturated but in this configuration H_i is not bound to G, all of the V_j must be bound together with G. So the configuration δ' which immediately precedes δ in the path must have all of the monomers in a single polymer. Since p is a saturated path, $H(\gamma_H) = H(\delta')$, and since p was arbitrary, $b(\gamma_H, \gamma_V) \geq h(p) \geq E(\delta') - E(\gamma_H) = S(\gamma_H) - S(\delta') = n$. The path which achieves height n proceeds by merging G_H and all the V_j's, and then splitting off all the H_i's, leaving G_V behind. By symmetry, $b(\gamma_H, \gamma_V) = b(\gamma_H, \gamma_V)$. □

We can generalize this basic result to a *copy-tolerant* setting, in which we assume the starting configuration has several copies of each G_H and G_V, as well as excess copies of each H_i and V_j, showing that the energy barrier of n remains for conversion of any G_H's into G_V's, and vice versa.

To make our result precise, we parameterize a set of configurations of the network based upon the count of polymers G_H and G_V and the number of horizontal and vertical monomers that are unbound. For $c_{G_H}, c_{G_V} \in \mathbb{N}$, define $\Gamma(c_{G_H}, c_{G_V})$ to be the set of configurations of the network $\mathcal{T}$ that contain precisely c_{G_H} polymers of type G_H and c_{G_V} polymers of type G_V, and in which all other monomers are in separate polymers by themselves. It turns out that the stable configurations of the network are exactly the ones in some $\Gamma(c_{G_H}, c_{G_V})$ (proof in the full version of this paper). The following theorem generalizes Claim 3.2 to the multi-copy context. Note that obtaining a similar copy-tolerant result for the translator cycle remains open.

Theorem 3.3. *Let* $\gamma \in \Gamma(c_{G_H}, c_{G_V})$. *If* $\gamma' \in \Gamma(c_{G_H} - \Delta, c_{G_V} + \Delta)$ *for some nonzero* $\Delta \in \mathbb{N}$, *then* $b(\gamma, \gamma') = n$.

The proof appears in the full version of this paper.

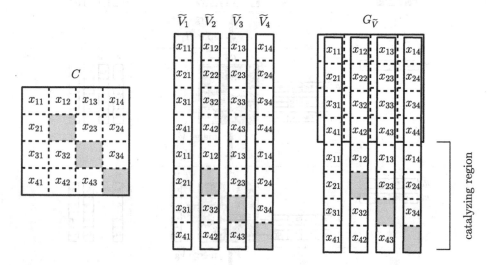

Fig. 7. Catalysts and autocatalysts in the grid gate TBN for the case $n = 4$. Left: C is a single monomer that acts as a catalyst to convert between G_H and G_V. Middle: Modified vertical monomers $\{\tilde{V}_j\}_{j=1}^n$ with extra sites. Right: After C converts G_H to G_V with modified vertical monomers, G_V has the same excess sites as C and acts as a catalyst itself.

3.2.1   Catalysis and Autocatalysis

The kinetic barrier shown for the grid gate can be disrupted by the presence of new monomer types. In fact, the model admits a *catalyst* monomer C that lowers the energy barrier from n to 1, i.e., in the presence of C, a G_H can be converted into a G_V, and vice versa, with a sequence of merge-split pairs. In the notation of chemical reaction networks, this binding network implements the net reaction

$$G_H + C \leftrightharpoons G_V + C$$

with energy barrier 1 but maintains a high energy barrier for the reaction $G_H \leftrightharpoons G_V$.

For the grid gate of size $n \times n$, we define a catalyst: $C := \{x_{ij} \mid 1 \leq i, j \leq n, i \neq j\} \cup \{x_{11}\}$, illustrated in Fig. 7 (left). C is a monomer containing all of the non-diagonal unstarred sites, while retaining exactly one of the diagonal sites. The mechanism by which C can transform G_H to G_V with merge-split pairs is by an alternating processes of merges and splits shown in Fig. 8. Intuitively, in step i of the catalyzed reaction $G_H + C \rightarrow G_V + C$, site x_{ii}^* on G switches its association from x_{ii} on H_i (left) to its counterpart on V_i (right) by merging the

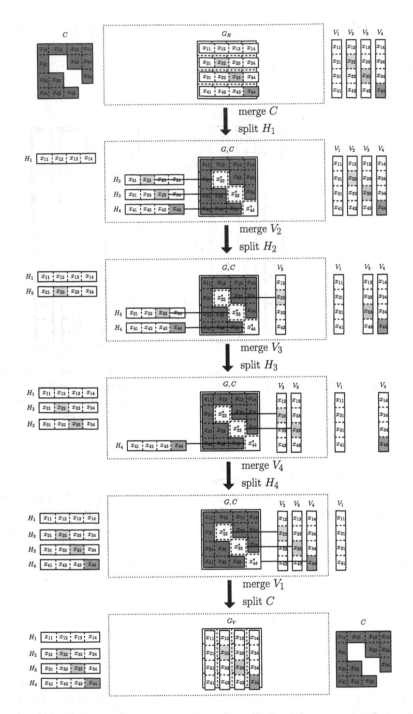

Fig. 8. Full pathway for reaction $G_H + C \rightarrow G_V + C$. In each stage, exactly one merge and one split occurs, and the center polymer remains saturated. (Color figure online).

evolving polymer (center) with V_i and then splitting off H_i. Only the diagonal sites need be considered in the intermediate steps since the catalyst C (blue, center, overlaid) balances the other sites of G.

Consider a network $T^C := \{G, C\} \cup \{H_i\}_{i=1}^n \cup \{V_j\}_{j=1}^n$ which includes one instance of every monomer type that has been introduced, as well as the catalyst. As before, we shall be interested in net transitions between G_H and G_V, and so for the network T^C we define the following configurations: $\gamma_H^C := \{G_H, C\} \cup \{V_j\}_{j=1}^n$ and $\gamma_V^C := \{G_V, C\} \cup \{H_i\}_{i=1}^n$.

Theorem 3.4 states that transitions that previously had arbitrarily large energy barriers are reduced in barrier to one.

Theorem 3.4. $b(\gamma_H^C, \gamma_V^C) = b(\gamma_V^C, \gamma_H^C) = 1$.

By adding to the network additional instances of $\{G\} \cup \{H_i\}_{i=1}^n \cup \{V_j\}_{j=1}^n$, these results can be extended to the more general case of $\gamma(c_{G_H}, c_{G_V}) \cup \{C\}$ by induction.

The grid gate can also be modified to act in an *autocatalytic* manner. By modifying the vertical monomers it is possible for G_V to have a set of exposed monomers acting as a "catalyzing region", which has the same structure and function as C (see Fig. 7, middle and right). We give the formal construction and related proofs in the full version of this paper.

4 Future Work

Single-molecule applications are rare compared with multi-copy situations. Indeed, using a catalyst or autocatalyst for amplification requires many copies of the components. Although we proved the correctness of the grid gate in the multi-copy context, we could not easily generalize our proof of the correctness of the translator cycle. Since the translator cycle has a straightforward strand displacement implementation, it is important to show that the uncatalyzed energy barrier is still n in bulk applications typical of strand displacement experiments. Success in this line of research provides a promising approach to making leakless strand displacement amplifiers [10]. More generally, it is important to develop general methods for extending single-copy TBN results to multi-copy contexts.

Ideally we not only want a large energy barrier to "bad" configurations, but we want to avoid getting stuck in local minima that keep us from getting to the good configurations. We can define "self-stabilizing" TBNs with the property that from *any* configuration, the TBN can reach *some* stable configuration with a low energy barrier path. As argued in the full version of this paper, this property is true for the grid gate, but it is not true for the translator cycle. This property is worth more general exploration.

Can we use the definition of energy in TBNs to bootstrap a reasonable notion of probability of configurations or paths? For instance, in statistical thermodynamics it is common to consider the Boltzmann distribution induced by energy E, where for each configuration γ, $\Pr[\gamma] = e^{-E(\gamma)}/(\sum_\beta e^{-E(\beta)})$. This is the probability of seeing γ at thermodynamic equilibrium. One can also use the relative energy between two states to predict the relative rates of transition between them, which might allow defining a notion of path probability in the kinetic theory of TBNs.

A useful chemical module consists of the reaction $X + X \rightarrow Y + Y$ (or more generally converting $k > 1$ copies of X to Y), which can act as a "threshold" to detect whether there are at least k copies of X. Analogous to a catalytic system, implementing the above reaction while forbidding $X \rightarrow Y$ requires control of the energy barrier, and cannot be done simply by varying the energies of X and Y. Can we construct TBNs with arbitrarily large energy barriers in this case?

What is the computational complexity of deciding whether $b(\gamma, \delta) \leq k$, given two configurations γ, δ and a threshold k? This problem is decidable in polynomial space in the number of monomers in γ: any configuration can be written down in polynomial space, and guessing merges and splits yields a nondeterministic polynomial space algorithm (placing it in PSPACE by Savitch's Theorem). However, low-height paths could be of exponential length, and thus it is not clear that the problem is in NP, since the obvious witness is a path with height $\leq k$. Is this problem possibly PSPACE-complete?

Can kinetic barriers be proven in geometric self-assembly models such as the abstract Tile Assembly Model (aTAM) [4,6,12]? One approach to providing thermodynamic arguments for correctness of self-assembly systems is by showing the stable configurations of the system as a TBN are exactly the desired terminal assemblies in the aTAM. It has been shown that the class of systems which have this notion of stability is limited [2]. An alternative approach would be to relax the requirement that the set of desired assemblies correspond exactly to stable configurations, and instead argue that there is a large kinetic barrier to reach undesired assemblies even if they are stable.

Acknowledgement. DD and DH were supported by NSF grant CCF-1619343. CC, KB, and DS were supported by NSF grants CCF-1618895 and CCF-1652824.

References

1. Cardelli, L.: Strand algebras for DNA computing. Nat. Comput. **10**(1), 407–428 (2011)
2. Chalk, C., Hendricks, J., Patitz, M.J., Sharp, M.: Thermodynamically favorable computation via tile self-assembly. In: Stepney, S., Verlan, S. (eds.) UCNC 2018. LNCS, vol. 10867, pp. 16–31. Springer, Cham (2018). https://doi.org/10.1007/978-3-319-92435-9_2
3. Chen, Y.-J., et al.: Programmable chemical controllers made from DNA. Nat. Nanotechnol. **8**(10), 755–762 (2013)
4. Doty, D.: Theory of algorithmic self-assembly. Commun. ACM **55**(12), 78–88 (2012)
5. Doty, D., Rogers, T.A., Soloveichik, D., Thachuk, C., Woods, D.: Thermodynamic binding networks. In: Brijder, R., Qian, L. (eds.) DNA 2017. LNCS, vol. 10467, pp. 249–266. Springer, Cham (2017). https://doi.org/10.1007/978-3-319-66799-7_16
6. Patitz, M.J.: An introduction to tile-based self-assembly and a survey of recent results. Nat. Comput. **13**(2), 195–224 (2014)

7. SantaLucia Jr., J., Hicks, D.: The thermodynamics of DNA structural motifs. Ann. Rev. Biophys. Biomol. Struct. **33**, 415–440 (2004)
8. Soloveichik, D., Seelig, G., Winfree, E.: DNA as a universal substrate for chemical kinetics. Proc. Nat. Acad. Sci. **107**(12), 5393–5398 (2010)
9. Srinivas, N., Parkin, J., Seelig, G., Winfree, E., Soloveichik, D.: Enzyme-free nucleic acid dynamical systems. Science **358**(6369), eaal2052 (2017)
10. Thachuk, C., Winfree, E., Soloveichik, D.: Leakless DNA strand displacement systems. In: Phillips, A., Yin, P. (eds.) DNA 2015. LNCS, vol. 9211, pp. 133–153. Springer, Cham (2015). https://doi.org/10.1007/978-3-319-21999-8_9
11. Wang, B., Thachuk, C., Ellington, A.D., Winfree, E., Soloveichik, D.: Effective design principles for leakless strand displacement systems. Submitted for publication
12. Winfree, E.: Simulations of computing by self-assembly. Technical report, CaltechCSTR: 1998.22, California Institute of Technology (1998)

A Trace Query Language for Rule-Based Models

Jonathan Laurent[1]([✉]), Hector F. Medina-Abarca[2], Pierre Boutillier[2],
Jean Yang[1], and Walter Fontana[2]

[1] Carnegie Mellon University, Pittsburgh, USA
jonathan.laurent@cs.cmu.edu
[2] Harvard Medical School, Boston, USA

Abstract. In this paper, we introduce a unified approach for querying
simulation traces of rule-based models about the statistical behavior of
individual agents. In our approach, a query consists in a trace pattern
along with an expression that depends on the variables captured by this
pattern. On a given trace, it evaluates to the multiset of all values of
the expression for every possible matching of the pattern. We illustrate
our proposed query language on a simple example, and then discuss
its semantics and implementation for the Kappa language. Finally, we
provide a detailed use case where we analyze the dynamics of β-catenin
degradation in Wnt signaling from an agent-centric perspective.

Keywords: Rule-based modeling · Query language · Kappa

1 Introduction

Rule-based modeling languages such as Kappa [4] and BioNetGen [9] can be
used to write mechanistic models of complex reaction systems. Models in these
languages consist of stochastic graph-rewriting rules that are equipped with rate
constants indicating their propensity to apply. Together with an initial mixture
graph, these rules constitute a dynamical system that can be simulated using
Gillespie's algorithm [2,5,8]. Each run of simulation results in a sequence of
transitions that we call a trace.

In practice, simulation traces are often discarded in favor of a limited number
of global features, such as the concentration curves of a set of observables. However, a more detailed analysis of their structure and statistical properties can
provide useful insights into a system's dynamics. For example, causal analysis
methods exist [4,6] that compress a large trace into a minimal subset of events
that are necessary and jointly sufficient to replicate an outcome of interest, and
then highlight causal influences between those remaining events. Queries about
the statistical behavior of individual agents can lead to complementary insights.
Examples include (i) measuring the average lifespan of a complex under different
conditions, (ii) computing a probability distribution over the states in which a

© Springer Nature Switzerland AG 2018
M. Češka and D. Šafránek (Eds.): CMSB 2018, LNBI 11095, pp. 220–237, 2018.
https://doi.org/10.1007/978-3-319-99429-1_13

particular type of agent can be when targeted by a given rule, and (iii) estimating how much of a certain kind of substrate getting phosphorylated is due to a particular pathway at different points in time.

In this paper, we propose a unifying language to express queries of this kind, that are concerned with statistical features of groups of molecular events that are related in specific motifs. These motifs are formalized using a notion of *trace pattern*. Then, evaluating a query comes down to computing the value of an expression for every matching of a pattern into a trace. We give a first illustration of this paradigm on a toy example in Sect. 2. After that, we introduce our query language in Sect. 3 and give it a formal semantics. We then characterize a natural subset of this language for which an efficient evaluation algorithm exists and discuss our implementation for the Kappa language (Sect. 4). Finally, we leverage our query engine to explore aspects of the dynamics of the Wnt signaling pathway in a detailed use case (Sect. 5).

2   A Starting Example

In order to illustrate our Trace Query Language, we introduce a toy Kappa model in Fig. 1. It is described using a rule notation that has been introduced in the latest release of the Kappa simulator and which we borrow in our query language. With this notation, a rule is described as a pattern that is annotated with rewriting instructions. The pattern denotes a precondition that is required for a rule to target a collection of agents. Rewriting instructions are specified by arrows that indicate the new state of a site after transformation.

The model of Fig. 1 features two types of agents: substrates S and kinases K. Both kinds of agents have two different sites, named x and d. In addition, x-sites can be in two different internal states: *unphosphorylated* and *phosphorylated*. We write those states u and p, respectively. Rule b expresses the fact that a substrate and a kinase with free d-sites can bind at rate λ_b. Rules u and u^* express the fact that the breaking of the resulting complex happens at different rates, depending on the phosphorylation state of the kinase involved. Finally, rule p expresses the fact that a substrate that is bound to a kinase can get phosphorylated at rate λ_p. In all our examples, we consider initial mixtures featuring free substrates and kinases in smiliar quantity. Substrates are initially unphosphorylated and kinases are present in both phosphorylation states.

By playing with this model a bit, one may notice that the concentration of phosphorylated substrate reaches its maximal value faster when the ratio of phosphorylated kinases is high (given the rules of our model, the latter quantity is invariant during the simulation). This phenomenon cannot be explained by looking at rule p alone. The query provided in (1) can be run to estimate the probability that a substrate is bound to a phosphorylated kinase when it gets phosphorylated:

$$\text{match } t : \big\{\, S\,(\,x_{u \to p},\; d^1\,),\; k{:}K\,(\,d^1\,)\,\big\} \;\text{ return }\; \text{int\_state}\,[\,{\bullet}\,t\,]\,(\,k,\,\texttt{"x"}\,) \qquad (1)$$

Given a trace, this query matches every transition where a substrate is getting phosphorylated and outputs the phosphorylation state of the attached kinase.

$$\lambda_u \gg \lambda_{u^*} \approx \lambda_p$$

$b\ :\ S\left(d^{\bullet \to 1}\right),\ K\left(d^{\bullet \to 1}\right) \qquad @\ \lambda_b$

$u\ :\ S\left(d^{1 \to \bullet}\right),\ K\left(d^{1 \to \bullet},\ x_u\right) \qquad @\ \lambda_u$

$u^*:\ S\left(d^{1 \to \bullet}\right),\ K\left(d^{1 \to \bullet},\ x_p\right) \qquad @\ \lambda_{u^*}$

$p\ :\ S\left(d^1,\ x_{u \to p}\right),\ K\left(d^1\right) \qquad @\ \lambda_p$

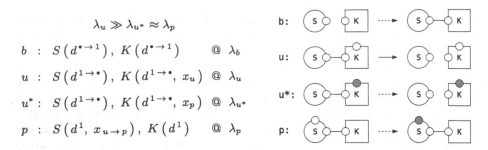

Fig. 1. An example Kappa model. On the left, it is described using the *edit notation* introduced in KaSim 4. Numbers in a rule expression correspond to local bond identifiers and • indicates a free site. Sites not mentioned in a rule are left unchanged by it. A graphical representation is provided on the right. Phosphorylated sites are indicated in grey. Dotted and solid arrows indicate *slow* and *fast* reactions, respectively.

The variables t and k denote a transition and an agent, respectively. Moreover, the expression int_state $[\ \bullet t\ ]\ (\ k,\ $"x"$\ )$ refers to the internal state of the site of agent k with name "x" in the mixture preceding transition t.

Running the previous query, we learn that substrates are much more likely to be phosphorylated by kinases that are phosphorylated themselves, even when such kinases are in minority in the mixture. This leads us to conjecture a causal link between the phosphorylation state of a kinase and its efficiency. After some thoughts, this link can be easily interpreted: because $\lambda_u \gg \lambda_{u^*}$, phosphorylated kinases form more stable complexes with substrates, leaving more chances for a phosphorylation interaction to happen. In fact, the average lifespan of a kinase-substrate complex is exactly $\lambda_{u^*}^{-1}$ when the kinase is phosphorylated and λ_u^{-1} when it is not. We can check these numbers experimentally by running the following query:

$$\text{match}\quad b:\left\{\,s{:}S\left(d^{\bullet \to 1}\right),\ K\left(d^{\bullet \to 1},\ x_p\right)\right\}$$

$$\text{and}\quad\ \text{first}\ u:\left\{\,s{:}S\left(d^{\to \bullet}\right)\right\}\ \text{after}\ b \qquad\qquad (2)$$

$$\text{return}\quad \text{time}[\,u\,] - \text{time}[\,b\,]$$

This query outputs a multiset of numbers, whose mean is the average lifespan of a complex formed by a substrate and a phosphorylated kinase. The same quantity can be computed for unphosphorylated kinases by replacing x_p by x_u in the first line of (2). The pattern in this query does not match single transitions but pairs of related transitions (b, u), where b is a binding transition and u the first unbinding transition to target the same substrate.

More generally, a query is defined by a pattern $P[t, a]$ and an expression $E[t, a]$, which feature a shared set t of transition variables and a shared set a of agent variables. The pattern P can be regarded as a predicate that takes as its arguments a trace τ and a *matching* ϕ mapping the variables in t and a to actual transitions and events in τ. The expression E can be regarded as a function that maps such (τ, ϕ) pairs to values. Then, the query evaluates on a trace τ to the multiset of all values of E, for every matching ϕ that satisfies P in τ.

3   The Core Query Language

In this section, we introduce the extensible core of our proposed query language and give it a formal semantics.

3.1   Meaning and Structure of Queries

As shown in Fig. 2, a query Q consists in a pattern P and an expression E. It can be interpreted as a function $[\![Q]\!]$ from traces to multisets[1] of values. The set of allowed values can grow larger as richer expressions are added to the language. Our current implementation defines a value as a tuple of base values and features the following types for base values: bool, int, float, string, agent, agent_set and snapshot.

A pattern P is interpreted as a function $[\![P]\!]$ that maps a trace to a set of matchings. A matching ϕ is defined by two functions ϕ_t and ϕ_a, which map variable names to transition identifiers and agent identifiers, respectively. We call ϕ_t a transition matching and ϕ_a an agent matching. Given a trace τ and a matching ϕ, the transition variable v denotes the transition $\tau[\phi_t(v)]$, where $\tau[i]$ is a notation for the ith transition of a trace. In addition, an expression E is interpreted as a function $[\![E]\!]$ that maps a pair of a trace and a matching to a value. The expression language is extensible and is discussed in Sect. 3.3. Its syntax is documented in Fig. 3. Then, the semantics of a query can be formally defined as follows:

$$[\![\text{match } P \text{ return } E]\!](\tau) \;=\; \{\!\!\{ [\![E]\!](\tau, \phi) \mid \phi \in [\![P]\!](\tau) \}\!\!\}.$$

Our language constraints the structure of possible patterns. As shown in Fig. 2, a pattern consists in a sequence of clauses, which can take one of three different forms: $(t : T)$, (first $t : T$ after t') and (last $t : T$ before t'). Here, t and t' are transition variables and T is a *transition pattern*. In all three cases, we say that t is *constrained* by the clause.

3.2   Transition Patterns

A transition pattern can be thought as a predicate that takes as its argument a pair (t, ϕ_a) of a transition and an agent matching. Our current implementation supports specifying transition patterns using KaSim's *edit notation*. Transition patterns defined this way are enclosed within curly brackets. For example, in query (1) of Sect. 2,

$$\{ S(x_{u \to p}, d^1), k : K(d^1) \}$$

---

[1] Note that multisets are indicated in Fig. 2 using Dijkstra's bag notation, whereas sets are indicated using the standard curly brackets notation.

is true for a transition t and a matching ϕ_{a} if and only if t has the effect of phosphorylating a substrate that is bound to the kinase with identifier $\phi_{\mathsf{a}}(k)$. Formally, a transition pattern T is interpreted as a function $[\![\,T\,]\!]$ that maps transitions into sets of agent matchings. Using the predicate terminology, one may say that $\phi_{\mathsf{a}} \in [\![\,T\,]\!](t)$ if and only if (t, ϕ_{a}) satisfies T.

Our query language can be instantiated with any choice of a language specifying transition patterns. The only requirement is that transition patterns should be decidable efficiently in the following sense. Given a transition pattern T and a transition t, one should be able to efficiently compute whether $[\![\,T\,]\!](t)$ is empty and generate an element of it in the case it is not. Our evaluation algorithm relies on this property.

3.3   Expression Language

We show in Fig. 3 the syntax of our expression language. An expression can consist of an agent variable, a constant, a parenthesized expression, a binary operation, a function[2] of an expression, a tuple of expressions or a *measure*.

Measures are the basic constructs through which information is extracted from a trace. They come in two different kinds: *state measures* and *event measures*. State measures are used to extract information about the state of the mixture at different points in the trace. They are parametered with *state expressions* that can take the form $\bullet t$ or $t \bullet$, denoting the states before and after transition t, respectively. For example, the int_state measure that is used in (1) is a state measure. In addition, event measures are used to extract information that is about a transition itself (in contrast to the states that it connects). They are parametered by transition variables. For example, the time measure that is used in (2) is an event measure.

The expression language can be easily extended with new operators, functions, measures and types. In the same way than the language for specifying transition patterns, it should be regarded as a parameter of our query language and not as a rigid component.

4   Evaluating Queries

In this section, we introduce a natural subset of the language described in Sect. 3, for which we provide an efficient implementation. Queries in this subset are said to be *regular*, and they display an interesting rigidity property.

4.1   Rigidity

Intuitively, a pattern is said to be rigid if its matchings are completely determined by the value of a single transition variable.

---

[2] Note that functions always take a single argument, which can be a tuple.

$$
\begin{array}{rcll}
\text{query} & Q & := & \text{match } P \text{ return } E & [\![Q]\!] \in \mathsf{Trace} \to \wr \mathsf{Value} \wr \\
\text{pattern} & P & := & C & [\![P]\!] \in \mathsf{Trace} \to \{\,\mathsf{Matching}\,\} \\
& & | & C \text{ and } C & \\
\text{clause} & C & := & t : T & [\![C]\!] \in \mathsf{Trace} \to \{\,\mathsf{Matching}\,\} \\
& & | & \text{first } t : T \text{ after } t' & \\
& & | & \text{last } t : T \text{ before } t' & \\
\text{transition pat.} & T & := & \cdots & [\![T]\!] \in \mathsf{Transition} \to \{\,\mathsf{Matching}_\mathsf{a}\,\} \\
\text{expression} & E & := & \cdots & [\![E]\!] \in \mathsf{Trace} \times \mathsf{Matching} \to \mathsf{Value}
\end{array}
$$

$$[\![\text{match } P \text{ return } E]\!]\,(\tau) \;=\; \wr\, [\![E]\!]\,(\tau,\phi) \mid \phi \in [\![P]\!]\,(\tau)\,\wr$$

$$[\![C \text{ and } C']\!]\,(\tau) \;=\; [\![C]\!]\,(\tau) \cap [\![C']\!]\,(\tau)$$

$$[\![t : T]\!]\,(\tau) \;=\; \{\,\phi \mid \phi_\mathsf{a} \in [\![T]\!]\,(\tau[\phi_\mathsf{t}(t)])\,\}$$

$$[\![\text{first } t : T \text{ after } t']\!]\,(\tau) \;=\; \left\{\,\phi \;\middle|\; \begin{array}{l} \phi_\mathsf{a} \in [\![T]\!]\,(\tau[\phi_\mathsf{t}(t)]),\;\; \phi_\mathsf{t}(t') < \phi_\mathsf{t}(t), \\ \forall i.\; \phi_\mathsf{t}(t') < i < \phi_\mathsf{t}(t) \implies \phi_\mathsf{a} \notin [\![T]\!]\,(\tau[i]) \end{array} \right\}$$

$$[\![\text{last } t : T \text{ before } t']\!]\,(\tau) \;=\; \left\{\,\phi \;\middle|\; \begin{array}{l} \phi_\mathsf{a} \in [\![T]\!]\,(\tau[\phi_\mathsf{t}(t)]),\;\; \phi_\mathsf{t}(t) < \phi_\mathsf{t}(t'), \\ \forall i.\; \phi_\mathsf{t}(t) < i < \phi_\mathsf{t}(t') \implies \phi_\mathsf{a} \notin [\![T]\!]\,(\tau[i]) \end{array} \right\}$$

Fig. 2. Syntax and semantics of the trace query language

$$
\begin{array}{rcll}
\text{expression} & E & := & a \mid C \mid (E) \mid E \bowtie E \mid f(E) \mid E, E \mid \\
& & & M_\mathsf{s}[S] \mid M_\mathsf{s}[S](E) \mid M_\mathsf{e}[t] \mid M_\mathsf{e}[t](E) \\
\text{constant} & C & := & 0 \mid 1 \mid \cdots \mid \texttt{"foo"} \mid \cdots \\
\text{binary operator} & \bowtie & := & + \mid - \mid = \mid < \mid \cdots \\
\text{function} & f & := & \mathsf{agent\_id} \mid \mathsf{size} \mid \mathsf{count} \mid \cdots \\
\text{state measure} & M_\mathsf{s} & := & \mathsf{int\_state} \mid \mathsf{component} \mid \mathsf{snapshot} \mid \cdots \\
\text{state expression} & S & := & \bullet t \mid t \bullet \\
\text{event measure} & M_\mathsf{e} & := & \mathsf{time} \mid \mathsf{rule} \mid \cdots
\end{array}
$$

(with a an agent variable and t a transition variable)

Fig. 3. Syntax of expressions

Definition 1. *Given a Kappa model, a pattern P is said to be* rigid *if and only if it features a transition variable r called* root variable *such that for any trace τ that is valid in the model, we have*

$$\forall \phi, \phi' \in [\![P]\!](\tau),\ \phi_t(r) = \phi'_t(r) \implies \phi = \phi'.$$

For example, the pattern P of query (2) is rigid, with root variable b. Indeed, suppose that b is matched to a specific transition t. Then, the agent variable s is determined by t as no more than one substrate can get bound during a single transition given the rules of our model (Fig. 1). Finally, u is uniquely determined as the first unbinding event that targets s after b.

An easy consequence of Definition 1 is that the number of matchings of a rigid pattern into a trace is bounded by the size of this trace.

4.2   Regular Queries

Our evaluation algorithm handles a subset of queries whose patterns admit a certain tree structure. For those patterns, rigidity is implied by a weaker notion of *local rigidity*.

Definition 2. *Given a Kappa model, a transition pattern T is said to be* rigid *if and only if for any agent variable a that appears in T and every valid transition t, we have*

$$\forall \phi_a, \phi'_a \in [\![T]\!](t),\ \phi_a(a) = \phi'_a(a).$$

Intuitively, a transition pattern is rigid if matching it to a transition determines all its agent variables.

Definition 3. *Given a model, a pattern P is said to be* locally rigid *if it features only rigid transition patterns. Then, a transition variable t is said to* determine *an agent variable a if there is a clause of P that constrains[3] t and features a.*

For patterns with a particular structure, local rigidity implies rigidity. This structural assumption can be expressed in terms of a pattern's *dependency graph*.

Definition 4. *The* dependency graph *of a pattern P is a graph whose nodes are the transition variables of P and for which there is an edge from t to t' if and only if P contains a clause of the form (first t' : T after t) or (last t' : T before t).*

We can now define the notion of a regular pattern, and thus of a regular query.

Definition 5. *A pattern is said to be* regular *if the following three conditions hold: (i) it is locally rigid (ii) its dependency graph is a tree (iii) whenever two of its transition variables determine a same agent variable, one of them has to be a descendent of the other in the dependency tree.*

---

[3] As defined in Sect. 3.1.

This structure enables an efficient enumeration of the matchings of a regular pattern into a trace. Moreover, the number of these matchings is bounded by the size of the trace, as regular patterns can be proven to be rigid.

Proposition 1. *Regular patterns are rigid.*

Finally, regular queries are defined as expected.

Definition 6. *A query is said to be* regular *if its pattern is regular.*

This notion of regularity may appear unintuitive at first, and we agree that its formal definition is somewhat involved. However, we argue that regular queries are exactly those queries that admit a natural operational interpretation. Therefore, experimentalists tend to think in terms of regular queries instinctively.

4.3 Evaluating Regular Queries Efficiently

When designing an algorithm for evaluating trace queries, one has to keep in mind that the corresponding sequence of state mixtures cannot fit in random-access memory all at once, even for small traces. In fact, even the most economic representation of a trace, which is specified by an initial mixture and a sequence of labeled rewriting events, may fail to fit in memory in some cases. Therefore, as often as possible, one should only be allowed to stream such a representation from disk, recomputing intermediate states dynamically and never keeping more than a small number of them in memory at once (two in our case).

Our algorithm for evaluating a regular query proceeds in two steps. First, it streams the trace to compute the set of all matchings of the pattern. Then, it streams the trace a second time to compute the value of the expression for all these matchings. The second step is quite simple to implement. Indeed, once the matchings are known, it is easy to compute the sequence of all measures that need to be performed and order them in increasing order of time. The first step attempts to match the root variable of the pattern to every transition in the trace. For each candidate matching, it uses rigidity to determine all other variables progressively as the trace is streamed, in an order that is determined by the dependency tree and with a minimal amount of caching. Overall, the algorithm runs in linear time with the length of the trace.

4.4 Our Implementation

We provide an implementation of our proposed trace query language, which relies on the algorithm that is mentioned in Sect. 4.3 for evaluating regular queries. Our query engine takes for inputs (i) a file that contains a list of queries written in the same syntax that we use in our examples and (ii) a trace file that has been generated by the Kappa simulator using the -trace option. It evaluates all queries at once and generates one output file per query, in comma-separated values (CSV) format.[4]

---

[4] Every line of an output file represents a single value. In our expression language, values are tuples of *base values*. These are separated by commas within a line.

Queries that are non-regular for structural reasons – i.e. that fail to meet criterion (ii) or (iii) of Definition 5 – are rejected immediately. As there is no easy static check for local rigidity, queries that do not meet this criterion are rejected at runtime.

We now introduce a use case in which we leverage our query engine to explore aspects of the dynamics of the Wnt signaling pathway.

5   A Use Case on Wnt Signaling

In this use case, we are focusing on a simplified model of the β-catenin destruction complex from canonical Wnt signaling. This complex is highly conserved in animals, and operates from humans to nematodes to insects to amphibians, regulating the establishment of the dorso-ventral axis. It is also heavily involved in colon cancer.

A source of complexity in our model is the fact that none of the enzymes involved in destroying β-catenin bind it directly. Instead they are loaded onto a scaffold. Moreover, the scaffold can head-to-tail homopolymerize, in addition to having three independent binding sites on a second scaffold, itself capable of dimerization. This allows a complex of scaffolds, where connection paths or stoichiometries are dynamic. It is this complex that acts as a super-scaffold to bring the substrate in contact with the enzymes. Considering both scaffolds contain large regions of disorder (i.e. chunks of unfolded peptide with high flexibility), it is sensible to believe an enzyme loaded on one scaffold could modify the substrate loaded on the neighboring scaffold. Lacking experimental evidence to suggest a ballpark limit for this reachable horizon, we leave it unconstrained: an enzyme will be able to modify any substrate loaded onto its complex.

Another source of complexity is that having kinases (i.e. enzymes that add a phosphate group) and phosphatases (i.e. enzymes that remove a phosphate group) loaded on the same complex will result in unimolecular do-undo loops. Conceivably the kinetics of complexes will vary heavily with the amount of kinases, phosphatases, and substrates loaded onto them.

We leverage our trace query engine to explore the dynamics of this system. More precisely, we develop queries to probe the agent-centric dephosphorylation dynamics, to measure the time it takes for an agent to navigate the modification steps, and to explore the complexes at which events happen.

Our results are relevant to other pathways in addition to Wnt, from NFκB to RAS/ERK to the most studied protein on the world, P53; the pathways these proteins regulate make heavy use of polymeric scaffold complexes, sequential modifications, and do-undo loops.

5.1   Experimental Protocol and Queries

To explore our system, we create a Kappa model with three parametrizations. The model contains the scaffold proteins Axin1 (Axn) and APC, the kinases CK1α (CK1) and GSK3β (GSK), the protein phosphatases PP1 and PP2, and the substrate of all these reactions, β-catenin (Cat). The destruction complex recruits Cat through Axn. It then gets phosphorylated at the Serine on position 45 (S45) by CK1. While S45-phosphorylated, it can be phosphorylated at the Threonine on position 41 (T41) by GSK. While T41-phosphorylated, it can be phosphorylated on both Serines on positions 37 and 33 (S37 and S33). Once S37- and S33-phosphorylated, Cat is degraded. Meanwhile, PP1 undoes the phosphorylations of CK1, while PP2 undoes those of GSK. Each kinase-phosphatase pair also compete against each other for binding sites on Axn.

The three parametrizations explore the relationship between phosphatase/kinase ratio and the distribution of do-undo events. The three parameter pairs are 50/10, 10/10, and 10/50, all in units of number of agents, and represent the number of kinases and phosphatases in the model (e.g. 10/50 presents 10 copies of PP1, 10 copies of PP2, 50 copies of CK1, and 50 copies of GSK). The scaffolds remain at an abundance of 100 each. The models begin with an initial amount of Cat of 500 agents, and the models are run for 500 simulated seconds. We use global stochastic rates for our reactions, a bi-molecular binding of 10^{-4} per second per agent, a uni-molecular binding of 10^{-2} per second, an unbinding of 10^{-2} per second, and a catalytic of 1.0 per second.

For all three parametrizations, we run the following queries on the resulting traces.

Undoing S45, T41, S37 and S33 Phosphorylation. Considering phosphatases undoing the phosphorylation of sites, does this happen to all agents? Does it happen to just a few agents? What is the distribution of dephosphorylation events per agent? (Fig. 4)

$$\text{match} \quad e : \{\, c\!:\!\text{Cat}\,(S45^{1}_{\,p \to u}\,)\,\}$$
$$\text{return} \quad \text{agent\_id}\,(\,c\,)\,,\, \text{time}\,[\,e\,]$$

$$\text{match} \quad e : \{\, c\!:\!\text{Cat}\,(T41^{1}_{\,p \to u}\,)\,\}$$
$$\text{return} \quad \text{agent\_id}\,(\,c\,)\,,\, \text{time}\,[\,e\,]$$

$$\text{match} \quad e : \{\, c\!:\!\text{Cat}\,(S37^{1}_{\,p \to u}\,)\,\}$$
$$\text{return} \quad \text{agent\_id}\,(\,c\,)\,,\, \text{time}\,[\,e\,]$$

$$\text{match} \quad e : \{\, c\!:\!\text{Cat}\,(S33^{1}_{\,p \to u}\,)\,\}$$
$$\text{return} \quad \text{agent\_id}\,(\,c\,)\,,\, \text{time}\,[\,e\,]$$

Wait Times. What is the distribution of times spent between the first phosphorylation on an agent, and the time it gets degraded? (Fig. 5)

$$
\begin{aligned}
&\text{match} \quad i : \{\, c : \text{Cat+} \,\} \\
&\text{and} \quad \text{first}\ p : \{\, c\!:\!\text{Cat}\,(S45_{\,u \to p}\,)\,\}\ \text{after } i \\
&\text{and} \quad \text{first}\ d : \{\, c : \text{Cat−} \,\}\ \text{after } p \\
&\text{return} \quad \text{time}\,[\,d\,] - \text{time}\,[\,p\,]
\end{aligned}
\tag{3}
$$

About this Query. Agent creation and destruction is expressed by suffixing agent names with $+$ and $-$, respectively.

Component Size and Enzyme Identity. Where do the phosphorylation steps that actually lead to degradation occur? Do they happen mostly on large complexes? What is the composition in units of Axn and APC of the complexes where the phosphorylation events leading to degradation took place? What is the distribution of kinase identifiers for the last phosphorylation events that lead to degradation? (Fig. 6)

$$
\begin{aligned}
\text{match} \quad & d : \{\, c : \mathrm{Cat}-\,\} \\
\text{and} \quad & \text{last } p_1 : \{\, c : \mathrm{Cat}\left(S45^{1}_{u \to p}\right),\ k_1 : \mathrm{CK1}\left(c^{1}\right)\,\} \text{ before } d \\
\text{and} \quad & \text{last } p_2 : \{\, c : \mathrm{Cat}\left(T41^{1}_{u \to p}\right),\ k_2 : \mathrm{GSK}\left(c^{1}\right)\,\} \text{ before } d \\
\text{and} \quad & \text{last } p_3 : \{\, c : \mathrm{Cat}\left(S37^{1}_{u \to p}\right),\ k_3 : \mathrm{GSK}\left(c^{1}\right)\,\} \text{ before } d \\
\text{and} \quad & \text{last } p_4 : \{\, c : \mathrm{Cat}\left(S33^{1}_{u \to p}\right),\ k_4 : \mathrm{GSK}\left(c^{1}\right)\,\} \text{ before } d \\
\text{return} \quad & \text{agent\_id}\left(k_1\right),\ \text{count}\left(\text{component}\left[\bullet p_1\right]\left(k_1\right),\ \text{"Axn"},\ \text{"APC"}\right), \\
& \text{agent\_id}\left(k_2\right),\ \text{count}\left(\text{component}\left[\bullet p_2\right]\left(k_2\right),\ \text{"Axn"},\ \text{"APC"}\right), \\
& \text{agent\_id}\left(k_3\right),\ \text{count}\left(\text{component}\left[\bullet p_3\right]\left(k_3\right),\ \text{"Axn"},\ \text{"APC"}\right), \\
& \text{agent\_id}\left(k_4\right),\ \text{count}\left(\text{component}\left[\bullet p_4\right]\left(k_4\right),\ \text{"Axn"},\ \text{"APC"}\right)
\end{aligned}
\tag{4}
$$

About this Query. The component state measure computes the connected component that contains an agent in a mixture. It returns a set of agents S. The count function takes such a set S along with n strings denoting agent types and returns an n-tuple of integers indicating how many agents of each type appear in S.

5.2   Results and Interpretation

Distribution of Undo Events per Agent. To study the effect of adding phosphatase, we look at the distribution of dephosphorylation events per agent in Fig. 4. S45 is the first residue to be modified in the causal chain leading to degradation; S37 is the last. Based on the 1:1 system, it is surprising to see increasing the phosphatase level five-fold maintains a similar total number of dephosphorylation events (compare curves' integrals). However, their distribution is quite different. Interestingly, increasing the amount of kinase to 1:5 led to decrease in dephosphorylation events, even though the dephosphorylation enzyme's abundance and rates were kept at the same levels. It is also worth noting, the 1:1 saw almost 30 thousand dephosphorylation events of S45, occurring on a shrinking pool of at most 500 copies of Cat. Clearly certain agents are caught in the do-undo loop; some specific agents are getting dephosphorylated almost 800 times. It is worth noting these levels of dephosphorylation imply a comparable number of phosphorylation events.

To answer the question that motivated this query, for S45 under 1:5 regime, most agents don't get sabotaged by the phosphatase: the blue line is quite flat. Decreasing the amount of kinase changes this, and under a 1:1 regime some agents get undone multiple times, a quarter seeing upwards of hundreds of undo events (e.g. from id 300 onward). Increasing the phosphatase to a 5:1 regime further exacerbates this, with over half the agents receiving undo events hundreds of times. The unavailability of phosphorylated S45 in turn inhibits the phosphorylation of T41, and so forth to S37 and S33. It is worth noting that, based on the 1:1 system, *increasing* the phosphatase five-fold *decreases* the number and extent of advanced dephosphorylation events, such as S33 and S37. Paradoxically, increasing the kinase five-fold has this same effect. We attribute the former to decreased availability of the intermediate phosphorylated states (i.e. if T41 is not phosphorylated, S37 can't be phosphorylated, ergo can't be dephosphorylated), and the latter to increased throughput to degradation (i.e. Cat is not around for long enough to get dephosphorylated, as once it gets fully phosphorylated it quickly proceeds to get degraded).

We call attention to the number of agents whose final sites got dephosphorylated (Fig. 4), vs. the number of agents who got degraded (Fig. 7, in Appendix A). The 1:5 or 1:1 systems both degraded over 450 agents each, but the former undid around 160 agents (Fig. 4 S37, domain of blue curve) while the latter undid over 350 (Fig. 4 S37, domain of red curve). For the 1:5 and 5:1 systems, both undid around 160 agents (Fig. 4 S37, domain of blue and yellow curves), but the former degraded over 450 agents (Fig. 7, blue curve) while the latter less than 50 (Fig. 7, yellow curve). This argues the notion of efficiency (e.g. minimizing the amount of undo steps) can't readily be inferred from the throughput of the system.

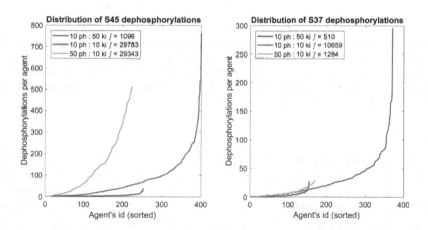

Fig. 4. Distribution of dephosphorylation events per agent. Each time an agent gets dephosphorylated, its ID is registered. After sorting, we plot the distribution of these IDs for two residues in the three parameter regimes. The area under the curve is also presented on each legend. S45 is the first residue to get phosphorylated, S37 (along with S33) is the last. (Color figure online)

Wait Times. Looking at the distribution of wait times (Fig. 5), from first phosphorylation to degradation, we note the bulk of degradation events occur rapidly, in less than 50 s. Worth noting that, from the 1:1 regime, increasing the amount of kinase five-fold marginally reduced wait times.

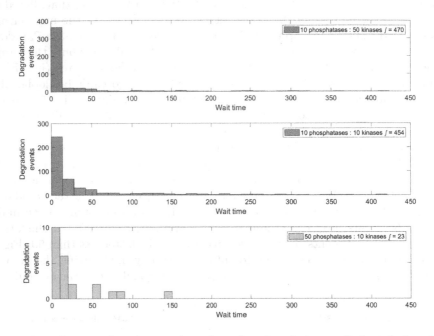

Fig. 5. Distribution of wait times from first phosphorylation until degradation. The sum of the bins is presented in the legend of each plot, and corresponds to the total number of degradation events, matching what is seen on Fig. 7. The height of each bin represents the number of agents that waited the bin's position (in seconds) since they were first modified until they were degraded.

Complex Composition. A way of looking at the question of complex contribution is to query the size of the complex at the last phosphorylation event before degradation. Taking S45 as representative of all the other residues, we plot the size of the complex, in terms of Axn and APC, at the time the final S45 occurred. Overall, we see a broad distribution of sizes, with some phosphorylation events occurring in large complexes (i.e. >80 Axn, >40 APC), but a significant number occurring in far smaller complexes (i.e. <10 Axn, <10 APC). Changing the parameter regime of kinase to phosphatase does not seem to alter this behavior significantly.

5.3   Summary of Findings

1. The number of undo events does not inform us of overall throughput (contrast Figs. 4 and 7).

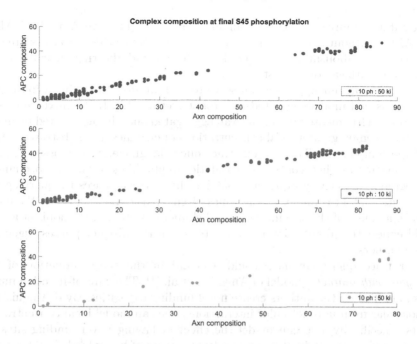

Fig. 6. Composition of the complex, in terms of Axn and APC components, at the last event where Cat got S45 phosphorylated before being degraded. The number of points corresponds to the number of degradation events. The points of this scatter plot have been nudged with a random noise factor of 0.2 to increase visual perception of discrete points where the data overlap.

2. How a step may be affected by changing abundances depends greatly on its upstream context (Fig. 4).
3. Entities that got degraded waited a short while since their first modification (Fig. 5), and yet most modifications were futile (Fig. 4).
4. We can't argue that giant complexes, nor small complexes, nor medium complexes, are the sole entities responsible for performing the effective (i.e. final) phosphorylation events (Fig. 6); the distribution of complexes is wide, and they all contribute to the kinetics.

The capacity of querying a simulation's trace offers a mechanistic description of the inner workings of our system. Since this description uses the vocabulary of molecular biology, it can greatly inform the search for drug targets.

For example, in our setup, complexes with over 60 copies of Axn and over 40 copies of APC contributed a large amount of degradation events (Fig. 6). Considering there were a total of 100 copies of each scaffold, these large complexes are giant components, having recruited the majority of scaffolds into a single entity. If a single entity is contributing an amount of degradation events comparable to the rest of the mixture, it means its *effective* catalytic rate is greater than that of smaller entities. One could therefore reduce overall degradation of Cat by destabilizing any of the three scaffold interactions (i.e. Axn-Axn, APC-APC, Axn-APC), without affecting the enzymes directly. Since these enzymes are also involved in metabolism, it would be desirable to avoid affecting their behaviors outside our pathway of interest.

Beyond Wnt, our approach can serve to add an analytical dimension to the phase-transition model suggested by Pronobis et al. [12]. If the mixture transitions from an fragmented regime into an aggregated one, the aggregated regime is expected to enjoy greater local concentrations of enzymes and substrates within the large components. Through our trace query language, we can assay the size of components as they contribute catalytic events. This allows us to quantify the degradation activity inside and outside the distinct phases by querying the size of the scaffold complex at the times of enzymatic modifications. Were one to make a model of the specific setup and chimeric protein of Pronobis et al., we would expect the distribution of events to reveal the distinct phases suggested by the authors.

Our trace query language could also aid in the characterization of the *aggregate-with-tentacles* model of Anvarian et al. [1]. The gain-of-function mutations identified by the authors create novel binding capacities. By distinguishing bi-molecular from uni-molecular interactions, one can model local concentration effects. Specifically, one can model the effect of having novel binding sites on the complex' *tentacles*, in close proximity by nature of being loaded on the same complex. One can therefore model the growth of the complex as additional binding sites become active (e.g. by setting a binding rate to greater than zero). Were one to make a model informed by the specific experimental setup used by the authors, one could quantify the change in unbound state as additional binding sites become active.

6   Related Works

Languages already exist to collect, filter and aggregate data from simulation. For example, the Kappa simulator [2] features an *intervention language* that allows taking repeated conditional measurements during simulation, possibly updating a global state every time. Chromar [11] proposes an *extended language* with similar capabilities, where queries can be defined in a more functional style from a selector and a fold operation. Finally, query languages have been proposed [10,13] that are inspired by the structured query language (SQL).

All these languages are similar in the sense that (i) queries can only evaluate expressions in the scope of a single state or transition and (ii) only population-level quantities can be measured. In contrast, queries in our proposed language can (i) match *trace patterns* that consist in multiple transitions acting on common agents and (ii) measure and compare the state of individual agents, possibly at different points in time. As a consequence, most of the example queries shown in this paper could not be expressed using preexisting query languages.

Thanks to its ability to match complex trace patterns, our query language may be especially interesting to use in combination with causal analysis [4,6].

Indeed, the pathways uncovered by causal analysis can be regarded as trace patterns and then matched using our query language. This may be useful to measure the relative frequency of pathways in different settings, but also to analyze how individual agents participating to a pathway evolve as this pathway unfolds. For example, query (4) in Sect. 5 compares the composition of the complex at which β-catenin is attached at different points of the pathway leading to its degradation.

7   Conclusion

How can one explore a question like "which complexes contribute to kinetics"? Experimental biologists have been using labeling techniques for decades, but implementing this in a modeling framework requires being able to track individual agents, and query particular events. Implementing a framework to query events on the trace of an agent and rule simulation seems a natural way of tackling these classes of problems.

Moreover, once a sufficiently rich *mechanistic* model is available, questions on *mechanism* arise. For a subset of these, a satisfying answer will require a change of vocabulary; the explanations desired use the individual's lexicon (e.g. it bound, it unbound, it got dephosphorylated 800 times), rather than a whole system lexicon (e.g. the abundance changed from 500 to 50). Thus, rather than tracking the whole model's behavior (akin to a top-down approach), one needs to focus on agents, and observe their individual experiences (akin to a bottom-up approach). These approaches are complementary, as they explore a model's intricacies from very different viewpoints. We hope that the query language proposed in this paper will contribute to make agent-centric analysis more widespread and accessible.

Acknowledgments. This work was sponsored by the Defense Advanced Research Projects Agency (DARPA) and the U.S. Army Research Office under grant numbers W911NF-14-1-0367 and W911NF-17-1-0073.

A   Use Case Appendix

Concentration Time Traces. From the output of the simulator, we get the evolution of the abundance of Cat through time. In Fig. 7, we can see that the systems with low phosphatase behave similarly, even though one has five times the amount of kinases than the other (blue vs red traces). In contrast, the system with high phosphatase shows markedly less degradation of Cat; where the other two systems degraded around 450 units, this one has only degraded 23. From this whole-system view, it would seem the amount of phosphatase is more critical than the amount of kinase: based on the 1:1 system, increasing the kinase five-fold has little effect, whereas increasing the phosphatase has a more dramatic effect.

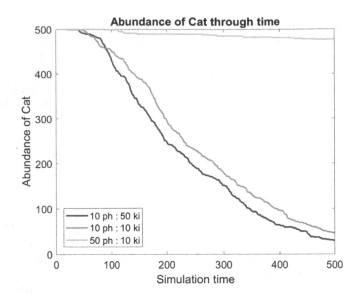

Fig. 7. Tracking the abundance of agent cat through the simulation. At time $T = 0$, the agents are introduced, all in monomeric form. The simulation was stopped after five hundred simulated seconds. In this legend and throughout the figures, "ph" stands for phosphatase, "ki" stands for kinase, and numbers indicate agent multiplicity. Thus "10 ph: 50 ki" means the system with 10 units of phosphatase and 50 of kinase. (Color figure online)

References

1. Anvarian, Z., et al.: Axin cancer mutants form nanoaggregates to rewire the wnt signaling network. Nat. Struct. Mol. Biol. **23**(4), 324 (2016)
2. Boutillier, P., Ehrhard, T., Krivine, J.: Incremental update for graph rewriting. In: Yang, H. (ed.) ESOP 2017. LNCS, vol. 10201, pp. 201–228. Springer, Heidelberg (2017). https://doi.org/10.1007/978-3-662-54434-1_8
3. Clarke, E.M., Faeder, J.R., Langmead, C.J., Harris, L.A., Jha, S.K., Legay, A.: Statistical model checking in *BioLab*: applications to the automated analysis of T-cell receptor signaling pathway. In: Heiner, M., Uhrmacher, A.M. (eds.) CMSB 2008. LNCS (LNAI), vol. 5307, pp. 231–250. Springer, Heidelberg (2008). https://doi.org/10.1007/978-3-540-88562-7_18
4. Danos, V., Feret, J., Fontana, W., Harmer, R., Krivine, J.: Rule-based modelling of cellular signalling. In: Caires, L., Vasconcelos, V.T. (eds.) CONCUR 2007. LNCS, vol. 4703, pp. 17–41. Springer, Heidelberg (2007). https://doi.org/10.1007/978-3-540-74407-8_3
5. Danos, V., Feret, J., Fontana, W., Krivine, J.: Scalable simulation of cellular signaling networks. In: Shao, Z. (ed.) APLAS 2007. LNCS, vol. 4807, pp. 139–157. Springer, Heidelberg (2007). https://doi.org/10.1007/978-3-540-76637-7_10
6. Danos, V., et al.: Graphs, rewriting and pathway reconstruction for rule-based models. In: IARCS Annual Conference on Foundations of Software Technology and Theoretical Computer Science, FSTTCS 2012, vol. 18, pp. 276–288 (2012)

7. Fages, F., Rizk, A.: On the analysis of numerical data time series in temporal logic. In: Calder, M., Gilmore, S. (eds.) CMSB 2007. LNCS, vol. 4695, pp. 48–63. Springer, Heidelberg (2007). https://doi.org/10.1007/978-3-540-75140-3_4
8. Gillespie, D.T.: Exact stochastic simulation of coupled chemical reactions. J. Phys. Chem. **81**(25), 2340–2361 (1977)
9. Harris, L.A., et al.: BioNetGen 2.2: advances in rule-based modeling. Bioinformatics **32**(21), 3366–3368 (2016)
10. Helms, T., Himmelspach, J., Maus, C., Röwer, O., Schützel, J., Uhrmacher, A.M.: Toward a language for the flexible observation of simulations. In: Proceedings of the Winter Simulation Conference, Winter Simulation Conference, p. 418 (2012)
11. Honorato-Zimmer, R., Millar, A.J., Plotkin, G.D., Zardilis, A.: Chromar, a language of parameterised agents. Theor. Comput. Sci. (2017)
12. Pronobis, M.I., Deuitch, N., Posham, V., Mimori-Kiyosue, Y., Peifer, M.: Reconstituting regulation of the canonical Wnt pathway by engineering a minimal β-catenin destruction machine. Mol. Biol. Cell **28**(1), 41–53 (2017)
13. Zehe, D., Viswanathan, V., Cai, W., Knoll, A.: Online data extraction for large-scale agent-based simulations. In: Proceedings of the 2016 ACM SIGSIM Conference on Principles of Advanced Discrete Simulation, pp. 69–78. ACM (2016)

Inferring Mechanism of Action of an Unknown Compound from Time Series Omics Data

Akos Vertes[1], Albert-Baskar Arul[1], Peter Avar[1], Andrew R. Korte[1], Hang Li[1], Peter Nemes[1], Lida Parvin[1], Sylwia Stopka[1], Sunil Hwang[1], Ziad J. Sahab[1], Linwen Zhang[1], Deborah I. Bunin[2], Merrill Knapp[2], Andrew Poggio[2], Mark-Oliver Stehr[2], Carolyn L. Talcott[2(✉)], Brian M. Davis[3], Sean R. Dinn[3], Christine A. Morton[3], Christopher J. Sevinsky[3], and Maria I. Zavodszky[3]

[1] Department of Chemistry, George Washington University,
Washington DC 20052, USA
[2] SRI International, Menlo Park, CA 94025, USA
clt@csl.sri.com
[3] GE Global Research, Niskayuna, NY 12309, USA

Abstract. Identifying the mechanism of action (MoA) of an unknown, possibly novel, substance (chemical, protein, or pathogen) is a significant challenge. Biologists typically spend years working out the MoA for known compounds. MoA determination is especially challenging if there is no prior knowledge and if there is an urgent need to understand the mechanism for rapid treatment and/or prevention of global health emergencies. In this paper, we describe a data analysis approach using Gaussian processes and machine learning techniques to infer components of the MoA of an unknown agent from time series transcriptomics, proteomics, and metabolomics data.

The work was performed as part of the DARPA Rapid Threat Assessment program, where the challenge was to identify the MoA of a potential threat agent in 30 days or less, using only project generated data, with no recourse to pre-existing databases or published literature.

1 Introduction

Lethal chemical and biological agents could be introduced through deliberate malevolent activity or inadvertent release into the environment. Technologies that enable rapid characterization of the mechanism of action (MoA) of these agents would facilitate the identification of countermeasures, such as a known antidote or antagonist for a specific cellular pathway. Likewise, characterizing the MoA of pathogenicity of a virus or bacterium could uncover host/microbial interactions that could be exploited to control the infection.

Sponsored by the US Army Research Office and the Defense Advanced Research Projects Agency; accomplished under Cooperative Agreement W911NF-14-2-0020.

M. Češka and D. Šafránek (Eds.): CMSB 2018, LNBI 11095, pp. 238–255, 2018.
https://doi.org/10.1007/978-3-319-99429-1_14

A simple representation of this problem is the exposure of immortalized or primary human cells to the agent in question. Although this model system does not possess the complexity of the response by an organ or an entire organism, it still retains much of the challenges associated with trying to infer the MoA from perturbing biochemical pathways in a living system. At the cellular level, there are many potential mechanisms affecting cellular processes, such as regulation of the cell cycle, transcription/translation, metabolism, and signaling. The molecules that modulate these processes are diverse and their interactions are complex. Thus, a systems biology approach, i.e., untargeted data collection from multiple data types (such as comparative analysis of quantitative changes in the transcriptome [8,28,30], proteome [11,14,23], and metabolome [4,5,22] is critical for elucidating the MoA [20]. Furthermore, mechanistic events induced by exposure to the agent may vary over a wide range of timescales following the challenge; some events occur in seconds to minutes, others take hours to days to manifest. Thus, it becomes important to gather timeseries data in order to observe the unfolding of complex molecular changes in the cells.

These considerations motivate a multi-omics timeseries approach to gather experimental data at the molecular level. Following data acquisition, robust analytical and network building methods [2,3,10] are used to identify components of the MoA and their relationships. A primary goal is to identify biomolecules with significantly altered abundances, as these may represent potential participants in the MoA. Another goal is to distinguish the agent induced MoA from normal metabolic activity in a population of cells. There are two main approaches to arrive at MoA candidates from comparative multi-omics data. Knowledge-based methods, e.g., Ingenuity® Pathway Analysis (Qiagen) and PANTHER, rely on combining extensive curated information extracted from biomedical literature with enrichment analysis of the transcriptomic, proteomic and metabolomic profile changes as a result of the experimental perturbation [13,17]. The algorithms operating on the knowledge base are used to infer the causal networks underlying the user data, potential upstream regulators, and potential downstream effects. Although these methods show strong promise for the identification of the MoA, they are less useful when the agent is novel because of the scarcity or absence of curated information. In the absence of reliable knowledge base content for the studied perturbing agent, de novo approaches can be followed. These can rely on the analysis of static interaction networks, such as the DeMAND method [29], or on cross-correlation analysis between the time courses of the differentially regulated compounds identified in the omics data, e.g., ProTINA [18]. After an array of potential MoAs is identified, downselection to the potential mechanism is performed through confirmatory experiments to quantify effects on specific pathways and/or cellular responses to the agent.

In this paper, we describe data analysis algorithms and workflows developed to address the need to identify MoA components based purely on comparative multi-omics time course data. We applied a broad untargeted approach quantifying changes in the transcriptome, proteome and metabolome, using challenge agents with well-characterized MoAs, to develop and verify the analytic tools and approaches that would enable rapid MoA discovery. We developed robust techniques for high throughput, high coverage multi-omics data generation and data centric analysis. The data analyses were carried out with team members blinded to agent identity during multiple testing periods. A description of experimental methods and early results have been reported in prior conference proceedings [24–27]. Here, we detail the approaches used for data analysis, specifically those approaches based on generating Gaussian process models from time series data. Multiple approaches for identifying MoA components allowed us to cover MoAs with diverse characteristics by looking at the data in multiple ways.

Plan. In Sect. 2 we describe the Gaussian process model and analysis workflows based on this model. Section 3 briefly describes the data to be analyzed. Section 4 illustrates application of the analysis workflow to identify candidate MoA participants. Section 5 summarizes, compares to related work, and discusses future directions.

2 Gaussian Process-Based Analysis Workflow

In the following we present some core parts of our data analysis workflow. The overall workflow has been represented and automated using *JupyterFlow*, an in-house tool developed at SRI International that is based on Petri nets (generalized dataflow graphs) [7] and integrates features of interactive computing (Jupyter/Python Notebooks) [12] with data-driven distributed and parallel execution on heterogenous clusters, typically with GPUs (Graphical Processing Units) on a subset of the nodes. JupyterFlow integrates with Tensor Flow [1], which we use both for estimating Gaussian Process Models [6] and for training Neural Networks (such as the Autoencoders discussed below).

JupyterFlow supports multiple modes of operations: the parallel execution of workflows on powerful GPU servers, the distributed execution on heterogeneous clusters and the cloud (where GPUs may be only available on some machines), and a mode of disconnected operation, where laptops may occasionally synchronize with the main workflow, but otherwise can work independently (without continuous network connectivity). Generally, the workflow executes automatically as a collection of computational notebooks with automatically inferred dataflow dependencies and resource constraints

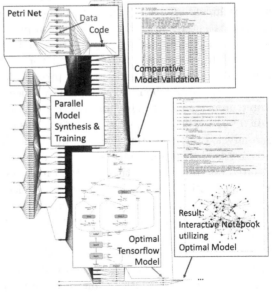

Fig. 1. Small excerpt of the RTA workflow in JupyterFlow. Each transition in the Petri net is defined by a Python/Jupyter notebook.

(e.g., memory, number of GPUs), which are essential to optimize the location of computations in the cluster. In addition, each notebook can also be executed interactively (e.g., for diagnosis and experimentation). Just as TensorFlow utilizes a form of dataflow graphs in the small, JupyterFlow utilizes dataflow graphs in the large. In the case of RTA, our current workflow (see Fig. 1 for a small excerpt) consists of ≈10,000 notebooks utilizing Python and TensorFlow that give rise to ≈270,000 metalevel dataflow dependencies (this is not counting dependencies at the TensorFlow level). We are currently executing this workflow on SRI's private cluster (with up to 4 GPUs per node) and in the Google Cloud (using large instances with up to 48 CPUs and 8 GPUs).

2.1 Gaussian Processes and What They Accomplish

A Gaussian process [21] represents a *probability distribution* over a *parameterized class of continuous functions*. The projection at each time point is Gaussian, i.e., normally distributed, with a *mean and variance that is explicitly represented* in the Gaussian process model. A time series of observations is defined by sampling at a finite number of arbitrary time points.

Gaussian process modeling allows us to address several challenges when facing high-dimensional, very noisy biological time series data. First, observations are only available at a small number of irregularly spaced time points. Also the number of observations at each time point may vary. By exploiting the continuity of the underlying functions, Gaussian processes allow us to benefit from the local structure to estimate the mean and to provide an explicit estimate of variance at each time point. The second challenge arises in the integration of multiple data sources, in our case we have transcriptomic, proteomic, and metabolomic data, each with different sampling time points, and different amounts of measurement noise. Again, the Gaussian processes allow us to integrate all data into a uniform continuous time scale and allow us to build specialized models for each data source to reflect the *biological variation* and the *measurement error*.

Specifically, we use *GPflow* [6], a library based on Tensor Flow, to estimate Gaussian processes separately for each type of data, for each compound (transcript, protein, or metabolite), and for treated and control condition. All observed quantities are uniformly represented in log2 scale. Furthermore, we use a log-based time scale to reflect the approximate spacing of observations with more observations closer to the beginning of the experiment (defined as Time 0). The class of functions is generated by a kernel that combines squared exponentials (reflecting biological variation) in the log-based time scale with additive white noise (to capture measurement error). The squared exponential time scale parameter is fixed at 2 to reflect the density of available observations. The squared exponential variance is fixed for each type of data by using an average of the optimal values over all compounds. The white noise variance is computed by Maximum Likelihood Estimation (MLE) for each specific compound.

The output of the Gaussian process workflow is a uniform time series for each compound with 100 time points for control and treated sample means and the corresponding standard deviations. In reality, the learned Gaussian process model is continuous. The 100 time points sample time dimension at sufficiently many discrete points to preserve the information. The *log-ratio time series* for each compound, which is the primary basis for subsequent stages of our workflow, is then computed by point-wise difference between treated and control. Figure 2 shows example plots.

In addition the confidence in the change (as measured by the log-ratio) is scored by computing the area between the standard deviation bands for control and treated samples. We perform computation for 1 and 2 standard deviations, the former to account for the higher degree of noise in the proteomics and metabolomics data. This provides a *ranking of the measured compounds* according to the area between the bands generated by the standard deviations in the visualization of Gaussian processes. It generalizes the biologist criteria that for a significant change the standard deviations of the control and treated samples should not overlap. This ranking can be used as a first filter to focus attention on compounds in which we have some confidence in the measured change.

2.2   PCA and the Resulting Ranking

Standard Principal Component Analysis (PCA) is carried out on the log-ratio timeseries generated by the Gaussian process model to determine the main sources of variance.[1] In the following we refer to the application of PCA along the time dimension, which yields principal components that are vectors over the 100 time points, assigning a weight to each timepoint corresponding to its contribution to the overall variance.[2]

Our results show that across all experiments, the top three components allow substantial *reduction in dimension* and hence in complexity of computations based on the PCA representation. As few as three components allow us to capture at least 90% of the variance (typically more than 95%). The output of this stage is a linear transformation of the time series data into a three dimensional vector space. As a by-product, the contribution of a compound to each PCA component provides another ranking of its response to a treatment.

2.3   Many Kinds of Clusterings

Clustering is a form of *abstraction*, which is essential for the biologist to cope with the overwhelming amount of observational data. However, since no single abstraction technique is superior in all applications, the clustering algorithm should be seen as an open-ended parameter in our approach. From the large number of clustering algorithms, we identified three general algorithm types that are sufficiently scalable to deal with the large number of observed compounds. Each type focuses on different features and they are all parameterizable to influence the degree of abstraction, giving the biologist a tunable magnifying glass.

---

[1] We use the machine learning framework [19] for PCA and basic clustering.

[2] Alternatively, PCA can be applied along the gene/compound dimension which we have done in another part of our RTA workflow.

Our current workflow includes well-known algorithms such as k-means clustering and two kinds of standard hierarchical clustering. It also includes a more recently proposed density-based hierarchical clustering algorithm (hdbscan) [16] that is more scalable than the standard ones, and supports partial clustering, meaning that not all compounds have to be assigned to the generated clusters.

Specifically, our k-means clustering workflow is based on the vector space arising from the PCA dimensionality-reduction (applied to the log-ratio time series data) equipped with the Euclidian metric. For tunability we allow the number of clusters k to range over $\{16, 32, 64, 128, 256, 512\}$. The other clustering algorithms are used with Pearson correlation ρ directly derived from the log-ratio time series data, using $1 - \rho$ as an affinity measure.[3] Standard hierarchical clustering is performed with two linkage methods (average distance and Ward's minimum variance method) that are very familiar for biologists working with gene expression data. A number of distance cutoffs are then used to generate clusterings of different granularity from the hierarchical clustering ($\{0.25, 0.5, 1, 2, 3, 4, 5, 10, 20, 30, 40, 50\}$ for average and $\{2, 3, 4, 5, 10, 20, 30, 40, 50, 100, 200, 300, 400, 500\}$ for Ward). Density-based clustering is performed with minimum cluster sizes $\{2, 3, 4, 5\}$ and an α-parameter ranging over $\{0.01, 0.1, 0.25, 0.5, 0.75, 0.9, 0.99\}$ to control the granularity.

2.4   Clustered Cross-Correlation Graphs

Our workflow includes several *graph synthesis methods* that are intended to support the biologist in the identification of the MoA. For example, Pearson correlation ρ can be naturally visualized as a family of *undirected* graphs parameterized by a correlation threshold. These graphs have compounds (or suitable subsets) as nodes and edges denote positive or negative correlation $\|\rho\|$ beyond the threshold, and they often result in scale-free networks characterized by a power law node degree distribution [3].

Slightly more generally, we can define cross-correlation graphs by replacing the symmetric notion of correlation by cross-correlation, which is defined as correlation between a time series and its shifted version and associated with a time direction and a time lag. Hence, the resulting graphs are *directed*. In our current workflow, we consider discrete time lags in the range 1–20 based on the Gaussian process representation with 100 time points. We use green (red) edges to indicate the existence of a positive (negative) cross-correlation with a time lag in the range, and always perform some pruning of the graph to eliminate orphan nodes, i.e., isolated compounds. A heuristic is used to keep the size of the graph manageable and amenable to automatic layout algorithms and to facilitate interpretability by the biologist.

In spite of these efforts, the resulting graphs are often too low-level or detailed. Hence, we developed a more abstract notion of *clustered cross-correlation graphs* (Fig. 2). Here, cross-correlation is lifted from pairs of compounds to pairs of clusters, as they arise from one of the algorithms in our

---

[3] Another affinity measure we have explored is based on correlated changes (using time series derivatives) but beyond the scope of this paper.

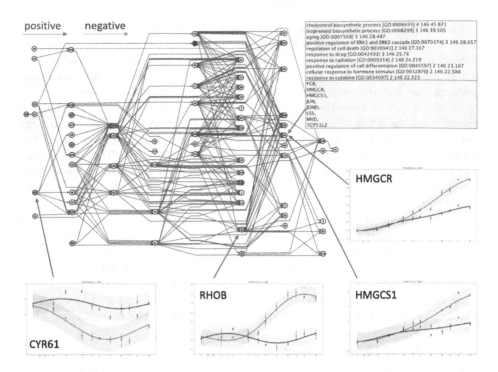

Fig. 2. Sample clustered cross-correlation graph for Unk5 Genes (see Sect. 4.3) using PCA-based clustering. Each node represents to a cluster of genes modeled as Gaussian processes with the color indicating the average log2-ratio (fold change) over time and over the genes in the cluster (green = increased, red = suppressed). For one cluster we show the full list of genes and ranked GO-process annotations. The Gaussian models for sample transcripts are shown in the timeseries plots: green is treated, black is control (surrounded by 2 standard deviation bands), dots with whiskers show the original data. (Color figure online)

clustering workflow. The lifting is done through averaging, i.e., cluster cross-correlation for a fixed lag from a source to a target cluster is defined as the average cross-correlation for the same fixed lag over all pairs with the first component in the source and the second component in the target cluster. Using this notion, clustered cross-correlation graphs have clusters of compounds as nodes and defined in complete analogy to the basic cross-correlation graphs. Similarly, orphan clusters are removed from the graph so that only non-trivial connected components remain.

The advantage of using cross-correlation as a basis for graph synthesis is that it is a well-studied concept from signal processing that is visually intuitive for the biologist. The mathematical simplicity of the concept can however be a disadvantage, since it will miss potential causal interactions of higher complexity that do not manifest themselves as (homogeneous) time-lagged correlation. Hence, our workflow contains other complementary techniques for graph synthesis (beyond the scope of this paper) based on neural networks which try to learn and exploit characteristic features in the time profiles.

2.5 Differential Anomaly Identification Using Autoencoders

It is expected that the observed time series data contains many anomalies due to measurement errors or other forms of noise. However, given that Gaussian processes allow us to filter most of the noise and smoothen the data to some extent, any *remaining anomalies* suggest a potential role in the MoA, and can give valuable clues in conjunction with the previous results and graphs.

To detect such anomalies we use a variation of *convolutional autoencoders* [15]. An autoencoder [9] is a neural network that is a sequential composition of an encoder and a decoder that together are trained so that the output reconstructs the original input as closely as possible. The encoder transforms the input (time series with 100 data points in our case) into a low-dimensional feature space (we found that as few as 4 dimensions are sufficient to reconstruct most inputs). In our specific application it is a convolutional neural network (using a window of size 41) without weight-sharing (reflecting heterogeneity in time) but with L1-regularization (favoring sparsity), followed by max-pooling, and exponential-linear activation functions. The decoder is a dense neural network with L2-regularization and linear activation. In order to prevent certain compounds from dominating the data set, all input time series are normalized. We use mean square error as the loss function for training, which uses the Adam-optimizer with a fixed number of 40000 epochs and a batch size of 1000 (500 for proteomics). A random subset of 10% of the compounds is used for validation. Autoencoder models are synthesized separately for transcriptomics and proteomics data sets as well as for the combination of transcriptomics, proteomics, and metabolomics data. When building models involving proteomics data we found that adding a dropout layer between encoder and decoder (with a low dropout rate of 0.01 and 0.0001 for the combined data set) reduced the risk of overfitting.

One of our anomaly-detection workflows aims at detecting what we call *differential anomalies*, that is behavior in the treated condition that is hard to predict by a model that only learns the behavior under the control condition. To detect differential anomalies, we use an autoencoder model trained on the control time series data and apply it to the treated time series data for all compounds to quantify the reconstruction error (mean-square error between original and predicted time series). We then generate a ranked list of compounds sorted by this reconstruction error (highest error first).[4]

2.6 Heuristic Annotation of Clusters

Our workflow implements a *generic heuristic method* to compute likely classifications of a cluster from classifications of its elements. It is used to annotate

---

[4] We explored some other metrics based on the original (non-normalized) time series data that we omit for brevity.

each cluster with a ranked lists of classifications based on GO terms (process or function) or HGNC families. This is the *only use of prior knowledge* in our approach, and serves mostly to interpret proposed MoA candidates.

The generic method assumes a set of classification concepts and a known relationship between compounds and concepts, which does not have to be one-to-one (i.e., overlapping concepts or hierarchies are allowed, as they occur for example in the GO hierarchy). We denote by p_c the probability that the concept c is associated with a random compound (assuming uniform distribution) covered by the relationship. By $f_c(S)$ we denote the number of compounds in a cluster S that are associated with the concept c, and by $f(S)$ we denote the corresponding sum over all concepts. For each cluster S we use the score $s_c = -f_c(S)\log_2(p_c)$, which is the logarithm of $1 - p_c^{f_c(S)}$, to rank all concepts for the given cluster.

In our workflow, we use GO processes, GO functions, and HGNC families as sources of classification concepts. In each case, we use the top 10 concepts c to label each cluster S with $(c, f_c(S), f(S), s_c)$, or a few more if there is a tie in the score. The label makes it explicit that the classification c is based on $f_c(S)$ out of $f(S)$ concepts. Intuitively, our scoring function reduces the impact of frequently occurring concepts (low information content) but amplifies the impact of co-occurring concepts (a form of confirmation) in the same cluster. It should be noted that the scoring function does not rely on expression levels of compounds and is not intended for comparisons between clusters.

3   The Omics Data

The data to be analysed was generated as follows. HepG2/C3A hepatocyte were exposed to five different challenge agents: forskolin, nocodazole, bendamustine, nexturastat A, and atorvastatin. Molecular responses were measured from three biological replicates at eight to ten time-points between 10 s and 48 h post-exposure by microarray-based transcriptomics, multiplexed high-resolution LC-MS shotgun proteomics, and novel untargeted metabolomics to quantify >67,000 transcripts, >5,000 proteins and >200 metabolites. Immuno-assays and qRT-PCR were performed to validate the proteomic and transcriptomic data, respectively. Biological assays were performed to down select and confirm the MoA candidates inferred from the various data analyses.

The challenge agents spanned a wide range of mechanisms: forskolin is a diterpene that activates the enzyme adenylyl cyclase and the cAMP signaling pathway; nocodazole is a microtubule-targeting agent that disrupts microtubule polymerization and induces cell cycle arrest in the prometaphase; bendamustine is an alkylating agent with a nitrogen mustard group which causes the formation of crosslinks between DNA bases; nexturastat A is an inhibitor of HMG-CoA reductase which is a key step in cholesterol biosynthesis.

4   Choosing MoA Candidates

In this section we illustrate the use of the Gaussian process based analyses of time series omics data to identify candidate participants in the MoA in response to exposure to a challenge agent. We begin with an overview of the process we developed and then illustrate the ideas using data from two of the challenge compounds studied.

4.1   Overview of Selection Process

In addition to the Gaussian process (GP) based analysis we carry out some standard time series analysis to provide a cross check. For each entity, the log2 fold change was computed at each of the measured time points as the difference of the log2 of the means of the control and treated replicates. We refer to these change vs time plots as *basic* time profiles. Time points for which the set of control measurements are either all above or all below the set of treated measurements are annotated as significant.[5]

The Shape of Response. Before looking for specific candidate MoA participants we look at the shape of the overall response in two ways: the level of up/down regulation at each timepoint; and the population level cell cycle state over time. For the levels of up/down regulation we make *udby* (Up/Down By) charts. These are bar charts plotting the relative number of entities that are first up/down regulated by a given threshold amount at each measured time point. This gives a visual representation of the overall shape of the response (when are events happening). The given threshold magnitude is typically between .5 and 1 log2 fold change, adjusted to ensure at least 3–5% of the measured data is included. Maps are also computed associating to each entity the maximum and minimun log2 fold change along with the times the extrema are reached. Figure 3 on the left shows the *udby* chart for Unk3 transcripts (based on the basic time profile). From the chart we observe activity at 4–6 h and 18–24 h, but there is a lull at 8 h.

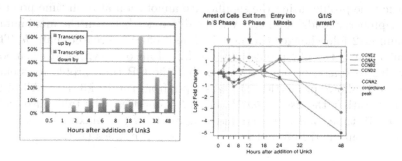

Fig. 3. Up/Dn By Chart and Cyclin gene plots for Unk3

Information about population cell cycle state is visualized by plotting the time series profiles of four key Cyclins. The shape of these plots contains information about whether or not the perturbing agent disrupts the cell cycle and if so at

---

[5] Currently we restrict analysis of transcriptomics data to protein coding genes.

which points. Figure 3 on the right shows the basic time profiles of Cyclin mRNA expression for Unk3 transcripts. From the plot we see a pattern of Cyclin mRNA expression that reflects a transient S phase arrest followed by a synchronized transition through the cell cycle and an apparent arrest at G1/S. The reasoning and the reason it is interesting to look at the cyclins is as follows. Cyclins are proteins whose protein and mRNA concentrations vary at different stages in the cell cycle. As long as the cells are growing exponentially, the proportion of cells in each cell cycle stage stays constant and the expression of each cyclin does not change over time. An increase in the expression of cyclin E2 (CCNE2) indicates that the proportion of cells in S-phase is increasing. If the arrest were permanent, the level of CCNE2 would stay high for the rest of the time course. The decrease in CCNE2 expression from 6–18 h indicates that the arrest is transient.

Cross Correlation and Visualization. To provide a visual summary of the potential relations among responding elements, we use the cross-correlation graphs on clusters (described in Sect. 2.4). Graph nodes are annotated by results from analyses including: classification, cluster members, average change, GO terms and family classifications (see Fig. 2). One can visually identify potentially interesting nodes by criteria such as being input, output, or highly connected, i.e., with high betweenness centrality, and use the annotations to drill down.

Filtering and Ranking. We consider several criteria for selecting or ranking transcripts, proteins and metabolites (aka entities). The *1sd* and *2sd* separation properties identify entities with statistically significant change by measuring the space between one or two standard deviation bands around the control and treated GP time profiles as explained in Sect. 2.1. The *udby.th* property holds for entities that have a significant log2 fold change of at least *th* at some measured time point, where *th* is typically between .5 and 1. This is computed using the basic time profile, using the significance annotation at each time point. The *delta.th* property holds of an entity if the difference between the maximum and minimum log2 fold changes (in the GP time profile) is greater than *th*. This is relevant because the GP profile may be shifted up or down, or over smoothed (likely due to sparse or noisy data) so that the GP time profile is essentially flat. Figure 4 shows properties for transcripts from the Unk3 data. All of these transcripts satisfy the *2sd*, *udby.75*, *delta.1* properties. The table also includes the *upby* .75 time, and the time and magnitude of the maximum fold change (*mxt,mxlf*) computed from both the basic and GP profiles (*L, G*).

name	pca-r.c	pca-r.r	pca-d.c	pca-d.r	delta1	2sd	ud.75	ubtL	ubtG	mxtL	mxlfL	mxtG	mxlfG	timem
CDKN1A	p0	18	p2	34	1	1	1	0.5	2	48	2.7497	48	2.933	530
FAS	p0	19	p2	9	1	1	1	6	4	24	2.9089	24	2.506	206
MDM2	p0	49	p2	6	1	1	1	6	4	24	2.2144	18	2.062	630
POLH	p0	17	p2	15	1	1	1	0.5	2	24	2.9415	24	2.464	144
PPM1D	p0	23	p2	4	1	1	1	4	4	24	2.7921	24	2.362	417
SFN	p0	32	p2	12	1	1	1	6	4	24	2.6263	24	2.216	966

Fig. 4. Examples of properties for Unk3 transcripts.

The *pca.r* and *pca.d* ranking functions assign rank n (smaller is better) to an entity if it appears as the nth element in the ranked list of one of the PCA components (positive or negative), where the PCA analysis is based on the GP ratio time profile or its derivative, respectively. From Fig. 4 we see that the example transcripts all have rank below 20 for at least one PCA analysis.

The *timem*, *time*, *diffm*, and *diff* ranking functions assign rank n to an entity if it is the nth element in the ranked list according to the corresponding anomaly calculation as described in Sect. 2.5. *timem* and *time* anomaly rankings are based on the GP ratio time profile where for *timem* the mean square error is computed after transforming the model back to the original data scale, while for *time* it is computed on the normalized model. Thus *timem* is biased towards larger changes. *diffm* and *diff* anomaly rankings are based on predicting the treated time series from the control time series. In the case of the example transcripts of Fig. 4, none have an anomaly rank below 100. Only *timem* is shown. Given these properties and ranking functions we combine them by conjunction and disjunction to select subsets of entities for further examination.

Feature Summary Spreadsheets. As an aid to exploring the candidate subsets we summarize the above properties/ranks and other information in a spreadsheet with rows corresponding to entities in some subset of interest. The columns record satisfaction of properties, ranks, and results from other computations. (Figure 4 is a selection from one such spread sheet.) The entities can then be grouped by sorting on different combinations of columns to hopefully reveal key players. The columns include the ten properties and rankings discussed above. There are also columns labeled *cl.r* and *cl.d* that record the number of the cluster containing the entity, using the PCA K-means clustering (with 128 clusters for transcriptomics and proteomics). *cl.r* similarity is based on the GP ratio time profile and *cl.d* similarity is based on the derivative of the GP ratio time profile. We also include columns *maxlf*, *minlf*, *maxt*, and *mint* recording the log2 fold change maximum, minimum, and corresponding times.

Picking Winners. With all the analyses, properties, and ranking functions in hand, the challenge is to identify the most significant responders to the treatment. The main criteria of success is whether the data somehow reveals the 'canonical' MoA. But there are generally many possibly important things going on that are not included in the canonical pathway. Perhaps the data will reveal something new.

We start by investigating two subsets. One subset is the entities that satisfy a separation property (*1sd* or *2sd*) or an udby property (*udby.th*). The choice of property within the two classes is determined by initially wanting to limit the number of candidates to ~1000 or less. The other subset contains the entities that rank in the top n of one of the six ranking functions. Generally, taking n to be 20 gives us a reasonable set to consider.

We make feature summary spreadsheets for each of the two subsets and look for groups that stand out according to some collection of features. When sorting by one of the cluster columns we can get additional information by looking at the annotations (GO term or family classifications) associated with clusters that have a relatively large number of elements from the subset under consideration. Of particular interest are the most specific annotations. GO terms such as signal transduction, cell cycle, or metabolic process don't tell you much although they can support conjectures derived from clusters annotated with more specific terms.

4.2 Challenge Compound Unk3

For Unknown 3 (henceforth Unk3) the transcriptomics data supported basic time profiles for 18208 transcripts and GP-based time profiles for 18834 transcripts. There were 4396/2803 transcripts that had a log2 fold change magnitude of at least 0.75 at some time point according to the basic/GP timeprofile. Of the 2803 transcripts identified by GP-based analysis, 1579 satisfied the *delta.75* property (defined in Sect. 4.1). The difference between basic and GP-based results has to do with the Gaussian process smoothing and integrating all the data, while the basic analysis computes changes point-wise. There were 4993 transcripts that passed the one standard deviation separation filter (satisfied *1sd*), while 1627 transcripts passed the two standard deviation separation filter (satisfied *2sd*).

The *udby* chart for Unk3 transcripts and the time profiles of Cyclin mRNA expression have already been presented in Sect. 4.1. Now we look at the top ranked transcripts according to PCA ranking and anomaly ranking. Thus we defined top20PCA-u3 to be the set of transcripts ranked in the top 20 of one of the three PCA components (positive or negative), using PCA based on the GP time profile or its derivative. We defined top20anom-u3 to be the set of transcripts ranked in the top 20 of one of four anomaly measures (timem, time, diffm, diff). We let top20anomPCA-u3 denote the union of top20PCA-u3 and top20anom-u3. To get a first impression of how the filters based on statistical significance relate to filters based on features we compared the PCA and anomaly ranked sets to those passing the one and two standard deviation separation filters (*1sd, 2sd* respectively) and to the filter selecting transcripts with log2 fold change magnitude at least 0.75 at some time point (*ud*). The numbers in each set and their intersections are shown in Fig. 5.

Transcript Set	Number in Set	Intersections	Number
top20anom-u3	177	top20anom-u3 & top20PCA-u3	16
top20PCA-u3	76	top20PCA-u3 & 1sd-u3	175
top20anomPCA-u3	237	top20PCA-u3 & 2sd-u3	153
1sd-u3	4993	top20PCA-u3 & ud-u3	177
2sd-u3	1627	top20anom-u3 & 1sd-u3	36
ud-u3	4376	top20anom-u3 & 2sd-u3	32
		top20anom-u3 & ud-u3	34

Fig. 5. Overview of rank-based selections for Unk3.

We see that the ranking according to PCA vs anomaly is quite different, only 16 transcripts pass both filters. Also, the statistical filters overlap substantially with the PCA based filters, but much less with the anomaly based filters.

For the next step we chose to focus on the top20anomPCA-u3 and produced a sortable feature table with one row for each transcript in top20anomPCA-u3. Sorting by feature (PCA or anomaly) doesn't give any obvious insights. So, we sort by cluster (ratio-based or derivative based). We start by looking at clusters that contain at least six elements from top20anomPCA-u3 and search the GO annotations associated with these clusters for the most specific annotations. The second annotation of Cluster 16 is "DNA damage response, signal transduction by p53 class mediator resulting in cell cycle arrest [GO:0006977]". This cluster contains CDKN1A, FAS, MDM2, PPM1D, SFN, POLH, (P53 responsive and DNA repair genes). With this clue we look further and find two more clusters with this annotation and several additional clusters with DNA damage/DNA repair annotations. The transcripts in these clusters are up-regulated early (by 2–6 h) and reach a maximum around 24 h. Consistent with the observation based on the plot of the time profiles of the Cyclins (Fig. 3), four of the larger clusters are annotated with GO terms related to G1-S transition. Two of these clusters are down regulated by 18 h and reach minimum at 48 h, while two have periods of up and down regulation. The highly ranked POLH, upregulated between 4 and 24 h, is a DNA polymerase involved in DNA repair.

We conclude that the treatment likely resulted in DNA damage leading to a P53 response in addition to the temporary cell cycle arrest. After the measurements and data analysis, Unk3 was revealed to be bendamustine, known to crosslink DNA strands and interfere with DNA damage repair mechanisms. This confirmed the results of the analyses.

4.3 Challenge Compound Unk5

For Unknown 5 (henceforth Unk5) the transcriptomics data supported basic time profiles for 15685 transcripts and GP-based time profiles for 17347 transcripts. There were 570/407 transcripts that had a log2 fold change magnitude of at least .75 at some time point according to the basic/GP time profile.

Of the 407 transcripts identified by GP-based analysis, 272 satisfy $delta.75$. There were 4442 transcripts that passed the one standard deviation separation filter, while 1287 transcripts passed the two standard deviation separation filter.

Figure 6 shows the $udby$ chart for Unk5 transcripts and the time profiles of Cyclin mRNA expression. From the chart we see that the bulk of the transcriptomics activity happens at 18 h or later. From the Cyclin plot we see a pattern of Cyclin mRNA expression that reflects a permanent arrest in G1, namely the steady increase in Cyclin D2 (CCND2) after 24 h.

As for Unk3, we next look at the top ranked transcripts according to PCA ranking and anomaly ranking. Thus we defined top20PCA-u5 to be the set of transcripts t ranked in the top 20 according to PCA ranking, that is $pca.r(t) < 20$ or $pca.d(t) < 20$. We defined top20anom-u5 to be the set of transcripts t ranked in the top 20 according to one of the four anomaly measures, i.e., $timem(t) < 20$

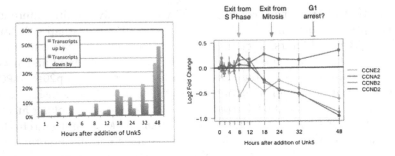

Fig. 6. Up/Dn By Chart and Cyclin plots for Unk5

or $time(t) < 20$ or $diffm(t) < 20$ or $diff(t) < 20$. We let `top20anomPCA-u5` denote the union of `top20PCA-u5` and `top20anom-u5`. As for Unk3, the ranking according to PCA vs. anomaly is quite different, only 24 transcripts pass both filters. We note that `top20anom-u5` and `top20PCA-u5` contain eight metallothioneins: MT1A, MT1B, MT1E, MT1F, MT1G, MT1H, MT1M, and MT1X. These transcripts appear in four different clusters, all are upregulated by 1 h and reach a maximum between 2 h and 8 h.

For the next step we chose to focus on the `top20anomPCA-u5` and produced a sortable feature table with one row for each transcript in `top20anomPCA-u5` and columns as described in Sect. 4.1. Sorting by $cl.r$ (ratio based clustering) we find three relatively large clusters annotated with Cholesterol Biosynthetic process GO terms. Two more clusters are annotated with metabolic process GO terms. Cluster 54 contains HMGCS1, HMGCR, MVD, LSS, FGB, JUN, JUND, and TCP11L2, the first four are enzymes in the canonical Cholesterol biosynthesis pathway (see Figs. 2, 7). All of the entities in the cluster are up-regulated midway and reach a maximum late. The Cholesterol enzymes pass the .75 threshold at 18 h and reach a maximum at 48h. Five clusters are annotated with metal ion and zinc binding GO terms. Sorting by $cl.d$ (derivative based clustering) we find two clusters annotated with Cholesterol Biosynthetic process GO terms and one with Cholesterol import. There are three clusters annotated with metal ion and zinc binding GO terms.

Although the proteomics data for this pathway is too sparse to be highly ranked by the GP analysis, once we have the clue, we plot the raw data for transcripts and proteins in the pathway and see a general trend for slow steady up-regulation and a remarkable agreement between transcripts and the proteins they code for. The metabolomic data plots for Cholesterol and Chenodeoxycholic indicate possible down regulation sometime after 8 h, adding evidence to a conjecture that the unknown is inhibiting something in that pathway and the cells are responding by trying to make more. This is illustrated in Fig. 7.

After the omics experiments and data analysis, Unk5 was revealed to be atorvastatin, an inhibitor of HMG-CoA reductase (HMGCR), as proposed by the analysis. The strong early response by the metallothioneins remains a mystery to be further investigated.

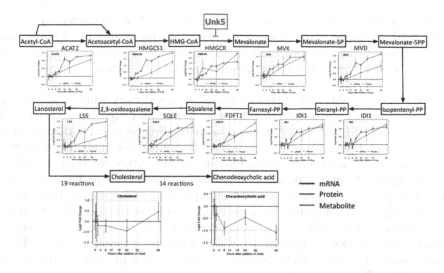

Fig. 7. Canonical Cholesterol pathway annotated with Unk5 data

5 Conclusions and Future Directions

We have presented a collection of algorithms and reproducible workflow for analysis of comparative multi-omics time series data. The goal is to be able to identify candidate participants in the MoA of chemical or biological agents using data analysis, without recourse to prior knowledge such as interaction and pathway databases. The algorithms are based on Gaussian process models that leverage a large number of data points to derive useful information from data that may be noisy or have missing data points. The Gaussian process models support integration of timeseries based on different sets of time points and different types of data. The workflows produce several functions for ranking candidates according to different features, functions for grouping entities according to different similarity criteria, and graph generation algorithms for visual representation of correlations and potential causality relations among entities or groups of entities. Each function and relation brings different insights into the cellular response to a perturbation. We illustrated the application of this computational analysis suite to identify key elements of the MoA of two unknowns, eventually revealed to be bendamustine and atorvastatin.

As mentioned in Sect. 1, existing approaches to determining the MoA of a cellular perturbation rely on existing knowledge, and may work with only single timepoints or a small number of timepoints (see [18] for an extensive overview). Two recent works [18,29] based on network analysis are perhaps the closest to our approach. DeMAND [29] scores candidate MoA proteins based on an assessment of the global dysregulation of their molecular interactions following perturbation, based on an input gene-protein interaction network. Multiple time points can be used, but the temporal information is not used. ProTina [18] creates a cell type specific protein-gene regulatory network (from existing knowledge bases) based on a dynamic model of the gene transcription. Protein targets are scored based on the enhancement/attenuation of protein-gene regulations. The dynamic model allows inclusion of temporal information in the scoring process. Note that in contrast, our approach does not require a regulation network as input.

Future Directions. We note that in contrast to usual work where a single approach is developed and validated against synthetic and experimental data, our workflows result in multiple approaches modeling diverse features that might be observed as part of an MoA, giving a flexible tool set for analyzing data from a truly unknown perturbing agent. One important future direction is validating the approach on a wide class of cellular perturbations and to generate a validated characterization of each of the approaches. The current tool suite is used by experts in computational modeling. Increasing the level of automation for identifying MoA candidates by combining the different approaches and integrating the visualization capabilities would make the capabilities accessible to more researchers and hopefully help to advance research in MoAs at the cellular response level.[6]

References

1. Abadi, M., et. al.: TensorFlow: a system for large-scale machine learning. In: Proceedings of the 12th USENIX Conference on Operating Systems Design and Implementation, pp. 265–283. USENIX Association (2016)
2. Barabasi, A.L., Gulbahce, N., Loscalzo, J.: Network medicine: a network-based approach to human disease. Nat. Rev. Genet. **12**(1), 56–68 (2011)
3. Barabasi, A.L., Oltvai, Z.N.: Network biology: understanding the cell's functional organization. Nat. Rev. Genet. **5**(2), 101–113 (2004)
4. Cajka, T., Fiehn, O.: Toward merging untargeted and targeted methods in mass spectrometry-based metabolomics and lipidomics. Anal. Chem. **88**(1), 524–545 (2016)
5. Dettmer, K., Aronov, P.A., Hammock, B.D.: Mass spectrometry-based metabolomics. Mass Spectrom. Rev. **26**(1), 51–78 (2007)
6. de Matthews, G., et al.: GPflow: a gaussian process library using tensorflow. J. Mach. Learn. Res. **18**, 40:1–40:6 (2017)
7. Girault, C., Valk, R.: Petri Nets for Systems Engineering: A Guide to Modeling, Verification, and Applications. Springer, Heidelberg (2010). https://doi.org/10.1007/978-3-662-05324-9
8. Goodwin, S., McPherson, J.D., McCombie, W.R.: Coming of age: ten years of next-generation sequencing technologies. Nat. Rev. Genet. **17**(6), 333–351 (2016)

[6] Disclaimer. Research was sponsored by the U.S. Army Research Office and the Defense Advanced Research Projects Agency and was accomplished under Cooperative Agreement Number W911NF-14-2-0020. The views and conclusions contained in this document are those of the authors and should not be interpreted as representing the official policies, either expressed or implied, of the Army Research Office, DARPA, or the U.S. Government. The U.S. Government is authorized to reproduce and distribute reprints for Government purposes notwithstanding any copyright notation hereon.

9. Hinton, G.E., Salakhutdinov, R.R.: Reducing the dimensionality of data with neural networks. Science **313**(5786), 504–507 (2006)
10. Jeong, H., Tombor, B., Albert, R., Oltvai, Z.N., Barabasi, A.L.: The large-scale organization of metabolic networks. Nature **407**(6804), 651–654 (2000)
11. Kim, M.S., et al.: A draft map of the human proteome. Nature **509**(7502), 575–581 (2014)
12. Kluyver, T., et. al.: Jupyter notebooks - a publishing format for reproducible computational workflows. In: Loizides, F., Schmidt, B. (eds.) Positioning and Power in Academic Publishing: Players, Agents and Agendas, pp. 87–90. IOS Press (2016)
13. Kramer, A., Green, J., Pollard, J., Tugendreich, S.: Causal analysis approaches in ingenuity pathway analysis. Bioinformatics **30**(4), 523–530 (2014)
14. Mann, M., Kulak, N.A., Nagaraj, N., Cox, J.: The coming age of complete, accurate, and ubiquitous proteomes. Mol. Cell **49**(4), 583–590 (2013)
15. Masci, J., Meier, U., Cireşan, D., Schmidhuber, J.: Stacked convolutional auto-encoders for hierarchical feature extraction. In: Honkela, T., Duch, W., Girolami, M., Kaski, S. (eds.) ICANN 2011 Part I. LNCS, vol. 6791, pp. 52–59. Springer, Heidelberg (2011). https://doi.org/10.1007/978-3-642-21735-7_7
16. McInnes, L., Healy, J., Astels, S.: HDBSCAN: hierarchical density based clustering. J. Open Sour. Softw. **2**(11) (2017)
17. Mi, H., et al.: Panther version 11: expanded annotation data from gene ontology and reactome pathways, and data analysis tool enhancements. Nucleic Acids Res. **D45**(1), D183–D189 (2017)
18. Noh, H., Shoemaker, J.E., Gunawan, R.: Network perturbation analysis of gene transcriptional profiles reveals protein targets and mechanism of action of drugs and influenza a viral infection. Nucleic Acids Res. **46**(6), e34 (2018)
19. Pedregosa, F., et al.: Scikit-learn: machine learning in Python. J. Mach. Learn. Res. **12**, 2825–2830 (2011)
20. Pujol, A., Mosca, R., Farres, J., Aloy, P.: Unveiling the role of network and systems biology in drug discovery. Trends Pharmacol. Sci. **31**(3), 115–123 (2010)
21. Rasmussen, C.E., Williams, C.K.I.: Gaussian Processes for Machine Learning (Adaptive Computation and Machine Learning). The MIT Press, Cambridge (2005)
22. Tautenhahn, R., et al.: An accelerated workflow for untargeted metabolomics using the METLIN database. Nat. Biotechnol. **30**(9), 826–828 (2012)
23. Uhlen, M., et al.: Tissue-based map of the human proteome. Science **347**(6220), 4 (2015)
24. Vertes, A., et. al.: Time-dependent metabolomics in systems biology context for mechanism of action studies. In: US HUPO Conference - Proteomics: From Genes to Function, San Diego, CA (2017)
25. Vertes, A., et. al.: Mechanism of action identification of threat agents within 30 days. In: Society of Toxicology 57th Annual Meeting, San Antonio, TX (2018)
26. Vertes, A., et. al.: Novel high-throughput metabolomic techniques and mainstream tools for the discovery of drug mechanism of action. In: US HUPO 14th Annual Conference - Technology Accelerating Discovery, Minneapolis, MN (2018)
27. Vertes, A., et. al.: Systems biology approach for mechanism of action identification in 30 days. In: ASMS Conference, San Diego, CA (2018)
28. Wang, Z., Gerstein, M., Snyder, M.: RNA-Seq: a revolutionary tool for transcriptomics. Nat. Rev. Genet. **10**(1), 57–63 (2009)
29. Woo, J.H., et al.: Elucidating compound mechanism of action by network Perturbation analysis. Cell **162**(2), 441–451 (2015)
30. Xu, W.H., et al.: Human transcriptome array for high-throughput clinical studies. Proc. Natl. Acad. Sci. U.S.A. **108**(9), 3707–3712 (2011)

Composable Rate-Independent Computation in Continuous Chemical Reaction Networks

Cameron Chalk[✉], Niels Kornerup, Wyatt Reeves, and David Soloveichik[✉]

The University of Texas at Austin, Austin, USA
{ctchalk,david.soloveichik}@utexas.edu

Abstract. Biological regulatory networks depend upon chemical interactions to process information. Engineering such molecular computing systems is a major challenge for synthetic biology and related fields. The chemical reaction network (CRN) model idealizes chemical interactions, abstracting away specifics of the molecular implementation, and allowing rigorous reasoning about the computational power of chemical kinetics. Here we focus on function computation with CRNs, where we think of the initial concentrations of some species as the input and the eventual steady-state concentration of another species as the output. Specifically, we are concerned with CRNs that are rate-independent (the computation must be correct independent of the reaction rate law) and composable ($f \circ g$ can be computed by concatenating the CRNs computing f and g). Rate independence and composability are important engineering desiderata, permitting implementations that violate mass-action kinetics, or even "well-mixedness", and allowing the systematic construction of complex computation via modular design. We show that to construct composable rate-independent CRNs, it is necessary and sufficient to ensure that the output species of a module is not a reactant in any reaction within the module. We then exactly characterize the functions computable by such CRNs as superadditive, positive-continuous, and piecewise rational linear. Our results show that composability severely limits rate-independent computation unless more sophisticated input/output encodings are used.

1 Introduction

A ubiquitous form of biological information processing occurs in complex chemical regulatory networks in cells. The formalism of chemical reaction networks (CRNs) has been widely used for modelling the interactions underlying such natural chemical computation. More recently CRNs have also become a useful model for designing synthetic molecular computation. In particular, DNA strand displacement cascades can in principle realize arbitrary CRNs, thus motivating the

These authors' work was supported in part by National Science Foundation grants CCF-1618895 and CCF-1652824.

M. Češka and D. Šafránek (Eds.): CMSB 2018, LNBI 11095, pp. 256–273, 2018.
https://doi.org/10.1007/978-3-319-99429-1_15

study of CRNs as a programming language [2,5,11]. The applications of synthetic chemical computation include reprogramming biological regulatory networks, as well as embedding control modules in environments that are inherently incompatible with traditional electronic controllers for biochemical, nanotechnological, or medical applications.

The study of information processing within biological CRNs, as well the engineering of CRN functionality in artificial systems, motivates the exploration of the computational power of CRNs. In general, CRNs are capable of Turing universal computation [7]; however, we are often interested in restricted classes of CRNs which may have certain desired properties. Previous work distinguished two programmable features of CRNs: the stoichiometry of the reactions and the rate laws governing the reaction speeds [4]. As an example of computation by stoichiometry alone, consider the reaction $2X \to Y$. We can think of the concentrations of species X and Y to be the input and output, respectively. Then this reaction effectively computes $f(X) = \frac{X}{2}$, as in the limit of time going to infinity, the system converges to producing one unit of Y for every two units of X initially present. The reason we are interested in computation via stoichimetry is that it is fundamentally *rate-independent*, requiring no assumptions on the rate law (e.g., that the reaction occurs at a rate proportional to the product of the concentrations of the reactants). This allows the computation to be correct independent of experimental conditions such as temperature, chemical background, or whether or not the solution is well-mixed.

Computation does not happen in isolation. In an embedded chemical controller, inputs would be produced by other chemical systems, and outputs would affect downstream chemical processes. Composition is easy in some systems (e.g. digital electronic circuits can be composed by wiring the outputs of one to the inputs of the other). However, in other contexts composition presents a host of problems. For example, the effect termed retroactivity, which results in insufficient isolation of modules, has been the subject of much research in synthetic biology [6]. In this paper, we attempt to capture a natural notion of composable rate-independent computation, and study whether composability restricts computational power.

$$X_1 + X_2 \to Y$$

(a)

$$X_1 \to Z_1 + Y$$
$$X_2 \to Z_2 + Y$$
$$Z_1 + Z_2 \to K$$
$$Y + K \to \varnothing$$

(b)

Above, we see two examples of rate-independent computation. Example (a) shows $y = \min(x_1, x_2)$. The amount of Y eventually produced will be the minimum of the initial amounts of X_1 and X_2, since the reaction will stop as soon as the first reactant runs out. Example (b) shows $y = \max(x_1, x_2)$. The amount of Y eventually produced in reactions 1 and 2 is the sum of the initial amounts of

X_1 and X_2. The amount of K eventually produced in reaction 3 is the minimum of the initial amounts of X_1 and X_2. Reaction 4 subtracts the minimum from the sum, yielding the maximum.

Now consider how rate-independent computation can be naturally composed. Suppose we want to compute $\min(\min(x_1, x_2), x_3)$. It is easy to see that simple concatenation of two min modules (with proper renaming of the species) correctly computes this function:

$$X_1 + X_2 \to Y$$
$$Y + X_3 \to Y'$$

where Y' represents the output of the composed computation. In contrast, suppose we want to compute $\min(\max(x_1, x_2), x_3)$. Concatenating the modules yields:

$$X_1 \to Z_1 + Y$$
$$X_2 \to Z_2 + Y$$
$$Z_1 + Z_2 \to K$$
$$Y + K \to \varnothing$$
$$Y + X_3 \to Y'$$

where Y' represents the output of the composed computation. Observe that depending on the relative rates of reactions 4 and 5, the eventual value of Y' will vary between $\min(\max(x_1, x_2), x_3)$ and $\min(x_1 + x_2, x_3)$, and the composition does not compute in a rate-independent manner.

Why is min composable, but max not? The problem arose because the output of the max module (Y) is consumed in both the max module and in the downstream module (min). This creates a competition between the consumption of the output within its own module and the downstream module.

Towards modularity, we assume the two CRNs to be composed do not share any species apart from the interface between them (i.e., a species Y representing the output of the first network is used as the species representing the input to the second network, and otherwise the two sets of species are disjoint). We prove that to construct composable rate-independent modules in this manner, it is necessary and sufficient to ensure that the output species of a module is not a reactant in any reaction of that module. We then exactly characterize the computational power of composable rate-independent computation.

Previously it was shown that without the composability restriction, rate-independent CRNs can compute arbitrary positive-continuous, piecewise rational linear functions [4]. Positive-continuity means that the only discontinuities occur when some input goes from 0 to positive, and piecewise rational linear means that the function can by defined by a finite number of linear pieces (with rational coefficients). Note that non-linear continuous functions can be approximated to arbitrary accuracy.[1] We show that requiring the CRN to be composable restricts the class of computable functions to be superadditive functions; i.e., functions that satisfy: for all input vectors a, b, $f(a) + f(b) \leq f(a + b)$. This strongly restricts computational power: for example, subtraction or max cannot be computed or approximated in any reasonable sense. In the positive direction, we show that any superadditive, positive-continuous, piecewise rational linear function can be computed by composable CRNs in a rate-independent manner. Our proof is constructive, and we further show that unimolecular and bimolecular reactions are sufficient.

We note that different input and output encodings can change the computational power of rate-independent, composable CRNs. For example, in the so-called *dual-rail* convention, input and output values are represented by differences in concentrations of two species (e.g., the output is equal to the concentration of species Y^+ minus the concentration of Y^-). Dual-rail simplifies composition—instead of consuming the output species to decrease the output value, a dual-rail CRN can produce Y^-—at the cost of greater system complexity. Dual-rail CRNs can compute the full class of continuous, piecewise rational linear functions while satisfying rate-independence and composability [4]. Note, however, that the dual-rail convention moves the non-superadditive subtraction operation to "outside" the system, and converting from a dual-rail output to a direct output must break composability.

2 Preliminaries

Let $\mathbb{N}$ and $\mathbb{R}$ denote the set of nonnegative integers and the set of real numbers, respectively. The set of the first n positive integers is denoted by $[n]$. If $x \in \mathbb{R}$, let $\mathbb{R}_{\geq x} = \{x' \in \mathbb{R} \mid x' \geq x\}$, and similarly for $\mathbb{R}_{>x}$. If Λ is a finite set (in this paper, of chemical species), we write $\mathbb{R}^{\Lambda}$ to denote the set of functions $f : \Lambda \to \mathbb{R}$, and similarly for $\mathbb{R}_{\geq 0}^{\Lambda}$, $\mathbb{N}^{\Lambda}$, etc. Equivalently, we view an element $c \in A^{\Lambda}$ as a vector of $|\Lambda|$ elements of A, each coordinate "labeled" by an element of Λ. Given a function $f : A \to B$, we use $f|_C$ to denote the restriction of f to the domain C. We also use the notation $c \upharpoonright \Delta$ to represent c projected onto $\mathbb{R}_{\geq 0}^{\Delta}$. Thus, $c \upharpoonright \Delta = 0$ iff $(\forall S \in \Delta) \, c(S) = 0$.

2.1 Chemical Reaction Networks

Given $S \in \Lambda$ and $c \in \mathbb{R}_{\geq 0}^{\Lambda}$, we refer to $c(S)$ as the *concentration of S in c*. For any $c \in \mathbb{R}_{\geq 0}^{\Lambda}$, let $[c] = \{S \in \Lambda \mid c(S) > 0\}$, the set of species *present* in c. If $\Delta \subseteq \Lambda$, we view a vector $c \in \mathbb{R}_{\geq 0}^{\Delta}$ equivalently as a vector $c \in \mathbb{R}_{\geq 0}^{\Lambda}$ by assuming $c(S) = 0$ for all $S \in \Lambda \backslash \Delta$.

---

[1] To approximate arbitrary continuous non-linear functions, piecewise linear functions are not sufficient, but rather we need piecewise affine functions (linear functions with offset). However, affine functions can be computed if we use an additional input fixed at 1.

Given a finite set of chemical species Λ, a *reaction* over Λ is a pair $\alpha = \langle \mathbf{r}, \mathbf{p} \rangle \in \mathbb{N}^\Lambda \times \mathbb{N}^\Lambda$, specifying the stoichiometry of the reactants and products, respectively.[2] In this paper, we assume that $\mathbf{r} \neq \mathbf{0}$, i.e., we have no reactions of the form $\varnothing \rightarrow \ldots$. For instance, given $\Lambda = \{A, B, C\}$, the reaction $A + 2B \rightarrow A + 3C$ is the pair $\langle (1, 2, 0), (1, 0, 3) \rangle$. A *(finite) chemical reaction network (CRN)* is a pair $\mathcal{C} = (\Lambda, R)$, where Λ is a finite set of chemical *species*, and R is a finite set of reactions over Λ. A *state* of a CRN $\mathcal{C} = (\Lambda, R)$ is a vector $\mathbf{c} \in \mathbb{R}_{\geq 0}^\Lambda$. Given a state $\mathbf{c}$ and reaction $\alpha = \langle \mathbf{r}, \mathbf{p} \rangle$, we say that α is *applicable* in $\mathbf{c}$ if $\lceil \mathbf{r} \rceil \subseteq \lceil \mathbf{c} \rceil$ (i.e., $\mathbf{c}$ contains positive concentration of all of the reactants).

2.2   Reachability and Stable Computation

We now follow [4] in defining rate-independent computation in terms of reachability between states (this treatment is in turn based on the notion of "stable computation" in distributed computing [1]). Intuitively, we say a state is "reachable" if some rate law can take the system to this state. For computation to be rate-independent, since unknown rate laws might take the system to any reachable state, the system must be able to reach the correct output from any such reachable state.

To define the notion of reachability, a key insight of [4] allows one to think of reachability via a sequence of straight line segments. This may be unintuitive, since mass-action[3] and other rate laws trace out smooth curves. However, a number of properties are shown which support straight-line reachability as an interpretation which includes mass-action reachability as well as reachability under other rate laws.

Let $m = |R|$ be the number of reactions in CRN $\mathcal{C}$, and let $n = |\Lambda|$ be the number of species in $\mathcal{C}$. The $n \times m$ *reaction stoichiometry matrix* $\mathbf{M}$ is such that $\mathbf{M}(i, j)$ is the net amount of the i'th species that is produced by the j'th reaction (negative if the species is consumed). We say state $\mathbf{d}$ is *straight-line reachable* from $\mathbf{c}$, written $\mathbf{c} \rightarrow^1 \mathbf{d}$, if $(\exists \mathbf{u} \in \mathbb{R}_{\geq 0}^m) \ \mathbf{c} + \mathbf{M}\mathbf{u} = \mathbf{d}$ and $\mathbf{u}(j) > 0$ only if reaction j is applicable at $\mathbf{c}$. Intuitively, a single segment means running the reactions applicable at $\mathbf{c}$ at a constant (possibly 0) rate to get from $\mathbf{c}$ to $\mathbf{d}$. We say state $\mathbf{d}$ is *l-segment reachable*, if $(\exists \mathbf{b}_1, \ldots, \mathbf{b}_{l+1}) \ \mathbf{c} = \mathbf{b}_1 \rightarrow^1 \mathbf{b}_2 \rightarrow^1 \mathbf{b}_3 \rightarrow^1 \ldots \rightarrow^1 \mathbf{b}_{l+1} = \mathbf{d}$. Generalizing to an arbitrary number of segments, we obtain our general notion of reachability below. Note that by the definition of straight-line reachability, only applicable reactions occur in each segment. The definition of reachability is closely related to exploring the "stoichiometric compatibility class" of the initial state [8].

---

[2] As we are studying CRNs whose output is independent of the reaction rates, we leave the rate constants out of the definition.

[3] Although the formal definition of mass-action kinetics is outside the scope of this paper, we remind the reader that a CRN with rate constants on each reaction define a system of ODEs under mass-action kinetics. For example, the two reactions $A + B \rightarrow A + C$ and $C + C \rightarrow B$ correspond to the following ODEs: $\dot{a} = 0$, $\dot{b} = k_2 c^2 - k_1 ab$, and $\dot{c} = k_1 ab - 2k_2 c^2$, where a, b, and c are the concentrations of species A, B, and C over time and k_1, k_2 are the rate constants of the reactions.

Definition 1. *State d is* reachable from c, *written $c \to d$, if $\exists l \in \mathbb{N}$ such that d is l-segment reachable from c for some $l \in \mathbb{N}$.*

We think of state d as being reachable from state c if there is a "reasonable" rate law that takes the system from c to d. Not surprisingly, previous work showed that if state d is reached from c via a mass-action trajectory, it is also segment-reachable.

Lemma 1 (Proven in [4]). *If d is mass-action reachable from c, then $c \to d$.*

We can now use reachability to formally define rate-independent computation. Formally, a *chemical reaction computer (CRC)* is a tuple $\mathcal{C} = (\Lambda, R, \Sigma, Y)$, where (Λ, R) is a CRN, $\Sigma \subset \Lambda$, written as $\Sigma = \{X_1, \ldots, X_n\}$, is the *set of input species*, and $Y \in \Lambda \backslash \Sigma$ is the *output species*. For simplicity, assume a canonical ordering of $\Sigma = \{X_1, \ldots, X_n\}$ so that a vector $\boldsymbol{x} \in \mathbb{R}_{\geq 0}^n$ (i.e., an input to f) can be viewed equivalently as a state $\boldsymbol{x} \in \mathbb{R}_{\geq 0}^\Sigma$ of $\mathcal{C}$ (i.e., an input to $\mathcal{C}$).

Definition 2. *A state $\boldsymbol{o} \in \mathbb{R}_{\geq 0}^\Lambda$ is* output stable *if, for all $\boldsymbol{o}'$ such that $\boldsymbol{o} \to \boldsymbol{o}'$, $\boldsymbol{o}(Y) = \boldsymbol{o}'(Y)$, i.e., once $\boldsymbol{o}$ is reached, no reactions can change the concentration of the output species Y.*

Definition 3. *Let $f : \mathbb{R}_{\geq 0}^n \to \mathbb{R}_{\geq 0}$ be a function and let $\mathcal{C}$ be a CRC. We say that $\mathcal{C}$* stably computes $\bar{f}$ *if, for all $\boldsymbol{x} \in \mathbb{R}_{\geq 0}^n$ and all $\boldsymbol{c}$ such that $\boldsymbol{x} \to \boldsymbol{c}$, there exists an output stable state $\boldsymbol{o}$ such that $\boldsymbol{c} \to \boldsymbol{o}$ and $\boldsymbol{o}(Y) = f(\boldsymbol{x})$.*

The results herein extend easily to functions $f : \mathbb{R}^n \to \mathbb{R}^l$, i.e., whose output is a vector of l real numbers. This is because such a function is equivalently l separate functions $f_i : \mathbb{R}^n \to \mathbb{R}$.

Also note that initial states contain only the input species Σ; other species must have initial concentration 0. We briefly discuss in the conclusion how allowing some initial concentration of non-input species affects computation.

2.3 Composability

We call a CRC $\mathcal{C} = (\Lambda, R, \Sigma, Y)$ *output oblivious* if the output species Y does not appear as a reactant. We now show that an output oblivious CRC is composable. For simplicity, in this section we focus on single-input, single-output CRCs, but our results can be easily generalized to multiple input and output settings.

First, we define the composition of two CRCs as the concatenation of their chemical reactions, such that the output species of the first is the input species of the second:

Definition 4. *Given two CRCs $\mathcal{C}_1 = (\Lambda_1, R_1, \Sigma_1, Y_1)$ and $\mathcal{C}_2 = (\Lambda_2, R_2, \Sigma_2, Y_2)$, consider $\mathcal{C}_2' = (\Lambda_2', R_2', \Sigma_2', Y_2')$ constructed by renaming species of $\mathcal{C}_2$ such that $\Lambda_1 \cap \Lambda_2' = \{Y_1\}$ and $Y_1 \in \Sigma_2'$. The composition of $\mathcal{C}_1$ and $\mathcal{C}_2$ is the CRC $\mathcal{C}_{2 \circ 1} = (\Lambda_1 \cup \Lambda_2', R_1 \cup R_2', \Sigma_1 \cup \Sigma_2' \backslash \{Y_1\}, Y_2')$. In other words, the composition is constructed by concatenating $\mathcal{C}_1$ and $\mathcal{C}_2$ such that their only interface is the output species of $\mathcal{C}_1$, used as the input for $\mathcal{C}_2$.*

Definition 5. *A CRC C_1 which stably computes f_1 is composable iff $\forall C_2$ stably computing f_2, $C_{2 \circ 1}$ stably computes $f_2 \circ f_1$.*

We first show that output oblivious CRCs are composable. Second, we show that if a CRC is composable then any reactions using the output species as a reactant can be removed without affecting functionality.

Lemma 2. *Output oblivious CRCs are composable.*

Proof. Consider the composition $C_{2 \circ 1}$ of two CRCs $C_1 = (\Lambda_1, R_1, \Sigma_1, Y_1)$ and $C_2 = (\Lambda_2, R_2, \Sigma_2, Y_2)$ that stably compute f_1 and f_2 respectively, and consider an input $x \in \mathbb{R}_{\geq 0}$. Consider some state c reached by $C_{2 \circ 1}$. We want to show that from c, we can reach an output stable configuration o s.t. $o(Y_2) = f_2 \circ f_1(x)$. From c, first produce the maximal amount of Y_1 possible from the reactions in C_1. Since Y_1 is the only species shared between Λ_1 and Λ_2 and Y_1 is not a reactant in any reaction of R_1, reactions from R_2 do not inhibit reactions from R_1. Thus, from c, there is a state c' that is reachable where the reactions in C_1 have produced in total $f_1(x)$ of Y_1. Again since Y_1 is the only species shared between C_1 and C_2, we can now consider $c' \upharpoonright \Lambda_2$, as C_1 cannot produce any more Y_1. Observe that $c' \upharpoonright \Lambda_2$ must have also been reachable in C_2 with an input of $f_1(x)$. We know this is true because if we undid all of the performed reactions in R_2, we would end up with a state where the only compound present in Λ_2 with a positive concentration is $f_1(x)$ of Y_1. Thus, by the definition of stable computation, there is an output stable state o s.t. $c' \upharpoonright \Lambda_2 \to o$ and $o(Y_2) = f_2(f_1(x)) = f_2 \circ f_1(x)$. Since the reactions in C_1 can no longer affect the output of $C_{2 \circ 1}$, this is also an output stable state reachable in $C_{2 \circ 1}$. Thus, $C_{2 \circ 1}$ stably computes $f_2 \circ f_1(x)$. $\square$

Lemma 3. *If a CRC C stably computes f and is composable, then we can remove all reactions where the output species appears as a reactant, and the resulting output oblivious CRC will still stably compute f.*

Proof. Assume that C_1 is a composable CRC stably computing f with the output species Y_1. Suppose we compose it with C_2, which contains a single reaction $Y_1 \to Y_2$ with Y_2 as the output species of C_2. Since C_2 stably computes the identity function, the resulting CRN $C_{2 \circ 1}$ must stably compute f. By the definition of stable computation, from any reachable state c in $C_{2 \circ 1}$, we know that there is an output stable state o such that $c \to o$ and $o(Y_2) = f(x)$. Thus we can consider any state c' reachable by running the reactions in C_1 that don't use Y_1 as a reactant until no series of those reactions can further increase the concentration of Y_1. From c', run the reaction in C_2 to convert all Y_1 into Y_2 to reach the state c. Since we have produced a maximal amount of Y_1 possible in C_1 without using the reactions that involve Y_1 as a reactant, there are no more reactions that can run to produce Y_1. Since the reactions output molecule Y_1 cannot occur (since it has a concentration of zero), we know that c must be an output stable state. Thus we know that $c(Y_2) = f(x)$. This means that any trajectory of C_1 that produces a maximal amount of Y_1 without using reactions that involve Y_1 as a reactant must produce an amount of Y_1 equal to $f(\boldsymbol{x})$. Therefore removing all such reactions from $\mathcal{C}_1$ gives us a CRN $\mathcal{C}_1'$ that also stably computes f. □

To allow composition of multiple downstream CRCs, we can use the reaction $Y \to Y_1 + \ldots + Y_n$ to generate n "copies" of the output species Y, such that each downstream module uses a different copy as input. Additionally, if the downstream module is output oblivious, then the composition is also output oblivious and thus the composition is composable. These observations allow complex compositions of modules, and will be used in our constructions in Sect. 3.2.

3   Functions Computable by Composable CRNs

Here we give a complete characterization of the functions computable by composable CRNs. First, we define exactly our notions of *superadditive*, *positive-continuous*, and *piecewise rational linear*.

Definition 6. *A function $f : \mathbb{R}^n \to \mathbb{R}^l$ is superadditive iff $\forall \boldsymbol{a}, \boldsymbol{b} \in \mathbb{R}^n, f(\boldsymbol{a}) + f(\boldsymbol{b}) \leq f(\boldsymbol{a} + \boldsymbol{b})$.*

Note that superadditivity implies monotonicity in our case, since the functions computed must be nonnegative. As an example, we show that the max function is not superadditive:

Lemma 4. *The function $\max(x_1, x_2)$ is not superadditive.*

Proof. Pick any $x_1, x_2 > 0$. Observe that $\max(x_1, 0) + \max(0, x_2) = x_1 + x_2$. But since x_1 and x_2 are both positive, we know that $x_1 + x_2 > \max(x_1, x_2)$. Thus max is not superadditive and by Lemma 6 there is no composable CRN which stably computes max. □

Definition 7. *A function $f : \mathbb{R}^n_{\geq 0} \to \mathbb{R}^l$ is positive-continuous if for all $U \subseteq [n]$, f is continuous on the domain $D_U = \{x \in \mathbb{R}^n_{\geq 0} \mid (\forall i \in [n]), x(i) > 0 \iff i \in U\}$. I.e., f is continuous on any subset $D \subset \mathbb{R}^n_{\geq 0}$ that does not have any coordinate $i \in [n]$ that takes both zero and positive values in D.*

Next we give our definition of piecewise rational linear. One may (and typically does) consider a restriction on the domains selected for the pieces, however this restriction is unnecessary in this work, particularly because the additional constraint of positive-continuity gives enough restriction.

Definition 8. *A function $f : \mathbb{R}^n \to \mathbb{R}$ is rational linear if there exists $a_1, \ldots, a_n \in \mathbb{Q}$ such that $f(x) = \sum_{i=1}^{n} a_i x(i)$. A function $f : \mathbb{R}^n \to \mathbb{R}$ is piecewise rational linear if there is a finite set of partial rational linear functions $f_1, \ldots, f_p : \mathbb{R}^n \to \mathbb{R}$ with $\bigcup_{j=1}^{p} \operatorname{dom} f_j = \mathbb{R}^n$, such that for all $j \in [p]$ and all $x \in \operatorname{dom} f_j$, $f(x) = f_j(x)$. We call $f_1, \ldots, f_p$ the components of f.*

The following is an example of a superadditive, positive-continuous, piecewise rational linear function:

$$f(x) = \begin{cases} x_1 + x_2 & x_3 > 0 \\ \min(x_1, x_2) & x_3 = 0 \end{cases} \tag{1}$$

The function is superadditive since for all input vectors $a = (a_1, a_2, a_3)$, $b = (b_1, b_2, b_3)$, there are three cases: **(1)** $a_3 = b_3 = 0$, in which case both input vectors compute min which is a superadditive function; **(2)** $a_3, b_3 \neq 0$, in which case both input vectors compute $x_1 + x_2$, which is a superadditive function; **(3)** without loss of generality, $a_3 = 0$ and $b_3 \neq 0$, in which case $f(a) + f(b) = \min(a_1, a_2) + b_1 + b_2 \leq a_1 + a_2 + b_1 + b_2 = f(a + b)$. The function is positive-continuous, since the only points of discontinuity are when x_3 changes from zero to positive. The function is piecewise rational linear, since min is piecewise rational linear.

Theorem 1. *A function $f : \mathbb{R}^n_{\geq 0} \to \mathbb{R}_{\geq 0}$ is computable by a composable CRC if and only if it is superadditive positive-continuous piecewise rational linear.*

We prove each direction of the theorem independently in Sects. 3.1 and 3.2.

3.1   Computable Functions are Superadditive Positive-Continuous Piecewise Rational Linear

Here, we prove that a stably computable function must be superadditive positive-continuous piecewise rational linear. The constraints of positive-continuity and piecewise rational linearity stem from previous work:

Lemma 5 (Proven in [4]). *If a function $f : \mathbb{R}^n_{\geq 0} \to \mathbb{R}_{\geq 0}$ is stably computable by a CRC, then f is positive-continuous piecewise rational linear.*

In addition to the constraints in the above lemma, we show in Lemma 6 that a function must be superadditive if it is stably computed by a CRC. To prove this, we first note a useful property of reachability in CRNs.

Claim. Given states a, b, c, if $a \to b$ then $a + c \to b + c$.

This claim comes from the fact that adding species cannot prevent reactions from occurring. Thus, we can consider the series of reactions where c doesn't react to reach the state $b + c$ from the state $a + c$. We now utilize this claim to prove that composably computable functions must be superadditive.

Lemma 6. *If a function $f : \mathbb{R}^n_{\geq 0} \to \mathbb{R}_{\geq 0}$ is stably computable by a composable CRC, then f is superadditive.*

Proof. Assume $\mathcal{C}$ stably computes f. By definition of $\mathcal{C}$ stably computing f, $\forall$ initial states $x_1, x_2, \exists o_1, o_2$ such that $x_1 \to o_1$ with $o_1(Y) = f(x_1)$ and $x_2 \to o_2$ with $o_2(Y) = f(x_2)$. Consider $\mathcal{C}$ on input $x_1 + x_2$. By the claim,

$x_1 + x_2 \to o_1 + x_2$, and again by the claim, $o_1 + x_2 \to o_1 + o_2$. Looking at the concentration of output species Y, we have $(o_1 + o_2)(Y) = f(x_1) + f(x_2)$. Since $\mathcal{C}$ stably computes f, there exists an output stable state o' reachable from initial state $x_1 + x_2$ and reachable from state $o_1 + o_2$, with $o'(Y) = f(x_1 + x_2)$. Since $\mathcal{C}$ is composable, species Y does not appear as a reactant and thus its concentration in any state reachable from state $o_1 + o_2$ cannot be reduced from $f(x_1) + f(x_2)$, implying $o'(Y) = f(x_1 + x_2) \geq f(x_1) + f(x_2)$. This holds for all input states x_1, x_2, and thus f is superadditive. $\qquad\square$

Corollary 1. *No composable CRC computes* $f(x_1, x_2) = \max(x_1, x_2)$.

3.2 Superadditive Positive-Continuous Piecewise Rational Linear Functions Are Computable

It was shown in [9] that every piecewise linear function can be written as a max of mins of linear functions. This fact was exploited in [4] to construct a CRN that dual-rail computed continuous piecewise rational linear functions. To directly compute a positive-continuous piecewise rational linear function, dual-rail networks were used to compute the function on each domain, take the appropriate max of mins, and then the reaction $Y^+ + Y^- \to \varnothing$ was used to convert the dual-rail output into a direct output where the output species is Y^+. However, this technique is not usable in our case: by Corollary 1, we cannot compute the max function, and the technique of converting dual-rail output to a direct output is not output oblivious. In fact, computing $f(Y^+, Y^-) = Y^+ - Y^-$ is not superadditive, and so by Lemma 6, there is no composable CRC which computes this conversion.

Since our functions are positive-continuous, we first consider domains where the function is continuous, and show that it can be computed by composing rational linear functions with min. Since rational linear functions and min can be computed without using the output species as a reactant, we achieve composability. We then extend this argument to handle discontinuities between domains.

Definition 9. *An open ray ℓ in $\mathbb{R}^n$ from the origin through a point x is the set $\ell = \{y \in \mathbb{R}^n \mid y = t \cdot x, \ t \in \mathbb{R}_{>0}\}$. Note that t is strictly positive, so the origin is not contained in ℓ.*

Definition 10. *We call a subset $D \subseteq \mathbb{R}^n$ a cone if for all $x \in \mathbb{R}^n$, we know that $x \in D$ implies the open ray from the origin through x is contained in D.*

Lemma 7. *Suppose we are given a continuous piecewise rational linear function $f : \mathbb{R}_{>0}^n \to \mathbb{R}_{\geq 0}$. Then we can choose domains for f which are cones which contain an open ball of non-zero radius.*

Intuitively, we can consider any open ray from the origin and look at the domains for f along this ray. If the ray traveled through different domains, then there must be boundary points where the function switches domains. But we know that f is continuous, so the domains must agree on their boundaries.

Since there is only one line that passes through the origin and any given point, the domains must share the same linear function to be continuous. Thus we can place the ray into one domain corresponding to its linear function. Applying this argument to all rays gives these domains as cones. This argument is formalized in the below proof.

Proof. Since f is piecewise rational linear, we can pick a finite set of domains $\mathcal{D} = \{D_i\}_{i=1}^N$ for f, such that $f|_{D_i} = g_i|_{D_i}$, where g_i is a rational linear function. Fix a domain D_k, and consider any point $\boldsymbol{x} \in D_k$. Since the open ray $\ell_{\boldsymbol{x}}$ from the origin passing through $\boldsymbol{x}$ is contained in $\mathbb{R}_{>0}^n$, it is covered by the domains in $\mathcal{D}$. If we write any point $\boldsymbol{y} \in \ell_{\boldsymbol{x}}$ in the form $t \cdot \boldsymbol{x}$, then, for each i, the restriction of g_i to $D_i \cap \ell_{\boldsymbol{x}}$ is of the form $g_i(t \cdot \boldsymbol{x}) = \alpha_i t$ for some $\alpha_i \in \mathbb{R}$. Since $\boldsymbol{x} \in D_k \cap \ell_{\boldsymbol{x}}$, we know that $f(1 \cdot \boldsymbol{x}) = \alpha_k \cdot 1 = \alpha_k$

Now suppose that for some $s \in \mathbb{R}_{>0}$ we know that $f(s \cdot \boldsymbol{x}) \neq \alpha_k s$. First consider the case where $s > 1$. Then define the set $A = \{t \in [1, s] \mid f(t \cdot \boldsymbol{x}) = \alpha_k t\}$ and define the set $B = \{t \in [1, s] \mid f(t \cdot \boldsymbol{x}) \neq \alpha_k t\}$. We know that A is non-empty since $1 \in A$, so $\sup A$ exists - call it t'. From the standard properties of the supremum, we know that there exists a sequence of points $\{t_j\}_{j=1}^\infty$ such that $t_j \in A$ for all j and $\lim_{j \to \infty} t_j = t'$. As a result, from the continuity of f, we see that:

$$f(t' \cdot \boldsymbol{x}) = \lim_{j \to \infty} f(t_j \cdot \boldsymbol{x}) = \lim_{j \to \infty} \alpha_k t_j = \alpha_k t'$$

So $t' \in A$. However, by assumption, $s \in B$, so that $t' < s$. Since t' is an upper bound on A, it must then be the case that $(t', s] \subseteq B$, so that there exists a sequence of points $\{s_j\}_{j=1}^\infty$ such that $s_j \in B$ for all j and $\lim_{j \to \infty} s_j = t'$. Since there are only finitely many domains in $\mathcal{D}$, but infinitely many s_j, by the pigeonhole principle infinitely many of the s_j must be contained in a single domain $D_{k'}$. Now write the subsequence of points contained in $D_{k'}$ as $\{s_{j'}\}_{j'=1}^\infty$. We still know that $\lim_{j' \to \infty} s_{j'} = t'$, so by the continuity of f and the fact that $s_{j'} \in D_{k'}$, we see that:

$$\alpha_k t' = f(t' \cdot \boldsymbol{x}) = \lim_{j' \to \infty} f(s_{j'} \cdot \boldsymbol{x}) = \lim_{j' \to \infty} \alpha_{k'} s_{j'} = \alpha_{k'} t'$$

Since $t' > 0$, this implies that $\alpha_{k'} = \alpha_k$, so that $f(s_{j'} \cdot \boldsymbol{x}) = \alpha_k s_{j'}$. However, this contradicts the fact that we were able to choose $s_{j'} \in B$. As a result, our assumption, that there is some $s > 1$ such that $f(s \cdot \boldsymbol{x}) \neq \alpha_k s$, must be false. A similar argument, using the infimum instead of the supremum, shows that there can be no $s < 1$ such that $f(s \cdot \boldsymbol{x}) \neq \alpha_k s$. As a result, for every point $t \in \ell_{\boldsymbol{x}}$, we know $f(t \cdot \boldsymbol{x}) = \alpha_k t$. In other words, $f|_{\ell_{\boldsymbol{x}}} = g_k|_{\ell_{\boldsymbol{x}}}$, so we can replace D_k with $D_k \cup \ell_{\boldsymbol{x}}$ without issue. Doing this for every $\boldsymbol{x} \in D_k$, we can replace D_k with a cone. By enlarging every domain in $\mathcal{D}$ in this way, we can choose domains for f which are cones.

Since f is continuous, we can replace each $D_i \in \mathcal{D}$ by its closure, which is again a cone. Suppose that for any $D_i \in \mathcal{D}$, there is a point $\boldsymbol{x} \in D_i$ is not in the interior of D_i. Then $\boldsymbol{x}$ is in the closure of the complement of D_i, so there exists a sequence $\{\boldsymbol{x}_k\}_{k=1}^\infty$ of points in the complement of D_i such that $\lim_{k \to \infty} \boldsymbol{x}_k = \boldsymbol{x}$.

Since the complement of D_i is covered by the $D_j \in \mathcal{D}$, where $j \neq i$, we know that each $\boldsymbol{x}_k$ lies in one of the D_j. Since there are only finitely many D_j but infinitely many $\boldsymbol{x}_k$, we know that infinitely many $\boldsymbol{x}_k$ must lie in at least one of the D_j. As a result, $\boldsymbol{x}$ is in the closure of this D_j, and since D_j is closed, we see that $\boldsymbol{x} \in D_j$. Because of this, if D_i has no interior points, then it is completely contained in the other D_j, so we can remove it from the set of domains. After doing this for every D_i which contains no interior points, we can ensure that the domains we have chosen for f all contain an open ball of non-zero radius.   □

Lemma 8. *Any superadditive continuous piecewise rational linear function $f : \mathbb{R}^n_{>0} \to \mathbb{R}_{\geq 0}$ can be written as the minimum of a finite number of rational linear functions g_i.*

Proof. Since f is a continuous piecewise rational linear function, by Lemma 7, we can choose domains $\{D_i\}_{i=1}^{N}$ for f which are cones and contain an open ball of non-zero radius, such that $f|_{D_i} = g_i|_{D_i}$, where g_i is a rational linear function. Now pick any $\boldsymbol{x} \in \mathbb{R}^n_{>0}$ and any g_j. Then because D_j is a cone containing an open ball of finite radius, it contains open balls with arbitrarily large radii. In particular, it contains a ball with radius greater than $|\boldsymbol{x}|$, so there exist points $\boldsymbol{y}, \boldsymbol{z} \in D_j$ such that $\boldsymbol{y} + \boldsymbol{x} = \boldsymbol{z}$. By the superadditivity of f, the linearity of g_j, and the fact that $\boldsymbol{y}, \boldsymbol{z} \in D_j$, we see:

$$g_j(\boldsymbol{y}) + f(\boldsymbol{x}) = f(\boldsymbol{y}) + f(\boldsymbol{x}) \leq f(\boldsymbol{z}) = g_j(\boldsymbol{x} + \boldsymbol{y}) = g_j(\boldsymbol{y}) + g_j(\boldsymbol{x})$$

So that $f(\boldsymbol{x}) \leq g_j(\boldsymbol{x})$. Since this is true for all g_j, and since we know that $f(\boldsymbol{x}) = g_i(\boldsymbol{x})$ for some i, we see that $f(\boldsymbol{x}) = \min_i g_i(\boldsymbol{x})$, as desired.   □

Lemma 8 is particularly useful for us since, as seen in the introduction, CRCs computing min are easy to construct, and rational linear functions are relatively straightforward as well. The next lemma gives details on constructing a CRC to compute f by piecing together CRCs which compute the components (rational linear functions) of f and then computing the min across their outputs. However, since Lemma 8 as given applies to continuous functions with domain $\mathbb{R}^n_{>0}$, so does this lemma; we handle the domain $\mathbb{R}^n_{\geq 0}$ later on.

Lemma 9. *We can construct a composable CRC that stably computes any superadditive continuous piecewise rational linear function $f : \mathbb{R}^n_{>0} \to \mathbb{R}_{\geq 0}$.*

Proof. By Lemma 8, we know that f can be written as the minimum of a finite number of rational linear functions g_i. Observe that a general rational linear function $g(\boldsymbol{x}) = a_1 x_1 + a_2 x_2 + \dots a_n x_n$ is stably computed by the reactions

$$\forall i, \ k_i X_i \to a_i k_i Y$$

where k_i is a positive integer such that $k_i a_i$ is also a positive integer. Since f is the minimum of a number of g_i's, we can make a chemical reaction network where we compute each g_i using a copy of the input species, calling the output Y_i (the reaction $X_1 \to X_1^1 + \dots + X_1^5$ produces five species with concentrations equal to

X_1's initial concentration, effectively copying the input species so that the input may be a reactant in several modules without those modules competing). Next, we use the chemical reaction

$$Y_1 + \ldots + Y_n \to Y$$

to get the minimum of the Y_i's. Since each Y_i obtains the count of the corresponding g_i, this CRN will produce the minimum of the g_i's quantity of Y's. Thus, according to Lemma 8, the described CRC stably computes f. Note that each sub-CRC described in this construction is output oblivious, and thus composable, so the composition of these modules maintains correctness.   □

The above construction only handles the domain $\mathbb{R}^n_{>0}$, where we know our functions are continuous by positive-continuity. However, when extended to the domain $\mathbb{R}^n_{\geq 0}$, positive-continuity of our functions allows discontinuity where inputs change from zero to positive. The challenge, then, is to compute the super-additive *continuous* piecewise rational linear function corresponding to which inputs are nonzero.

Surprisingly, Lemma 11 below shows that we can express a superadditive positive-continuous piecewise rational linear function as a min of superadditive *continuous piecewise rational linear functions*. The first step towards this expression is to see that, given two subspaces of inputs wherein the species present in one subspace A are a superset of the species present in a subspace B, the function as defined on the subspace A must be greater than the function as defined on the subspace B; otherwise, the function would disobey monotonicity and thus superadditivity, as proven below:

Lemma 10. *Consider any superadditive positive-continuous piecewise rational linear function $f : \mathbb{R}^n_{\geq 0} \to \mathbb{R}_{\geq 0}$. Write $N = [n]$, and for each $S \subseteq N$, let $g_S(\boldsymbol{x})$ be the superadditive continuous piecewise rational linear function that is equal to f on D_S. If $S, T \subseteq N$ and $S \subseteq T$, then for all $\boldsymbol{x} \in D_S$ we know $g_S(\boldsymbol{x}) \leq g_T(\boldsymbol{x})$.*

Proof. Write $\boldsymbol{e}_i$ for the vector of length 1 pointing in the positive direction of the ith coordinate axis. Define the vector $\boldsymbol{v} = \sum_{i \in T \setminus S} \boldsymbol{e}_i$. Then for any $\boldsymbol{x} \in D_S$ and any $\epsilon \in \mathbb{R}_{>0}$, we know that $\boldsymbol{x} + \epsilon \boldsymbol{v} \in D_T$. Since f is superadditive, it is also monotonic. Suppose that $g_T(\boldsymbol{x}) < g_S(\boldsymbol{x})$. Because g_T is continuous, taking $\delta = g_S(\boldsymbol{x}) - g_T(\boldsymbol{x}) > 0$, there is some small enough $\epsilon > 0$ such that

$$f(\boldsymbol{x} + \epsilon \boldsymbol{v}) = g_T(\boldsymbol{x} + \epsilon \boldsymbol{v}) < g_T(\boldsymbol{x}) + \delta = g_S(\boldsymbol{x}) = f(\boldsymbol{x})$$

contradicting the monotonicity of f. Our assumption must be false, so $g_S(\boldsymbol{x}) \leq g_T(\boldsymbol{x})$.   □

Next we define a predicate for each subset of inputs which is true if all inputs in that subset are nonzero. Intuitively, in the CRC construction to follow, this predicate is used by the CRC to determine which inputs are present:

Definition 11. *For any set $S \subseteq [n]$, define the S-predicate $P_S : \mathbb{R}^n_{\geq 0} \to \{0,1\}$ to be the function given by:*

$$P_S(x) = \begin{cases} 1 & x(i) > 0 \ \forall i \in S \\ 0 & \text{otherwise} \end{cases}$$

A naïve approach might be the following: for each subdomain D_S, the function is continuous, so compute it by CRC according to Lemma 9, producing an output Y_S. Then compute the P_S predicate by CRC, and if the predicate is true (e.g., a species representing P_S has nonzero concentration), use that species to catalyze a reaction which changes the Y_S to Y, the final output of the system. However, note that if T is a subset of S, P_S and and P_T are both true, so this technique will overproduce Y.

The following technique solves this issue by identifying a min which can be taken over the intermediate outputs Y_S. In particular, for each S, we compute $g_S(x) + \sum_{K \not\subseteq S} P_K(x) g_K(x)$, and then take the min of these terms. When S corresponds to the set of input species with initially nonzero concentrations, then the summation term in this expression is 0, since $P_K(x) = 0$ for all $K \not\subseteq S$. When S does not correspond to the set of input species with initially nonzero concentration, then either **(1)** it is a superset of the correct set I, in which case Lemma 10 says that $g_S(x) \geq g_I(x)$ (thus the min of these is $g_I(x)$) or **(2)** the summation term added to $g_S(x)$ contains at least $g_I(x)$, and since $g_S(x) + g_I(x) \geq g_I(x)$, the min of these is $g_I(x)$. Thus taking the min for all S of $g_S(x) + \sum_{K \not\subseteq S} P_K(x) g_K(x)$ is exactly $g_I(x)$, where I is the correct set of initially present input species.

Lemma 11. *Consider any superadditive positive-continuous piecewise rational linear function $f : \mathbb{R}^n_{\geq 0} \to \mathbb{R}_{\geq 0}$. Write $N = [n]$, and for each $S \subseteq N$, let $g_S(x)$ be the superadditive continuous piecewise rational linear function that is equal to f on D_S. Then, $f(x) = \min\limits_{S \subseteq N} [g_S(x) + \sum\limits_{K \not\subseteq S} P_K(x) g_K(x)]$.*

Proof. For $S \subseteq N$, let $h_S : \mathbb{R}^n_{\geq 0} \to \mathbb{R}_{\geq 0}$ be given by

$$h_S(x) = g_S(x) + \sum_{K \not\subseteq S} P_K(x) g_K(x)$$

We want to show that $f(x) = \min_{S \subseteq N} h_S(x)$. To do this, fix $x \in \mathbb{R}^n_{\geq 0}$ and define the set $I = \{i \in N \mid x(i) > 0\}$. First, let's show that $h_I(x) = f(x)$. By the definition of I, for all $K \not\subseteq I$, we know $P_K(x) = 0$. Thus, $\sum_{K \not\subseteq I} P_K(x) g_K(x) = 0$, so $h_I(x) = g_I(x) = f(x)$. Now we must show that $h_S(x) \geq f(x)$ for all $S \subseteq N$. There are two cases to consider:

Case 1: $S \not\supseteq I$

In this case,

$$h_S(x) = g_S(x) + \sum_{K \not\subseteq S} P_K(x) g_K(x) \geq g_S(x) + P_I(x) g_I(x) \geq P_I(x) g_I(x)$$

By the definition of I, we know $P_I(x) = 1$, so $P_I(x)g_I(x) = g_I(x) = f(x)$. Thus we get that $h_S(x) \geq f(x)$.

Case 2: $S \supseteq I$

By Lemma 10, $g_S(x) \geq g_I(x)$. As a result,

$$h_S(x) = g_S(x) + \sum_{K \not\subseteq S} P_K(x)g_K(x) \geq g_S(x) \geq g_I(x) = f(x)$$

Since for all $x \in \mathbb{R}^n_{\geq 0}$, we know $h_S(x) \geq f(x)$ for all $S \subseteq N$ and $h_I(x) = f(x)$ for some $I \subseteq N$, it follows that $f(x) = \min_{S \subseteq N} h_S(x)$.   □

Lemma 12 takes the above Lemma 11 along with the construction which stably computes on strictly continuous domains from Lemma 9 to construct a CRC which stably computes on positive-continuous domains.

Lemma 12. *Given any superadditive positive-continuous piecewise rational linear function $f : \mathbb{R}^n_{\geq 0} \to \mathbb{R}_{\geq 0}$, there exists a composable CRC which stably computes f.*

Proof. The proof follows by identifying that the function can be expressed as a composition of functions (via Lemma 11) which are computable by output oblivious CRCs and are thus composable by Lemma 2. By Lemma 11, we know that $f(x) = \min_{S \subseteq N}[g_S(x) + \sum_{K \not\subseteq S} P_K(x)g_K(x)]$. The first subroutine copies the input species, e.g. $X_1 \to X_1^1 + \ldots + X_1^5$, in order for each sub-CRC to not compete for input species. This copying is output oblivious. Then for any $Q \subseteq [n]$, $P_Q(x)$ is computed using one set of copies via the reaction:

$$\sum_{i \in Q} X_i \to P_Q$$

noting that although the predicate $P_Q(x)$ is defined to be 0 or 1, it is sufficient in this construction for the concentration of the species representing $P_Q(x)$ to be zero or nonzero. This CRC is output oblivious.

We can also compute each $g_Q(x)$ (via Lemma 9) using copies of the input molecules. This construction is output oblivious. To compute $P_Q(x)g_Q(x)$ given the concentration species P_Q as nonzero iff $P_Q(x) = 1$ as shown above, we simply compute the following (assuming Y_Q is the output of the module computing $g_Q(x)$):

$$f(P_Q, Y_Q) = \begin{cases} Y_Q & P_Q \neq 0 \\ 0 & P_Q = 0 \end{cases}$$

which is computed by this output oblivious CRC:

$$Y_Q + P_Q \to Y + P_Q$$

The CRC computing min is output oblivious, as seen in the introduction. The CRC computing the sum of its inputs is output oblivious (e.g., $X_1 \to Y, X_2 \to$

Y computes $X_1 + X_2$). Since each CRC shown is output oblivious and thus composable, we can compose the modules described to construct a CRC stably computing $\min_{S \subseteq N}[g_S(\boldsymbol{x}) + \sum_{K \nsubseteq S} P_K(\boldsymbol{x})g_K(\boldsymbol{x})]$, which is equal to $f(\boldsymbol{x})$ by Lemma 11.

$\square$

Corollary 2. *Given any superadditive positive-continuous piecewise rational linear function* $f : \mathbb{R}^n_{\geq 0} \to \mathbb{R}_{\geq 0}$, *there exists a composable CRC with reactions with at most two reactants and at most two products which stably computes* f.

To deduce this corollary, note that the reactions with more than two reactants and/or products are used to compute the following functions: computation of a rational linear function, copying inputs, min, and predicate computation. We can decompose such reactions into a set of bimolecular reactions. For example, a reaction $X_1 + \ldots + X_n \to Y_1 + \ldots + Y_n$ can be decomposed into the reactions $X_1 + X_2 \to X_{12}, X_{12} + X_3 \to X_{123}, \ldots, X_{123\ldots n-1} + X_n \to Y_{12\ldots n-1} + Y_n, Y_{12\ldots n-1} \to Y_{12\ldots n-2} + Y_{n-1}, \ldots, Y_{12} \to Y_1 + Y_2$. We can verify that each affected module stably computes correctly with these expanded systems of reactions, and remains composable.

4 Example

In this section, we demonstrate the construction presented in the previous section through an example. Consider the function shown in Eq. 1 in Sect. 3. As shown in that section, the function is superadditive, positive-continuous, and piecewise rational linear. Thus, we can apply our construction to generate a composable CRN stably computing this function. Note that while this CRN is generated from our methodology, we have removed irrelevant species and reactions.

Making copies of input:

$$X_1 \to X_1'' + X_1'''$$
$$X_2 \to X_2'' + X_2'''$$
$$X_3 \to X_3'$$

Using X_3' to make P_3, which catalyzes reactions for the domain $X_3 > 0$:

$$X_3' \to P_{\{3\}}$$

Computing the sum in $Y_{\{3\}}$:

$$X_1'' \to Y_{\{3\}}$$
$$X_2'' \to Y_{\{3\}}$$

Computing the min in $Y_\varnothing$:

$$X_1''' + X_2''' \to Y_\varnothing$$

Making a copy of $Y_{\{3\}}$ for use in increasing $Y_\varnothing'$:

$$Y_{\{3\}} \to Y_{\{3\}}' + Y_{\{3\},\varnothing}$$

Increase $Y_\varnothing'$ so that it won't be the min when x_3 is present:

$$Y_{\{3\},\varnothing} + P_{\{3\}} \to Y_\varnothing' + P_{\{3\}}$$

Rename $Y_\varnothing$ to $Y_\varnothing'$ so that it will be summed with the term created by the previous reaction:

$$Y_\varnothing \to Y_\varnothing'$$
$$Y_\varnothing' + Y_{\{3\}}' \to Y$$

5 Future Work

Instead of continuous concentrations of species, one may consider discrete counts. This changes which functions are stably computed by CRNs. Without our composability constraint, [3] shows in the discrete model that a function $f : \mathbb{N}^n \to \mathbb{N}$ is stably computable by a direct CRN if and only if it is semilinear; i.e., its graph $\{(x, y) \in \mathbb{N}^n \times \mathbb{N} \mid f(x) = y\}$ is a semilinear subset of $\mathbb{N}^n \times \mathbb{N}$. The proof that composably computable functions must be superadditive (Lemma 6) holds for the discrete model as well. So, it is now known that functions computable in the discrete model by a direct, composable CRN must be superadditive and semilinear. Surprisingly, however, there exists a function which is superadditive and semilinear which is not computable by such CRNs (the proof is non-trivial and is omitted):

$$ f(x_1, x_2) = \begin{cases} x_1 - 1 & x_1 > x_2 \\ x_1 & x_1 \le x_2. \end{cases} $$

Therefore the exact characterization of the class of computable functions for the discrete, composable case remains an open question.

In our model of a chemical reaction computer, we restrict the concentrations of non-input species in the initial state to be zero. One may consider some (constant) initial concentration of non-input species, called *initial context*, and how that may affect computation. In the non-composable case, this allows the components of the piecewise functions to be rational *affine* functions of the form $f(\boldsymbol{x}) = \sum_{i=1}^{n} a_i \boldsymbol{x}(i) + b$, where the additional b concentration of output species is produced from the initial concentration of non-output species. However, in the composable case, a function as simple as $f(x) = x - 1$ is not computable, even though it is superadditive and affine. Additionally, since $f(x) = x + 1$ is not superadditive, it also cannot be computed. Uncovering what, if any, additional power is gained from initial context in the composable case remains open.

Our negative and positive results are proven with respect to stable computation, which formalizes our intuitive notion of rate-independent computation. It is possible to strengthen our positive results to further show that our CRNs converge (as time $t \to \infty$) to the correct output from any reachable state under *mass-action kinetics* (proof omitted). It is interesting to characterize the exact class of rate laws that guarantee similar convergence.

Apart from the dual-rail convention discussed in the introduction, other input/output conventions for computation by CRNs have been studied. For example, [10] considers *fractional encoding* in the context of rate-dependent computation. As shown by dual-rail, different input and output conventions can affect the class of functions stably computable by CRNs. While using any superadditive positive continuous piecewise rational linear output convention gives us no extra computational power—since the construction in this paper shows how to compute such an output convention directly—it is unclear how these conventions change the power of rate-independent CRNs in general.

Finally we can ask what insights the study of composition of rate-independent modules gives for the more general case of rate-dependent computation. Is there a similar tradeoff between ease of composition and expressiveness for other classes of CRNs?

References

1. Angluin, D., Aspnes, J., Eisenstat, D., Ruppert, E.: The computational power of population protocols. Distrib. Comput. **20**(4), 279–304 (2007)
2. Cardelli, L.: Strand algebras for DNA computing. Nat. Comput. **10**(1), 407–428 (2011)
3. Chen, H.-L., Doty, D., Soloveichik, D.: Deterministic function computation with chemical reaction networks. Nat. Comput. **13**(4), 517–534 (2014)
4. Chen, H.-L., Doty, D., Soloveichik, D.: Rate-independent computation in continuous chemical reaction networks. In: Proceedings of the 5th Conference on Innovations in Theoretical Computer Science, pp. 313–326. ACM (2014)
5. Chen, Y.-J., et al.: Programmable chemical controllers made from DNA. Nat. Nanotechnol. **8**(10), 755 (2013)
6. Del Vecchio, D., Ninfa, A.J., Sontag, E.D.: Modular cell biology: retroactivity and insulation. Mol. Syst. Biol. **4**(1), 161 (2008)
7. Fages, F., Le Guludec, G., Bournez, O., Pouly, A.: Strong turing completeness of continuous chemical reaction networks and compilation of mixed analog-digital programs. In: Feret, J., Koeppl, H. (eds.) CMSB 2017. LNCS, vol. 10545, pp. 108–127. Springer, Cham (2017). https://doi.org/10.1007/978-3-319-67471-1_7
8. Feinberg, M., Horn, F.J.M.: Dynamics of open chemical systems and the algebraic structure of the underlying reaction network. Chem. Eng. Sci. **29**(3), 775–787 (1974)
9. Ovchinnikov, S.: Max-min representation of piecewise linear functions. Contrib. Algebra Geom. **43**(1), 297–302 (2002)
10. Salehi, S.A., Liu, X., Riedel, M.D., Parhi, K.K.: Computing mathematical functions using DNA via fractional coding. Sci. Rep. **8**(8312) 2018
11. Soloveichik, D., Seelig, G., Winfree, E.: DNA as a universal substrate for chemical kinetics. Proc. Nat. Acad. Sci. **107**(12), 5393–5398 (2010)

Tool Papers

ASSA-PBN 3.0: Analysing Context-Sensitive Probabilistic Boolean Networks

Andrzej Mizera[3,4], Jun Pang[1,2]([✉]), Hongyang Qu[5], and Qixia Yuan[1]

[1] Faculty of Science, Technology and Communication, University of Luxembourg,
Esch-sur-Alzette, Luxembourg
jun.pang@uni.lu
[2] Interdisciplinary Centre for Security, Reliability and Trust,
University of Luxembourg, Esch-sur-Alzette, Luxembourg
[3] Luxembourg Centre for Systems Biomedicine, University of Luxembourg,
Esch-sur-Alzette, Luxembourg
[4] Department of Infection and Immunity, Luxembourg Institute of Health,
Esch-sur-Alzette, Luxembourg
[5] Department of Automatic Control and Systems Engineering,
University of Sheffield, Sheffield, UK

Abstract. We present a major new release of ASSA-PBN, a software tool for modelling, simulation, and analysis of probabilistic Boolean networks (PBNs). The new version enables the support for context-sensitive PBNs (CPBNs), which can well balance the uncertainty and stability of the modelled biological systems. It contributes mainly in three aspects. Firstly, it designs a high-level language for specifying CPBNs. Secondly, it implements various simulation-based methods for simulating CPBNs and analysing their long-run dynamics. Last but not least, it provides an efficient method to identify all the attractors of a CPBN. Thanks to its divide and conquer strategy, the implemented detection algorithm can deal with large and realistic biological networks under both synchronous and asynchronous updating schemes.

1 Introduction

Probabilistic Boolean networks (PBNs) [1,2] is a modelling framework widely used to model gene regulatory networks (GRNs). It was introduced as an extension of Boolean networks (BNs) to cope with the uncertainties related to the system under study. Instead of providing a fixed Boolean function for each node, a PBN associates a set of functions for each node and, at any time point, one of the functions in the set is selected in accordance with a probability distribution to determine the value of the corresponding node. This is the original definition of a PBN, which is known as an *instantaneously random* PBN (IPBN). This definition, however, does not consider the perspective that information might come from distinct sources, each representing a "context" of a cell. To capture this perspective, another type of PBNs, called *context-sensitive* PBNs (CPBN), is introduced [3]. A CPBN represents each of the contexts as a BN, and switches

M. Češka and D. Šafránek (Eds.): CMSB 2018, LNBI 11095, pp. 277–284, 2018.
https://doi.org/10.1007/978-3-319-99429-1_16

between contexts in a stochastic manner. At each step, a CPBN can either remain in a current context, or switch to a new context with a pre-defined probability and evolve accordingly. When a CPBN is within a certain context, it will eventually settle into a set of state cycles called *attractors*. The CPBN can only leave an attractor when the context is switched or perturbations are incorporated. The attractors characterise the long-run behaviours of a CPBN and are very important for understanding the underlying networks. Comparing to IPBNs, CPBNs can better reflect the stability of biological systems. The study of attractors of a biological system are therefore more performed in terms of CPBNs instead of IPBNs. Example studies of CPBNs can be found in [4–6].

Being able to capture both the attractor information and the uncertainties of the modelled networks, CPBNs is an important and useful framework for analysing and understanding the long-run behaviours of biological networks. A few tools [7–11] have been developed for analysing PBNs or BNs. However, none of them provides intrinsic support for CPBNs. Moreover, the attractor detection functionality provided for BNs is largely restricted by the network size in existing tools. A user-friendly tool supporting attractor detection of large CPBNs is still required.

Motivations of ASSA-PBN 3.0. The previous versions of ASSA-PBN [12,13] have provided rich functionalities for modelling, simulation and analysis of IPBNs. To support efficient analysis of CPBNs, we release this new version of ASSA-PBN. The most important novelty of this version is that it supports the modelling and analysis of CPBNs and includes an important function of *attractor detection*. This functionality is implemented using efficient binary decision diagram (BDD) operations and can identify all the attractors of a given BN. Compared to similar functionalities in other tools, a significant novelty of ASSA-PBN in this respect is that it uses a strongly connected component (SCC)-based decomposition method for dealing with large CPBNs; hence it is able to handle large and realistic PBNs efficiently.

In addition, this new version contributes in another two major aspects. Firstly, it designs a high-level language to specify a CPBN in a straightforward way. Secondly, it extends the existing simulation-based functionalities for analysing the long-run dynamics of CPBNs. This includes computation of steady-state probabilities of CPBNs and parameter estimation for CPBNs. Due to the differences in the semantics of IPBNs and CPBNs, the extension not only adapts the codes with the new semantics, but also applies a few new methods to optimise the performance of the analysis. We give below an overview of the major functionalities of ASSA-PBN and highlight with bold font the new functionalities provided in the newly released version:

- a high-level modelling language for specifying both IPBNs and **CPBNs**;
- random generation of PBNs (both instantaneously random and **context-sensitive**);
- efficient simulation of PBNs (both instantaneously random and **context-sensitive**);

- computation of steady-state probabilities (for both instantaneously random and **context-sensitive** PBNs);
- parameter estimation for both instantaneously random and **context-sensitive** PBNs;
- **attractor detection for constituent BNs (both under the synchronous and the asynchronous update scheme).**

2 Preliminaries

We briefly introduce the concept of BNs and PBNs in this section. A BN models the elements of a system with binary-valued nodes. Each node is assigned with a Boolean function to reflect the relationships among the nodes. In the literature, there are various updating schemes for BNs. We introduce here two most frequently used ones, i.e. the *synchronous* and the *asynchronous* schemes.[1] In a synchronous BN, the values of all the nodes are updated according to their Boolean functions at each time step. In an asynchronous BN, the value of a single, randomly selected node is updated according to its Boolean function at each time step. The dynamics of a BN can be viewed as a discrete-time Markov chain (DTMC). Given an initial state, a BN will eventually evolve into a set of states called an attractor, inside which the BN will stay forever.

A PBN extends a BN by assigning more than one Boolean function to each node. Formally, a PBN $G(V, \mathscr{F})$ consists of a set of binary-valued nodes (also referred to as genes) $V = \{v_1, v_2, \ldots, v_n\}$ and a list of vector-valued function sets $\mathscr{F} = (F_0, F_1, \ldots, F_k)$. Each vector-valued function $F_\ell = (F_\ell^{(1)}, F_\ell^{(2)}, \cdots, F_\ell^{(n)})$ determines a constituent Boolean network, or *context*, of the PBN. The function $F_\ell^{(i)}$ is a predictor of node i whenever the context ℓ is selected. At each time step, a decision is made whether to switch context based on the switching probability q. If $q < 1$, the PBN is said to be context-sensitive; if $q = 1$, the PBN is said to be instantaneously random. If the decision is to switch context, then a context is selected according to the selection probability distribution $D = (D_0, D_1, \ldots, D_k)$, where $D_i (i \in [0, k])$ is the probability that context i is selected. The newly selected context can be the same as the current context. As long as a CPBN remains in one context, it will eventually evolve into an attractor of this context and remain within it. When the context is switched, the CPBN can move out of this attractor. Given a small switch probability, the CPBN can remain in one attractor for a long time. Therefore, it is meaningful to identify the attractors of each context. The *attractors of a CPBN* are defined to be all the attractors of its constituent BNs. We refer to [3] and [1, p. 4] for a formal definition and more detailed description of PBNs.

---

[1] For other schemes, we refer to [14].

3   Usage and New Features

Similar to its previous versions, the newly released ASSA-PBN consists of three main modules, i.e. a modeller, a simulator, and an analyser, which interact with each other to provide a full support for PBN analysis. The modeller provides support to use a high-level language for constructing a PBN model of a real-life biological system, e.g. a GRN, and supports visu-

```
1  contextDistribution = 0.1, 0.2, 0.7
2  switch = 0.001
3  node x_0
4  0,1: (!x_2)&(x_1|x_0)
5  2: (!x_0)&x_1&x_2
6  endNode
```

Fig. 1. Specifying Boolean functions for node x_0.

alisation and interactive editing of PBN predictor function values. The simulator takes the PBN constructed in the modeller and performs simulation to produce trajectory samples. Based on the constructed model and generated samples, the analyser performs basic and in-depth analysis of the PBN. The analysis results can be used either to obtain insights into the original system or to optimise the fitting of the system to experimental measurements. We proceed with describing the three main newly added features.

A High-Level CPBN Specification Language. We design a simple high-level language for specifying CPBNs. There are three main parts of a definition of a CPBN in this language: the probability distribution on contexts, the probability for making a switch at each time step, and the Boolean functions in each context. The first line in Fig. 1 shows an example for specifying the probability distribution for three contexts, and the second line shows the specification of the switch probability. We describe the Boolean functions in all contexts node by node for a node at one time and specify which context a Boolean function belongs to. The contexts of a CPBN are referred to as numbers starting from 0. Lines 3–6 of Fig. 1 show an example for specifying the Boolean functions for node x_0. In this example, node x_0 has two Boolean functions. The first function, specified in the fourth line, is shared by contexts 0 and 1; and the second function, specified in the fifth line, is only assigned to context 2. To the best of our knowledge, this language for describing context-sensitive PBNs is the first high-level language to be supported by a software tool.

An Efficient Simulator and Simulation-Based Analyses for CPBNs. Statistical approaches are practically the only viable option for the analysis of huge PBNs [15]. Applications of such methods necessitate generation of trajectories of significant length. In order to make the analysis to execute in a reasonable amount of time, ASSA-PBN has applied a number of techniques like multiple CPU/GPU core parallel generation [16,17]. In this newly released version, the simulation of a CPBN is implemented. To make the simulation as efficient as possible, we combine s, the switch probability, with $D = (D_0, D_1, \ldots, D_k)$, the probability distribution over all the contexts, to form a united distribution $\mathscr{D} = (\mathscr{D}_0, \mathscr{D}_1, \ldots, \mathscr{D}_k, \mathscr{D}_{k+1})$, where $\mathscr{D}_i = D_i * s$ for $i \in [0, k]$ and $\mathscr{D}_{k+1} = 1 - s$. The value of $\mathscr{D}_i$ ($i \in [0, k]$) represents the probability that a switch of context

Table 1. Evaluation results on five real-life biological systems. *Singleton* refers to an attractor with only one state while *cyclic* refers to an attractor with more than one state.

Networks	References	# nodes	# attractors		Time (second)		Speedup
			singleton	cyclic	HyTarjan	Decomp	
Tumour	[25]	32	9	0	0.514	0.419	**1.23**
PC12	[26]	33	7	0	0.268	0.125	**2.14**
MAPK_r3	[27]	53	20	0	5.387	2.599	**2.07**
MAPK_r4	[27]	53	16	8	11.314	1.866	**6.06**
HGF	[28]	66	18	0	17.959	3.604	**4.98**
apoptosis	[29]	97	512	512	1524.09	84.426	**18.05**
apoptosis*	[29]	97	1536	512	7926.26	183.22	**43.26**

is made and the new context is the ith one; and the value of $\mathscr{D}_{k+1}$ represents the probability that no switch is made. The switching of a context can then be performed by sampling only once from this new distribution $\mathscr{D}$, instead of sampling twice when performed based on the original distribution D and s. The simulation of a CPBN provides the foundation for further analysis of it with simulation-based methods, e.g., steady-state probability computation, long-run influence analysis, long-run sensitivity analysis, and parameter estimation for CPBNs.

Decomposition-Based Methods for Attractor Detection. Attractor detection is a completely new functionality in this version. It is specially designed for large CPBNs and BNs. Attractor detection of a CPBN or a BN is non-trivial since attractors are determined based on the network's states, the number of which is exponential in the number of nodes. A lot of efforts have been put in the development of attractor detection algorithms and tools, e.g. [7,9,10,18–20]. However, these tools are prohibited in large BNs due to the well-known state-space explosion problem, especially for asynchronous networks where there are potentially multiple outgoing transitions in each state.

In order to deal with large BNs, we apply the SCC-based decompositional methods as described in [21] (for synchronous networks) and [22] (for asynchronous networks). The idea of the methods is as follows: a BN is decomposed into sub-networks called blocks according to the SCCs in the *network structure*; then, attractor detection is performed on each of the blocks; finally, the attractors of the original network are recovered with the attractors in the blocks. Since the attractor detection is performed on blocks, the detection speed is much faster than that on the original large network. We have proved the correctness of the decomposition-based methods in [21,22], and the implementation of these methods is made publicly available through this newly released version.

The decomposition-based attractor detection methods are implemented based on BDDs, which is an efficient compressed symbolic data structure. We encode the transitions of the states in a BN, known as the transition relations, with BDDs, and implement the decomposition-based methods with efficient BDD operations. For synchronous networks, the attractor detection in each block is performed using either the monolithic or enumerative algorithms as detailed in [23]. For asynchronous networks, it is performed with an adapted hybrid Tarjan algorithm, which is the state-of-the-art BDD based SCC detecion algorithm described in Algorithm 7 of [24].

4   Case Studies

The evaluation focuses on the new functionality: attractor detection. We have evaluated our decomposition-based method for detecting attractors on both randomly generated networks and real-life biological networks in [21,22]. It has been shown that our tool is much faster than genYsis [7] in attractor detection for both synchronous and asynchronous networks. In this section, we perform an evaluation of our method on five real-life biological networks to further demonstrate its efficiency. The evaluation results are shown in Table 1. The evaluation results shown here are all based on asynchronous networks. We refer to [21] for evaluation on synchronous networks.

Due to the two variants for the MAPK model and apoptosis model, there are in total seven networks. All the seven networks are analysed under the asynchronous updating scheme. We compute the attractors for all the networks with two different methods, i.e., the hybrid Tarjan method described in Algorithm 7 of [24], and our decomposition-based method. All the experiments are conducted on a computer with an Intel(R) Xeon(R) X5675 @ 3.07 GHz CPU. We show the results of the two methods ('HyTarjan' for the hybrid Tarjan method and 'Decomp' for our decomposition-based method) in Table 1. Note that the decomposition-based method needs to perform attractor detection in each subnetwork and this is performed using the hybrid Tarjan method. The results in Table 1 clearly demonstrates that the decomposition-based method outperforms the hybrid Tarjan method. In particular, the decomposition-based method is more likely to gain a large speedup when the network size is big and the number of cyclic attractors increases.

5   Future Developments

There are a few research directions that we want to pursuit in the future. We plan to apply the technique of the satisfiability (SAT) solver to further improve the performance of attractor detection in large networks, especially for synchronous BNs. We also plan to implement a few optimisations based on network topologies to further improve the performance of attractor detection. Meanwhile, we will apply our tool for the analysis of large realistic biological networks modelled as PBNs or BNs.

Acknowledgement. Qixia Yuan was supported by the National Research Fund, Luxembourg (grant 7814267). This work was also partially supported by by the research project SEC-PBN funded by the University of Luxembourg and the ANR-FNR project AlgoReCell (INTER/ANR/15/11191283).

References

1. Trairatphisan, P., Mizera, A., Pang, J., Tantar, A.A., Schneider, J., Sauter, T.: Recent development and biomedical applications of probabilistic Boolean networks. Cell Commun. Sig. **11**, 46 (2013)
2. Shmulevich, I., Dougherty, E., Zhang, W.: From Boolean to probabilistic Boolean networks as models of genetic regulatory networks. Proc. IEEE **90**(11), 1778–1792 (2002)
3. Shmulevich, I., Dougherty, E.R.: Probabilistic Boolean Networks: The Modeling and Control of Gene Regulatory Networks. SIAM Press, Auckland (2010)
4. Pal, R., Datta, A., Bittner, M.L., Dougherty, E.R.: Intervention in context-sensitive probabilistic Boolean networks. Bioinformatics **21**(7), 1211–1218 (2004)
5. Brun, M., Dougherty, E.R., Shmulevich, I.: Steady-state probabilities for attractors in probabilistic Boolean networks. Sig. Process. **85**(10), 1993–2013 (2005)
6. Hashimoto, R.F., Stagni, H., Higa, C.H.A.: Budding yeast cell cycle modeled by context-sensitive probabilistic Boolean network. In: Proceedings of 2009 IEEE International Workshop on Genomic Signal Processing and Statistics, pp. 1–4. IEEE (2009)
7. Garg, A., Xenarios, I., Mendoza, L., DeMicheli, G.: An efficient method for dynamic analysis of gene regulatory networks and in silico gene perturbation experiments. In: Speed, T., Huang, H. (eds.) RECOMB 2007. LNCS, vol. 4453, pp. 62–76. Springer, Heidelberg (2007). https://doi.org/10.1007/978-3-540-71681-5_5
8. Müssel, C., Hopfensitz, M., Kestler, H.A.: Boolnet-an R package for generation, reconstruction and analysis of Boolean networks. Bioinformatics **26**(10), 1378–1380 (2010)
9. Dubrova, E., Teslenko, M.: A SAT-based algorithm for finding attractors in synchronous Boolean networks. IEEE/ACM Trans. Comput. Biol. Bioinform. **8**(5), 1393–1399 (2011)
10. Chaouiya, C., Naldi, A., Thieffry, D.: Logical modelling of gene regulatory networks with GINsim. In: van Helden, J., Toussaint, A., Thieffry, D. (eds.) Bacterial Molecular Networks. Methods in Molecular Biology (Methods and Protocols), vol. 804, pp. 463–479. Springer, Heidelberg (2012). https://doi.org/10.1007/978-1-61779-361-5_23
11. Paulevé, L.: Pint: a static analyzer for transient dynamics of qualitative networks with IPython interface. In: Feret, J., Koeppl, H. (eds.) CMSB 2017. LNCS, vol. 10545, pp. 309–316. Springer, Cham (2017). https://doi.org/10.1007/978-3-319-67471-1_20
12. Mizera, A., Pang, J., Yuan, Q.: ASSA-PBN: an approximate steady-state analyser of probabilistic Boolean networks. In: Finkbeiner, B., Pu, G., Zhang, L. (eds.) ATVA 2015. LNCS, vol. 9364, pp. 214–220. Springer, Cham (2015). https://doi.org/10.1007/978-3-319-24953-7_16
13. Mizera, A., Pang, J., Yuan, Q.: ASSA-PBN 2.0: a software tool for probabilistic Boolean networks. In: Bartocci, E., Lio, P., Paoletti, N. (eds.) CMSB 2016. LNCS, vol. 9859, pp. 309–315. Springer, Cham (2016). https://doi.org/10.1007/978-3-319-45177-0_19

14. Zhu, P., Han, J.: Asynchronous stochastic Boolean networks as gene network models. J. Comput. Biol. **21**(10), 771–783 (2014)
15. Mizera, A., Pang, J., Yuan, Q.: Reviving the two-state Markov chain approach. IEEE/ACM Trans. Comput. Biol. Bioinform. (2018)
16. Mizera, A., Pang, J., Yuan, Q.: GPU-accelerated steady-state computation of large probabilistic Boolean networks. In: Fränzle, M., Kapur, D., Zhan, N. (eds.) SETTA 2016. LNCS, vol. 9984, pp. 50–66. Springer, Cham (2016). https://doi.org/10.1007/978-3-319-47677-3_4
17. Mizera, A., Pang, J., Yuan, Q.: Fast simulation of probabilistic Boolean networks. In: Bartocci, E., Lio, P., Paoletti, N. (eds.) CMSB 2016. LNCS, vol. 9859, pp. 216–231. Springer, Cham (2016). https://doi.org/10.1007/978-3-319-45177-0_14
18. Naldi, A., Thieffry, D., Chaouiya, C.: Decision diagrams for the representation and analysis of logical models of genetic networks. In: Calder, M., Gilmore, S. (eds.) CMSB 2007. LNCS, vol. 4695, pp. 233–247. Springer, Heidelberg (2007). https://doi.org/10.1007/978-3-540-75140-3_16
19. Zañudo, J.G., Albert, R.: An effective network reduction approach to find the dynamical repertoire of discrete dynamic networks. Chaos Interdisc. J. Nonlinear Sci. **23**(2), 025111 (2013)
20. Klarner, H., Bockmayr, A., Siebert, H.: Computing maximal and minimal trap spaces of Boolean networks. Nat. Comput. **14**(4), 535–544 (2015)
21. Mizera, A., Pang, J., Qu, H., Yuan, Q.: A new decomposition method for attractor detection in large synchronous Boolean networks. In: Larsen, K.G., Sokolsky, O., Wang, J. (eds.) SETTA 2017. LNCS, vol. 10606, pp. 232–249. Springer, Cham (2017). https://doi.org/10.1007/978-3-319-69483-2_14
22. Mizera, A., Pang, J., Qu, H., Yuan, Q.: Taming asynchrony for attractor detection in large Boolean networks, Technical report (2017). https://arxiv.org/abs/1704.06530
23. Yuan, Q., Qu, H., Pang, J., Mizera, A.: Improving BDD-based attractor detection for synchronous Boolean networks. Sci. Chin. Inf. Sci. **59**(8), 080101 (2016)
24. Kwiatkowska, M., Parker, D., Qu, H.: Incremental quantitative verification for Markov decision processes. In: Proceedings of 41st IEEE/IFIP International Conference on Dependable Systems & Networks, pp. 359–370. IEEE (2011)
25. Cohen, D.P.A., Martignetti, L., Robine, S., Barillot, E., Zinovyev, A., Calzone, L.: Mathematical modelling of molecular pathways enabling tumour cell invasion and migration. PLoS Comput. Biol. **11**(11), e1004571 (2015)
26. Offermann, B., et al.: Boolean modeling reveals the necessity of transcriptional regulation for bistability in PC12 cell differentiation. Front. Genet. **7**, 44 (2016)
27. Grieco, L., Calzone, L., Bernard-Pierrot, I., Radvanyi, F., Kahn-Perles, B., Thieffry, D.: Integrative modelling of the influence of MAPK network on cancer cell fate decision. PLOS Comput. Biol. **9**(10), e1003286 (2013)
28. Singh, A., Nascimento, J.M., Kowar, S., Busch, H., Boerries, M.: Boolean approach to signalling pathway modelling in HGF-induced keratinocyte migration. Bioinformatics **28**(18), 495–501 (2012)
29. Schlatter, R., et al.: ON/OFF and beyond - a Boolean model of apoptosis. PLOS Comput. Biol. **5**(12), e1000595 (2009)

KaSa: A Static Analyzer for Kappa

Pierre Boutillier[1], Ferdinanda Camporesi[2], Jean Coquet[3], Jérôme Feret[2($\boxtimes$)],
Kim Quyên Lý[4], Nathalie Theret[5,6], and Pierre Vignet[6]

[1] Harvard Medical School, Boston, USA
Pierre_Boutillier@hms.harvard.edu
[2] DI-ÉNS (INRIA/ÉNS/CNRS/PSL*), Paris, France
{camporesi,feret}@ens.fr
[3] Department of Biomedical Data Science, Stanford University, Stanford, CA, USA
coquet.jean@gmail.com
[4] Tezos France, Paris, France
lykimq@gmail.com
[5] Univ Rennes, Inserm, EHESP, Irset - UMR_S1085, 35043 Rennes, France
nathalie.theret@univ-rennes1.fr
[6] Univ Rennes, Inria, CNRS, IRISA, 35000 Rennes, France
pierre.vignet@inria.fr

Abstract. KaSa is a static analyzer for Kappa models. Its goal is two-fold. Firstly, KaSa assists the modeler by warning about potential issues in the model. Secondly, KaSa may provide useful properties to check that what is implemented is what the modeler has in mind and to provide a quick overview of the model for the people who have not written it.

The cornerstone of KaSa is a fix-point engine which detects some patterns that may never occur whatever the evolution of the system may be. From this, many useful information may be collected: KaSa warns about rules that may never be applied, about potential irreversible transformations of proteins (that may not be reverted even thanks to an arbitrary number of computation steps) and about the potential formation of unbounded molecular compounds. Lastly, KaSa detects potential influences (activation/inhibition relation) between rules.

In this paper, we illustrate the main features of KaSa on a model of the extracellular activation of the transforming growth factor, TGF-b.

1 Introduction

Kappa may be used to describe systems of mechanistic interactions between proteins by the means of site-graph rewriting rules. Each node in graphs denotes

This material is based upon works partially sponsored by ANR (Chair of Excellence AbstractCell), the Defense Advanced Research Projects Agency (DARPA) and the U. S. Army Research Office under grant number W911NF-14-1-0367, and by the ITMO Plan Cancer 2014 (TGFSysBio project). The views, opinions, and/or findings contained in this article are those of the authors and should not be interpreted as representing the official views or policies, either expressed or implied, of ANR, DARPA, the U. S. Department of Defense, or ITMO.

© Springer Nature Switzerland AG 2018
M. Češka and D. Šafránek (Eds.): CMSB 2018, LNBI 11095, pp. 285–291, 2018.
https://doi.org/10.1007/978-3-319-99429-1_17

Fig. 1. Two rules. (left) Two proteins may bind. (right) The protein on the left may phosphorylate the right site of the protein on the right.

an instance of a protein equipped with a kind and a finite set of identified sites. Rules may bind/unbind sites pair-wisely to establish/break links between proteins. Some sites may also have an internal state in order to specify if they are phosphorylated, methylated, and so on, so forth. We give, in Fig. 1, two examples of rules in Kappa.

Kappa is context-free: only the information that matters for a given interaction to happen has to be mentioned in rules. This feature is crucial to scale up to the size of large models. Thus Kappa provides the opportunity to design arbitrarily sophisticated models. These models may involve proteins with multiple phosphorylation sites, scaffolds, concurrency for shared resources, different time- and concentration scales, large variabilities in the kinds of molecular compounds, and non-linear feed-back loops. In general, we want to understand how the collective behavior of proteins may emerge from the mechanistic interactions between individual proteins. Yet, there are no modeling wizards and before investigating the long term behavior of a model, it is worth wondering whether the implementation matches faithfully our modeling assumptions. In the case of models written by others, extracting quickly some basic properties about models is also helpful to understand what the models are doing.

This motivates the use of formal methods. KaSa is a static analyzer that abstracts the set of reachable states of models, and then uses this information to collect insightful properties. In particular, KaSa may warn about rules that may never be applied, about potential definitive transformations of proteins and about the potential formation of unbounded molecular compounds. Lastly, KaSa detects the potential influences (activation/inhibition relation) between rules.

In this paper, we illustrate the main features of KaSa on a model of the extracellular activation of the transforming growth factor TGF-b, a protein which controls cell homeostasis in normal tissue, but promotes the development of fibrosis and cancer [6]. There has been a nice interplay between the design of the static analysis and the one of this model. On the first hand, KaSa has been helpful to curate the model, on the second hand, we have extended KaSa to cope with new properties of interest that we have identified during the modeling process.

2   Technical Description

Development. The development of KaSa has started in 2006, as a follow up of Complex, a static analyzer that had been designed by Plectix BioSystems (Cambridge, MA, USA). KaSa is now around 68,000 lines of OCaml [7] (excluding the front-end). It offers 53 command-line options. Jérôme Feret (2010-present) and Kim Quyên Lý (2015–2017) have been being the main developers.

Distribution. KaSa belongs to the Kappa modeling platform, which is completely open source www.kappalanguage.org. KaSa is partially integrated within the Kappa user interface, In particular, all the functionalities that are described in this paper, but local traces, are available on the fly while editing a model.

The development of the modeling platform is hosted on github https://github.com/Kappa-Dev/KaSim. An app is provided for MacOs and Windows. The nightly-builds of the development version may be downloaded at https://tools.kappalanguage.org/nightly-builds/. The modeling platform is also available as an opam package. With a properly installed opam, the instruction `opam pin add --dev KaSim` will compile all necessary dependencies as well as the current master branch of the git repository.

The manual may be consulted online at: https://tools.kappalanguage.org/docs/KaSim-manual-master/KaSim_manual.htm (see Chap. 6).

3 Main Functionalities

Now we browse the main functionalities of KaSa.

Note that the results computed by KaSa depend all on the choice of the initial state, or more precisely on the set of the proteins and molecular compounds that may be present in the initial state independently of their concentration. KaSa is purely qualitative: its results depend neither on rule rates, nor on initial concentrations.

Reachability Analysis. The cornerstone of KaSa is its reachability analysis. KaSa performs a mutual induction over some families of patterns, so as to prove that some of them may never occur in reachable states. Three families of patterns are considered [3]. The first one detects relations among the state of sites within each protein instance. The second one targets the relations between the state of sites in the proteins that are directly linked. The third one focuses on detecting whether or not a protein may be bound twice to the same instance of a protein. KaSa outputs a list of refinement lemmas. Each one consists in a precondition, that is a pattern, and a post-condition, that is a list of refinements of this pattern. The formal meaning of a refinement lemma is that whenever an instance of the precondition is found in a reachable state, this instance may be extended to an instance of a pattern in the post-condition.

In Fig. 2, we give some of the properties that are found in our case study. The analysis infers that in its latent form, *TGFB1* has always its site *a* (in pink) free. KaSa also detects that *TGFB1* may be bound twice to the same instance of the protein *THBS1*, but never to different instances simultaneously.

Dead Rule Detection. A rule the left hand side of which is in contradiction with the refinement lemmas cannot be applied whatever the evolution of the system is. There may be various reasons for this. Sometimes, several names have been used to denote the same protein. Sometimes, proteins have structural invariants that prevents the application of a rule. In our case study, dead rules have helped in

Fig. 2. Two refinement lemmas. (left) When TGFB1 is in its latent form, its site a is necessarily free. (right) When *TGFB1* has its two sites bound, it is bound twice to the same instance of the protein *THSBS1*. In each of these refinement lemmas, the refinement list is made of a single element. In more complicated cases, there maybe a choice of several patterns for refining the precondition.

identifying some missing parts in models, hence blocking the signaling pathways. The model has been completed after having consulted the literature.

Influences Among Rules. Rules may have a positive or a negative influence on each others. There is a positive (resp. negative) influence when an application of a given rule may potentially create (resp. remove) an instance of the left hand side of another rule. Influences provide an overview of the causality of the model.

We give an example in Fig. 3. We consider a protein with two phosphorylation sites. The left site may be freely phosphorylated and dephosphorylated, whereas the right site may get phosphorylated only when the left one is already phosphorylated. Thus the phosphorylation of the left site has a positive influence on the phosphorylation of the right one, while the dephosphorylation of the left site has a negative influence on the phosphorylation of the right one, as indicated in the influence map. This notion of influence is similar to the one that is used in Gene regulatory tools such as GinSIM [8] or reaction networks tools such as Biocham [2], except that, in Kappa, influences describe to which extent rules may influence each other, and not whether the variation of concentration of each molecular compound may influence the concentration of the other ones.

Since there are many rules, we use a hierarchy of abstractions to avoid the brute force approach which may not scale to large models. Firstly, we compute indirect influences. Indirect influences focus on the states of sites independently. There is a positive indirect influence whenever a rule may take a site into a state that is required by another rule to apply. Secondly, we compute direct influences. Direct influences are obtained by filtering indirect influences by checking that both rules have compatible requirements about their context of application. Thirdly, we refine direct influences further, by checking that the unifying context of both rules cannot be proved unreachable by our reachability analysis.

Local Transition Systems. It is sometimes useful to understand how a protein may go from one configuration to another. Thus KaSa computes a transition system for each kind of proteins of the model [4]. This abstraction completely ignores the context of the protein: the behavior of each protein is described independently without considering the state of the proteins it is attached to. Local traces do not intend to provide information about the collective behavior of proteins (i. e. their concentration): instead it focus on each protein individually.

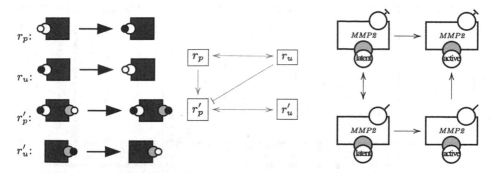

Fig. 3. Four rules (left) and the corresponding influence map (right). In rule r'_p, the right site may be phosphorylated only if the left site is. As a consequence the rule r_u inhibits the rule r'_p, and the rule r_p activates the rule r'_p.

Fig. 4. A local trace. In protein *MMP2*, the passage in the active form is definitive and it prevents the site *CDB* to bind the protein *COL*.

Non Weakly Reversible Transitions. Most mechanisms may be reverted in one or more steps of computation. This is crucial so that resources may be used several times (for instance an enzyme is expected to activate several instances of its substrate, thus it has to detach from it). However modelers often see signaling as cascades of interactions that push forward the signal.

Tarjan's strongly connected components decomposition algorithm is useful to detect which computation steps will never be reverted. Often, non weakly reversible transitions come from missing mechanisms and the model has to be completed. Sometimes, they come from a definitive degradation of a protein. In this case, the property is helpful to understand the behavior of this protein.

In the first versions of our case study, most rules about unbinding were missing. Our analysis has detected that corresponding binding steps were definitive and the model had to be completed accordingly. Once this done, all the remaining definitive transitions are related to the activation of the proteins *TGFB1*, *MMP2*, and *MMP14*, which is an irreversible process. In Fig. 4, we give a local trace associated to the protein *MMP2*.

Detection of Unbounded Polymers. Knowing which complexes may grow arbitrarily is important. Some models may assemble macro-molecules. But sometimes the presence of unbounded polymers is a side-effect of the lack of specification of the potential conflicts between protein interactions.

Unbounded polymers may only arise whenever a sequence of proteins may be repeated indefinitely in a reachable molecular compound. Such a sequence necessarily matches with a cycle in the oriented graph in which nodes are the different kinds of bonds between proteins (each kind bond is considered twice, one for each direction) and the edges connect two (oriented) bonds if the target of the first bond and the source of the second one are two different sites in a same kind of protein. We also use Tarjan's algorithm to detect these cycles.

model	rules	constraints	dead rules	non weakly reversible transitions	rules with non weakly reversible transitions	indirect influences	direct influences	realisable influences	strongly connected components (Syntactic)	strongly connected components (Realisable)	edges per scc in average (Syntactic)	edges per scc in average (Realisable)	reachability analysis	influence map (Indirect)	influence map (Direct)	influence map (Realisable)	polymer detection (Syntactic)	polymer detection (Realisable)	non weakly reversible transitions	local traces	all-in-one (at maximal accuracy)
egfr_net	39	17	0	0	0	764	280	280	0	0	N/A	N/A	0.02	0.04	0.02	0.04	0.02	0.02	0.03	0.02	0.05
fceri_fyn	46	21	0	8	8	758	304	304	1	0	8	N/A	0.04	0.05	0.03	0.10	0.04	0.04	0.06	0.04	0.10
fceri_fyn_lig	48	21	0	8	8	760	306	306	1	0	8	N/A	0.04	0.05	0.03	0.09	0.04	0.05	0.06	0.05	0.10
fceri_fyn_trimer	362	22	36	96	96	59971	7405	6763	1	0	10	N/A	0.53	4.63	0.66	2.15	0.54	0.54	0.58	0.54	2.24
fceri_fyn_gamma2	59	21	0	0	0	1464	518	518	1	0	8	N/A	0.06	0.10	0.04	0.15	0.06	0.06	0.08	0.06	0.16
fceri_fyn_ji	36	16	0	0	0	536	231	231	1	0	8	N/A	0.03	0.02	0.01	0.06	0.03	0.03	0.05	0.03	0.07
fceri_fyn_lyn_745	40	18	2	2	2	620	255	243	1	0	8	N/A	0.04	0.04	0.02	0.07	0.04	0.04	0.05	0.04	0.08
fceri_fyn_trimer	192	19	0	0	0	21557	2536	2536	1	0	10	N/A	0.24	1.60	0.24	0.81	0.24	0.24	0.27	0.24	0.86
machine	220	72	7	17	10	5319	2873	2735	0	0	N/A	N/A	0.77	0.13	0.10	1.05	0.76	0.77	0.97	0.77	1.22
ensemble	233	86	0	1	1	4841	2936	2936	0	0	N/A	N/A	0.62	0.15	0.13	0.82	0.61	0.63	0.91	0.62	1.14
korkut (2017/01/13)	3916	1289	1610	2016	2016	75563	75563	39280	1	1	131	131	14	2.49	2.71	16	14	14	14	14	18
korkut (2017/02/06)	5750	2571	884	1397	1397	81412	75472	55101	1	1	2693	2687	94	4.01	4.16	94	96	115	99	114	119
TGF (V19)	97	107	10	153	53	3471	3009	2631	1	1	78	74	0.24	0.09	0.09	0.47	0.23	0.25	0.37	0.25	0.63
TGF (2018/04/19)	292	112	0	314	28	6040	5504	5504	1	1	108	108	0.89	0.18	0.19	1.36	0.86	0.90	1.25	0.92	1.73
BigWnt (2015/12/28)	356	134	1	833	14	5974	5271	5264	1	1	49	49	3.99	0.16	0.16	4.47	3.98	3.96	125	4.00	127
BigWnt (2017/03/22)	1486	182	12	61	16	1091187	38110	37958	1	1	84	80	15	26	5.15	25	15	15	260	15	286

Fig. 5. Benchmarks (performed on a MacBook Pro, 3.3 Ghz intel Core). For each model, we provide the number of rules, information about what has been discovered by KaSa and about analysis time.

This feature is available at two accuracy levels. At syntactic level, every kind of bonds occurring in the initial state or in the right hand side of a rule is considered. A more precise analysis is obtained by filtering out the pairs of bonds for which the corresponding pattern is proved unreachable.

In our case study, KaSa detects a large strongly collected component related to the formation of the Fibronectin matrix (which may indeed grow arbitrarily).

4   Benchmarks

We apply KaSa to several Kappa models (e. g. see Fig. 5). The first eight models are translations in Kappa of some of the models which are provided with the BNGL distribution [1]. The models 'machine' and 'ensemble' are two versions of the MAPK signaling pathways published by Eric Deeds and Ryan Suderman [9]. Both versions of the model 'korkut' describe the Ras signaling pathways. They have been assembled by John Bachman and Benjamin Gyori (Sorger lab, BigMechanisms DARPA Project), following a three steps procedure [5]: automatic natural language processing, automatic assembling into Kappa, and human curation. We analyze two versions of the model of the extracellular matrix of the protein TGF-b that we have used to illustrate the different functionalities of KaSa. The assembling has been done by hand, by inspection of the literature and its curation has been assisted by KaSa. Lastly, we analyze two versions of the Wnt signaling pathway, written by Héctor F. Medina

Abarca (Fontana Lab, Big-Mechanisms DARPA Project). This model has also been assembled by hand by inspection of the literature. Some scripts have been used to refine the kinetics of rules according to some contextual information about the proteins.

For each model, we give the number of constraints, of detected dead rules, of detected non weakly reversible transitions, of the rules that are involved in those transitions, of potential influence relation (in each accuracy mode), of strongly connected components that may occur in polymers (in both modes). Then, we give the CPU time used for each of these functionalities. The last column gives the CPU time of the whole analysis, with all functionalities set to the maximal accuracy level. It is worth noting that the CPU time required to compute the direct influence map, is sometimes longer than the one to compute the indirect one. Indeed, dumping an imprecise result may take longer than filtering this result thanks to a more costly but more accurate analysis.

Contribution. Jérôme Feret (2010-present) and Kim Quyên Lý (2015–2017) are the main contributors of KaSa. KaSa is integrated within the Kappa modeling platform whose main architect is Pierre Boutillier. In particular, Pierre Boutillier has integrated KaSa in the user interface of Kappa which may be used either online, or locally. The model of the extracellular activation of the transforming growth factor, TGF-b, has been assembled by Jean Coquet (2012–2017), Nathalie Théret (2012-present), Pierre Vignet (2016-present), and Ferdinanda Camporesi (2016-present). Jérôme Feret has written the paper.

References

1. Blinov, M., et al.: BioNetGen: software for rule-based modeling of signal transduction based on the interactions of molecular domains. Bioinformatics **20**(17), 3289–3291 (2004)
2. Fages, F., Soliman, S.: From reaction models to influence graphs and back: a theorem. In: Fisher, J. (ed.) FMSB 2008. LNCS, vol. 5054, pp. 90–102. Springer, Heidelberg (2008). https://doi.org/10.1007/978-3-540-68413-8_7
3. Feret, J., Lý, K.: Reachability analysis via orthogonal sets of patterns. Electron. Notes Theor. Comput. Sci. **335**, 27–48 (2018)
4. Feret, J., Lý, K.: Local traces: an over-approximation of the behaviour of the proteins in rule-based models. IEEE/ACM TCBB (2018)
5. Gyori, B., et al.: From word models to executable models of signaling networks using automated assembly. bioRxiv (2017)
6. Horiguchi, M., Ota, M., Rifkin, D.: Matrix control of transforming growth factor-β function. J. Biochemistry **152**(4), 321–329 (2012)
7. Leroy, X., Doligez, D., Frisch, A., Garrigue, J., Rémy, D., Vouillon, J.: The OCaml system (2017). Release 4.06
8. Naldi, A., Berenguier, D., Fauré, A., Lopez, F., Thieffry, D., Chaouiya, C.: Logical modelling of regulatory networks with ginsim 2.3. Biosystems **97**(2), 134–139 (2009)
9. Suderman, R., Deeds, E.: Machines vs. ensembles: effective MAPK signaling through heterogeneous sets of protein complexes. PLoS Comput. Biol. **9**(10), e1003278 (2013)

On Robustness Computation
and Optimization in BIOCHAM-4

François Fages[✉] and Sylvain Soliman

Inria Saclay - Ile de France, Lifeware Group, Palaiseau, France
{Francois.Fages,Sylvain.Soliman}@inria.fr

Abstract. BIOCHAM-4 is a tool for modeling, analyzing and synthesizing biochemical reaction networks with respect to some formal, yet
possibly imprecise, specification of their behavior. We focus here on one
new capability of this tool to optimize the robustness of a parametric
model with respect to a specification of its dynamics in quantitative
temporal logic. More precisely, we present two complementary notions
of robustness: the statistical notion of model robustness to parameter
perturbations, defined as its mean functionality, and a metric notion of
formula satisfaction robustness, defined as the penetration depth in the
validity domain of the temporal logic constraints. We show how the formula robustness can be used in BIOCHAM-4 with no extra cost as an
objective function in the parameter optimization procedure, to actually
improve the model robustness. We illustrate these unique features with
a classical example of the hybrid systems community and provide some
performance figures on a model of MAPK signalling with 37 parameters.

1 Introduction

Computational systems biology aims at gaining a system level understanding
of high-level biological processes from their biochemical realm. Formal methods
from Computer Science have been soon introduced at the heart of this effort to
go beyond mathematical modeling and master the complexity of cell processes.
In particular, model-checking techniques have been used to analyze Boolean
gene regulatory networks and signaling or control protein reaction networks.
They have also been generalized, in particular in the hybrid systems community, to quantitative temporal logics such as MTL, MITL, STL, or in our case
FO-LTL($\mathbb{R}_{\text{lin}}$) [14] for dealing with numerical constraints, fitting quantitative
models to the time series data observed with increasing accuracy in biological
experiments [9,16], controling cell processes in real-time, or designing circuits in
synthetic biology [1,13].

One striking feature of natural biological processes is their robustness with
respect to both external perturbations and their intrinsic stochasticity. Measuring and optimizing robustness is thus an important topic in biological modeling
and a useful yet rare feature of modeling tools in this domain. In [11], Kitano
gives a definition of robustness of a biological system with respect to a dynamical property and a parameter perturbation law, as the mean functionality of the

© Springer Nature Switzerland AG 2018
M. Češka and D. Šafránek (Eds.): CMSB 2018, LNBI 11095, pp. 292–299, 2018.
https://doi.org/10.1007/978-3-319-99429-1_18

system. Such a statistical definition of robustness can be evaluated in a computational model by sampling the parameters' space according to their distribution laws, and checking the property on simulation traces, i.e. by runtime verification or monitoring. Since the capability of generating simulation traces is the only requirement, this can be done in a very general setting of non-linear hybrid systems.

In [7,14], different notions of robustness are proposed with the definition of a continuous degree of satisfaction of a temporal logic property on a trace. Such a satisfaction degree gives quantitative information on the satisfaction of the formula, as the distance to, and penetration depth within, the validity domain of the formula. In the simplest example of a threshold formula $x < c$, that satisfaction degree can be defined for instance as the value $c - x$, with negative values representing distance to satisfaction, and positive values the margins achieved for robust satisfaction.

In this paper, we present some unique features of BIOCHAM-4[1] for defining hybrid systems by reaction networks with rates and events, and focus on the use of quantitative temporal logic language to specify the dynamical properties of the system, observed or wished, verify them in the model, compute parameter sensitivity indices, measure robustness, synthesize model parameters and optimize robustness. We deal with the two complementary notions of robustness mentioned above, i.e. the statistical notion of model robustness to parameter change, and the formula satisfaction robustness. We show how the formula robustness can be used in BIOCHAM-4 with no extra cost as an objective function in the parameter optimization procedure to actually improve the robustness of the model with respect to parameter perturbations. We illustrate these features with a classical example of non-linear hybrid system, the bouncing ball example, and provide some performance figures on the Huang and Ferrell's model of MAPK signaling [10] with 37 parameters.

2   BIOCHAM Models

A BIOCHAM model is composed of a (multi)set of reactions with rate functions, and/or influences with forces, plus possibly events. Such models can be interpreted in a hierarchy of differential, stochastic, Petri net and Boolean semantics [4]. We will focus here on the differential semantics. They can be imported from model repositories such as BioModels using the BIOCHAM interface to SBML (SBML-qual for influence models), or from ODE models [3] through an interface to the XPP format.

This article comes with a set of examples (MAPK signaling, cell cycle, bouncing ball) available online under the form of BIOCHAM-4 Jupyter notebooks[2].

---

[1] http://lifeware.inria.fr/biocham4.
[2] https://lifeware.inria.fr/wiki/Main/Software#CMSB18.

3   Behavior Specifications in FO-LTL($\mathbb{R}_{\text{lin}}$)

3.1   Validity Domains of FO-LTL($\mathbb{R}_{\text{lin}}$) Constraints

In BIOCHAM, quantitative temporal properties of the behavior of a system can be formally specified in a first-order version of Linear Time Logic with free variables, linear constraints over $\mathbb{R}$, and quantifiers, named FO-LTL($\mathbb{R}_{\text{lin}}$). The grammar is:

$$\phi ::= c \mid \neg\phi \mid \phi \Rightarrow \psi \mid \phi \wedge \phi \mid \phi \vee \phi \mid \exists x\ \phi \mid \forall x\ \phi \mid \mathbf{X}\phi \mid \mathbf{F}\phi \mid \mathbf{G}\phi \mid \phi\mathbf{U}\phi \mid \phi\mathbf{W}\phi$$

where c denotes linear constraints over state variables (including *Time*) and free variables. The *validity domain* $\mathcal{D}_{\pi,\phi} \in \mathbb{R}^k$ of an FO-LTL($\mathbb{R}_{\text{lin}}$) formula ϕ containing $k \geq 1$ variables on a finite trace $\pi = (s_0, \ldots, s_n)$ can be defined by recursively bottom-up from the validity domain of the basic constraints to the complex FO-LTL($\mathbb{R}_{\text{lin}}$) formulae by intersection, union and complementation [5]. An FO-LTL($\mathbb{R}_{\text{lin}}$) formula is false if one validity domain is empty, valid if the validity domains of all variables are $\mathbb{R}$, and satisfiable otherwise.

For instance, the formula $\mathbf{F}(A \geq 0.2)$ where A is a state variable expresses that the concentration of molecule A gets greater than 0.2 at some time point in the trace ($\mathbf{F}$). If needed, the precise time values where the concentration of A gets greater than the threshold value can be expressed by introducing a free variable t with an equality constraint to the real time variable, $\mathbf{F}(A \geq 0.2 \wedge t = \text{Time})$. Constraints between time variables can then relate the time of different events. The maximum value of a state variable A can be specified and set in a variable v by the formula $\mathbf{G}(A \leq v) \wedge \mathbf{F}(A \geq v)$, a local maximum (or plateau) by $\mathbf{F}(A \leq v \wedge \mathbf{X}(A = v \wedge \mathbf{X}A \leq v))$. FO-LTL($\mathbb{R}_{\text{lin}}$) formulae are very expressive and can be used to define complex oscillation properties, with pseudo-period constraints defined by delay constraints between the local maxima, and phase constraints defined by delays between the peaks of different state variables [6].

3.2   Implementation

The recursive definition of the validity domain of an FO-LTL($\mathbb{R}_{\text{lin}}$) formula on a finite trace is implemented in BIOCHAM-4 by generating a C program that implements the necessary loops, starting from the last time point to the first, and considering the subformulae in the bottom-up order, i.e. first from the linear constraints at the leaves, to the root of the syntactic tree. The call to the C compiler is responsible for slower response time than the previous interpreted implementation on small examples, but faster on large examples. The Parma Polyhedra Library (PPL)[3] is used to solve the linear constraints and simplify the representation of validity domains by finite lists of polyhedra.

Bound constraints, i.e. constraints of the form $x \leq c$ or $x \geq c$ where x is a variable and c a constant, define boxes as a particular kind of polyhedra. In that case, the validity domains are finite union domains of boxes, since they

---

[3] http://www.bugseng.com/ppl.

are obtained by intersection, union, complementation and projection of boxes. However, it is worth noticing that even in the case of bound constraints, the validity domain of a temporal formula can contain an exponential number of polyhedra in the number of free variables in the FO-LTL($\mathbb{R}_{\text{lin}}$) formula [5]. The issue of trace simplification, such as keeping in the trace only the time points that are local extrema for one state variable, is also important to reduce computation time and justified in a number of practical cases [15].

4   Formula Robustness and Satisfaction Degree

In [13], the continuous satisfaction degree in the interval $[0, 1]$, of an FO-LTL($\mathbb{R}_{\text{lin}}$) formula ϕ on a trace π, was defined as the distance between the validity domain of the free variables $x_1, \ldots, x_k$ in ϕ and some objective values in $\mathbb{R}^k$ given as a valuation σ of the variables $x_1, \ldots, x_k$, i.e. a vector $\boldsymbol{v}_\sigma = (v_1, \ldots, v_k)$. This definition enforces that the violation degree is in $[0, +\infty)$ (0 when the formula $\sigma(\phi)$ is satisfied) and the satisfaction degree in $(0, 1]$ (1 when the formula is satisfied). Those notions can be generalized in order to take into account how *robustly* a formula is satisfied, by taking into account the penetration depth of the objective values in the validity domain, similarly to the space-time robustness defined in [2] for STL:

Definition 1. *The* violation degree $vd(\pi, \phi, \sigma) \in (-1, +\infty)$ *of an FO-LTL($\mathbb{R}_{lin}$) formula ϕ in a numerical trace π with respect to an objective valuation σ is*

$$\min_{v \in \mathcal{D}_{\pi, \phi}} d(v, v_\sigma) \text{ if } v_\sigma \notin \mathcal{D}_{\pi, \phi}, \quad \frac{1}{1 + \min_{v \notin \mathcal{D}_{\pi, \phi}} d(v, v_\sigma)} - 1 \text{ if } v_\sigma \in \mathcal{D}_{\pi, \phi}.$$

The satisfaction degree $sd(\pi, \phi, \sigma) \in (0, +\infty)$ *of ϕ in π w.r.t. σ is* $\dfrac{1}{1 + vd(\pi, \phi, \sigma)}$

The first case is the same as [14], i.e. the distance in $[0, +\infty)$ between $\boldsymbol{v}_\sigma$ and the domain $\mathcal{D}_{\pi, \phi}$. In the second case, we get a negative number related to the distance between $\boldsymbol{v}_\sigma$ and the outside of the domain $\mathcal{D}_{\pi, \phi}$. The satisfaction degree is defined as in [14] but now provides a notion of formula robustness when its value is greater than 1. Indeed $sd(\pi, \phi, \sigma)$ is bounded by the radius of $\mathcal{D}_{\pi, \phi}$ and describes in the space of the variables $x_1, \ldots, x_k$ by how much one can change the objective $\boldsymbol{v}_\sigma$ while keeping $\sigma(\phi)$ satisfied on π.

In our implementation in BIOCHAM-4, the distance to, and penetration depth within, validity domains are not computed exactly, but approximated using the notions of generators and constraints in PPL. This is indeed sufficient to guide the search of parameter values and optimize formula robustness, using the Covariance Matrix Adaptive Evolutionary Strategy CMA-ES [8] as black-box continuous optimization tool with satisfaction degree as objective function.

5   Model Robustness and Parameter Sensitivity

The more classical notion of *model robustness* defined in [11] as the mean functionality with respect to a set of parameter perturbations P can be instanciated in our setting for a FO-LTL($\mathbb{R}_{lin}$) formula ϕ and some objective σ as

$$R_{P,\phi,\sigma} = \int_{p \in P} \min(1, sd(\pi(p), \phi, \sigma))prob(p)dp$$

In BIOCHAM this integral is evaluated by sampling (log-)normally-distributed parameter values given by the user. This sampling stops when a user-given number of samples is reached, or when the relative sample-standard-deviation becomes smaller than a given threshold. However, the computation may be computationally expensive with tens of numerical simulations needed and is generally not usable inside a search for model parameters.

6   BIOCHAM Commands

The bouncing ball is a classical example of the hybrid systems community with which we can illustrate our approach. One single simulation is already quite time consuming (about 2 s) because of the checking of the bouncing event condition. The FO-LTL($\mathbb{R}_{lin}$) formula F(x<h / F(x>h)) gives the height of the first bounce in the free variable h ($\sim$4 in the companion notebook). Computing the validity domain with the command

```
validity_domain(F(x<h / F(x>h)))
```

and satisfaction degree

```
satisfaction_degree(F(x<h / F(x>h)), [h -> 3.5])
```

on such a simulation trace is slightly faster (about 0.1 s). Now, the command

```
robustness(F(x<h / F(x>h)), [x0, K, D], [h->3.5]).
```

computes a value (0.884 in 160 s) that quantifies how robustly is satisfied the formula with an objective for h of 3.5 when parameters x_0, K, D are perturbed (with default Gaussian distribution). On the other hand, the formula robustness necessitates a single simulation to get the value $sd(\pi, \phi, \sigma) = 1.4664$. This low computational cost makes it possible to use formula robustness as an optimization criterion within the parameter search procedure. One can ask BIOCHAM-4 to search for parameter values such that the violation degree vd is below -0.5 (i.e., $sd \geq 2$):

```
search_parameters(F(x<h /\ F(x>h)), [5<=x0<=10, 0.5<=K<=1.0, -0.1<=D<=0.0], [h
->3.5], cmaes_stop_fitness: -0.5).
```

BIOCHAM-4 finds in $2.5\,s$ a solution with $x_0 = 9.555, K = 0.801, D = -0.0585$ such that $vd = -0.572$ and $sd = 2.339$. Computing again the model robustness as before we now get (in $101\,s$) an improved model robustness of 0.930.

It is worth noting however that, in theory, increasing the formula robustness does not necessarily improve the model robustness. Indeed, the formula robustness may increase with a parameter set that approaches a frontier where the formula is no longer satisfied, in which case the model robustness with respect to parameter perturbation may decrease. In many practical cases however, though the systems we tackle are highly non-linear, improving the formula robustness also did improve the model robustness.

7 Evaluation on MAPK Signaling Network

In [10], the authors present the dynamics of a signaling network, namely the Mitogen Activated Protein Kinase (MAPK) cascade, with a model encompassing 22 species and fully relying on Mass-Action kinetics with 37 parameters. This model remains today a reference for the MAPK cascade as it properly describes the 3-level global structure and the 2-step nature of each level. Though it was supposed that explicit feedback reactions were necessary for obtaining oscillations in this model, it was demonstrated by Qiao et al. in [12] that this is actually not necessary. The authors explored blindly the space of 36 parameters and then established a bifurcation diagram in the 37th parameter in order to find oscillations.

The FO-LTL($\mathbb{R}_{\mathrm{lin}}$) formula used in [14] to measure model robustness is $\mathbf{F}(\mathrm{PPK} \geq up \wedge \mathbf{F}(\mathrm{PPK} \leq up - amplitude))$, where PPK represents the output of the signaling cascade, i.e. the concentration of the doubly-phosphorylated kinase at the third level. By using that formula with an objective of for instance 0.5 for *amplitude*, one can actually search parameter values in a directed way, guided by continuous satisfaction degree. In order to ignore simple overshoots, one can consider the refined formula $\mathbf{F}(\mathrm{PPK} \geq up \wedge \mathbf{F}(\mathrm{PPK} \leq up - a_1 \wedge \mathbf{F}(\mathrm{PPK} \geq up - a_1 + 0.5 * a_2)))$, with objectives of 0.5 and 1 for respectively a_1 and a_2. BIOCHAM-4 is able to find a solution for the 37 parameters in a few minutes (averaged to $5\,min$ over 100 runs on a $3.6\,GHz$ Intel Core i7 machine) and to optimize formula robustness with no extra cost. This is also shown on a cell cycle model in the companion notebook mentioned above.

8 Conclusion

Building models that are robust to parameter variations is a key issue in systems biology and synthetic biology, because of both their fluctuating environment and the stochastic nature of biochemical reactions. BIOCHAM-4 is one of the very few tools to implement a metric notion of robustness for dynamical properties and optimize it by the parameter search procedure with no extra cost.

Although not true in all generality for non-linear hybrid systems, optimizing formula robustness does improve in practice the statistical notion of model robustness with respect to parameter variations. The integration of formula robustness in the notion of satisfaction degree thus provides a simple and effective approach to the design of robust parameterization of reaction networks.

Acknowledgment. This work benefited from partial support from the ANR project HYCLOCK contract DS0401.

References

1. Courbet, A., Amar, P., Fages, F., Renard, E., Molina, F.: Computer-aided biochemical programming of synthetic microreactors as diagnostic devices. Mol. Syst. Biol. **14**(4), e7845 (2018)
2. Donzé, A., Maler, O.: Robust satisfaction of temporal logic over real-valued signals. In: Chatterjee, K., Henzinger, T.A. (eds.) FORMATS 2010. LNCS, vol. 6246, pp. 92–106. Springer, Heidelberg (2010). https://doi.org/10.1007/978-3-642-15297-9_9
3. Fages, F., Gay, S., Soliman, S.: Inferring reaction systems from ordinary differential equations. Theor. Comput. Sci. **599**, 64–78 (2015)
4. Fages, F., Martinez, T., Rosenblueth, D., Soliman, S.: Influence networks compared with reaction networks: semantics, expressivity and attractors. IEEE/ACM Trans. Comput. Biol. Bioinform. (2018). https://doi.org/10.1109/TCBB.2018.2805686
5. Fages, F., Rizk, A.: On temporal logic constraint solving for the analysis of numerical data time series. Theor. Comput. Sci. **408**(1), 55–65 (2008)
6. Fages, F., Traynard, P.: Temporal logic modeling of dynamical behaviors: first-order patterns and solvers. In: del Cerro, L.F., Inoue, K. (eds.) Logical Modeling of Biological Systems, Chap. 8, pp. 291–323. Wiley, Hoboken (2014)
7. Fainekos, G.E., Pappas, G.J.: Robustness of temporal logic specifications. In: Havelund, K., Núñez, M., Roşu, G., Wolff, B. (eds.) FATES/RV 2006. LNCS, vol. 4262, pp. 178–192. Springer, Heidelberg (2006). https://doi.org/10.1007/11940197_12
8. Hansen, N., Ostermeier, A.: Completely derandomized self-adaptation in evolution strategies. Evol. Comput. **9**(2), 159–195 (2001)
9. Heitzler, D., et al.: Competing G protein-coupled receptor kinases balance G protein and β-arrestin signaling. Mol. Syst. Biol. **8**, 590 (2012)
10. Huang, C.Y., Ferrell, J.E.: Ultrasensitivity in the mitogen-activated protein kinase cascade. PNAS **93**(19), 10078–10083 (1996)
11. Kitano, H.: Towards a theory of biological robustness. Mol. Syst. Biol. **3**, 137 (2007)
12. Qiao, L., Nachbar, R.B., Kevrekidis, I.G., Shvartsman, S.Y.: Bistability and oscillations in the Huang-Ferrell model of MAPK signaling. PLoS Comput. Biol. **3**(9), 1819–1826 (2007)
13. Rizk, A., Batt, G., Fages, F., Soliman, S.: A general computational method for robustness analysis with applications to synthetic gene networks. Bioinformatics **12**(25), i169–i178 (2009)
14. Rizk, A., Batt, G., Fages, F., Soliman, S.: Continuous valuations of temporal logic specifications with applications to parameter optimization and robustness measures. Theor. Comput. Sci. **412**(26), 2827–2839 (2011)

15. Traynard, P., Fages, F., Soliman, S.: Trace simplifications preserving temporal logic formulae with case study in a coupled model of the cell cycle and the circadian clock. In: Mendes, P., Dada, J.O., Smallbone, K. (eds.) CMSB 2014. LNCS, vol. 8859, pp. 114–128. Springer, Cham (2014). https://doi.org/10.1007/978-3-319-12982-2_9

16. Traynard, P., Feillet, C., Soliman, S., Delaunay, F., Fages, F.: Model-based investigation of the circadian clock and cell cycle coupling in mouse embryonic fibroblasts: prediction of RevErb-α up-regulation during mitosis. Biosystems **149**, 59–69 (2016)

LNA++: Linear Noise Approximation with First and Second Order Sensitivities

Justin Feigelman[1,2], Daniel Weindl[1], Fabian J. Theis[1,2], Carsten Marr[1(✉)], and Jan Hasenauer[1,2(✉)]

[1] Helmholtz Zentrum München - German Research Center for Environmental Health, Institute of Computational Biology, Oberschleißheim, Germany
{carsten.marr,jan.hasenauer}@helmholtz-muenchen.de
[2] Center for Mathematics, Technische Universität München, Munich, Germany

Abstract. The linear noise approximation (LNA) provides an approximate description of the statistical moments of stochastic chemical reaction networks (CRNs). LNA is a commonly used modeling paradigm describing the probability distribution of systems of biochemical species in the intracellular environment. Unlike exact formulations, the LNA remains computationally feasible even for CRNs with many reactions. The tractability of the LNA makes it a common choice for inference of unknown chemical reaction parameters. However, this task is impeded by a lack of suitable inference tools for arbitrary CRN models. In particular, no available tool provides temporal cross-correlations, parameter sensitivities and efficient numerical integration. In this manuscript we present LNA++, which allows for fast derivation and simulation of the LNA including the computation of means, covariances, and temporal cross-covariances. For efficient parameter estimation and uncertainty analysis, LNA++ implements first and second order sensitivity equations. Interfaces are provided for easy integration with Matlab and Python.

Implementation and availability: LNA++ is implemented as a combination of C/C++, Matlab and Python scripts. Code base and the release used for this publication are available on GitHub (https://github.com/ICB-DCM/LNAplusplus) and Zenodo (https://doi.org/10.5281/zenodo.1287771).

Keywords: Linear noise approximation · Automatic construction
Numerical simulation · Sensitivity analysis · MATLAB · Python

1 Introduction

The intracellular environment consists of biological molecules undergoing continual change due to chemical reactions. The time-dependent behavior of such molecular systems is often modelled using chemical reaction networks (CRNs), which describe interactions between chemical species with rates determined by reaction constants. The time-dependent probability of the CRN is described by

© Springer Nature Switzerland AG 2018
M. Češka and D. Šafránek (Eds.): CMSB 2018, LNBI 11095, pp. 300–306, 2018.
https://doi.org/10.1007/978-3-319-99429-1_19

the chemical master equation (CME) [4]. The CME is a system of ordinary differential equations (ODE), which is usually infinite-dimensional. Although the CME is only solvable in the simplest of cases, many approximate solutions have been proposed [1,5,12,16].

The linear noise approximation (LNA) [16] provides a convenient approximation to the statistical moments of the solution of the CME, an ODE system for the means and temporal cross-covariances of each species. Moreover, the LNA has been successfully used for parameter inference in CRNs [2,10,14]. However, computational tools for numerical simulation, sensitivity analysis, and parameter inference using the LNA are still lacking. Tools such as the intrinsic noise analyzer (iNA) [15] implement the LNA but lack the ability to directly invoke the LNA without redefining the model via cumbersome creation of input files. LNAR [3] for the R language allows simulation of the LNA but does not provide sensitivities. CERENA [8] for Matlab provides the means to simulate the LNA, but does not provide information about temporal correlations. StochSens for MATLAB implements first order sensitivities but not second order sensitivities and does not provide a computationally efficient simulation routine. A detailed overview of the properties of individual tools is provided in Table 1.

In this work we present LNA++, an efficient implementation of the LNA for arbitrary CRNs, using C/C++ and CVODES [6] for numerical integration, and Blitz++ [17] for fast tensor operations. We provide Python and Matlab interfaces for generating and computing the LNA for arbitrary CRNs, and demonstrate the utility of the tool by performing parameter inference of a simple CRN. While LNA is an established technique, this is the first framework to make an efficient numerical simulation of temporal cross-covariances and associated sensitivities easily accessible to the research community.

In the following, we provide a short introduction to LNA, outline the implementation of LNA++, and describe its application. Further details are provided in the documentation of LNA++ (https://doi.org/10.5281/zenodo.1287771).

2   Background

2.1   Chemical Reaction Network and Linear Noise Approximation

LNA++ is designed to study CRNs described by a set of reactions

$$\mathcal{R}_j : \sum_i \mathbb{S}_{ij}^- X_i \to \sum_i \mathbb{S}_{ij}^+ X_i,$$

where $\mathbb{S}_{ij}^-$ is the number of molecules of species X_i consumed by reaction $\mathcal{R}_j$, and $\mathbb{S}_{ij}^+$ the number of produced molecules. The net change is described by the stoichiometric matrix, $\mathbb{S}_{ij} = \mathbb{S}_{ij}^+ - \mathbb{S}_{ij}^-$. The reaction propensities depend on the copy numbers of the reactants and kinetic constants Θ.

Table 1. Features of LNA++ and available toolboxes for linear noise approximation. Availability and lack of a feature is indicated by ✓ and ✗, respectively.

Feature	iNA [15]	LNAR [3]	CERENA [8]	StochSens [11]	LNA++
MATLAB compatible	✗	✗	✓	✓	✓
Python compatible	✗	✗	✗	✗	✓
R compatible	✗	✓	✗	✗	✗
Mean/covariance	✓	✓	✓	✓	✓
+ 1st order sensitivities	✗	✗	✓	✓	✓
+ 2nd order sensitivities	✗	✗	✗	✗	✓
Cross-covariance	✗	✗	✗	✓	✓
+ 1st order sensitivities	✗	✗	✗	✓	✓
+ 2nd order sensitivities	✗	✗	✗	✗	✓
Steady state calculation	✓	✗	✓	✗	✓
Fast numerical simulation using C code	✗	✓	✓	✗	✓
Integration with parameter estimation	✗	✓	✓	✗	✓
Robustness/ identifiability analysis	✗	✗	✗	✓	✗
SBML import	✗	✓	✓	✗	✓

The exact dynamics of the stochastic processes are governed by the CME [4]. The LNA approximates these processes with the macroscopic mean $\phi(t)$ and stochastic fluctuations $\boldsymbol{\xi}(t)$ [9]:

$$x(t) = \phi(t) + \boldsymbol{\xi}(t).$$

The macroscopic mean $\phi(t)$ satisfies the macroscopic rate equations:

$$\dot{\phi}(t) = \mathbb{S}F(\phi, \boldsymbol{\Theta}, t), \tag{1}$$

with reaction flux vector $F(\phi, \boldsymbol{\Theta}, t)$. The stochastic fluctuation $\boldsymbol{\xi}(t)$ obeys the stochastic differential equation

$$d\boldsymbol{\xi} = \mathbb{A}(\phi, \boldsymbol{\Theta}, t)\boldsymbol{\xi}dt + \mathbb{E}(\phi, \boldsymbol{\Theta}, t)d\boldsymbol{W}$$

with $\mathbb{A}(\phi, \boldsymbol{\Theta}, t) = \mathbb{S}\frac{\partial}{\partial \phi}F(\phi, \boldsymbol{\Theta}, t)$, $\mathbb{E}(\phi, \boldsymbol{\Theta}, t) = \mathbb{S}\sqrt{\mathrm{diag}(F(\phi, \boldsymbol{\Theta}, t))}$, and Wiener process $\boldsymbol{W}$. This yields the governing equation

$$\dot{\mathbb{V}}(t) = \mathbb{A}\mathbb{V} + \mathbb{V}\mathbb{A}^T + \mathbb{E}\mathbb{E}^T \tag{2}$$

for the covariance $\mathbb{V}(t) = \text{cov}(\boldsymbol{x}(t), \boldsymbol{x}(t))$ of the state $\boldsymbol{x}$ at time t. In addition, temporal cross-covariance $\text{cov}(\boldsymbol{x}(s), \boldsymbol{x}(t))$ between times s and t, with $t \geq s$, is

$$
\begin{aligned}
\text{cov}(\boldsymbol{x}(s), \boldsymbol{x}(t)) &= \mathbb{V}(s)\mathbb{G}(s,t)^T \\
\dot{\mathbb{G}}(s,t) &= \mathbb{A}(\boldsymbol{\phi}, \boldsymbol{\Theta}, t)\mathbb{G}(s,t), \qquad \mathbb{G}(s,s) = \mathbb{I}.
\end{aligned}
\tag{3}
$$

In addition to state variables $\boldsymbol{x}(t)$, LNA++ facilitates the calculation of mean, covariance and temporal cross-covariance of observables $\boldsymbol{y}(t)$. The observables $\boldsymbol{y}$ are a subset of the state variables $\boldsymbol{x}$. The measurement noise is assumed to be additive with mean zero. In this case, the mean of the measured output is the mean of the observable. The covariance of the measured output is the sum of the covariance of the observable and the covariance of the measurement noise, and the same holds true for the temporal cross-covariance.

For details on the LNA, including the derivation of the temporal cross-covariance and the equations for the measured outputs, we refer to the work of Komorowski et al. [9] and the LNA++ Documentation.

2.2 1st and 2nd Order Sensitivity Equations

The statistical moments of the state variables $\boldsymbol{x}(t)$ and the observables $\boldsymbol{y}(t)$ depend on the kinetic constants $\boldsymbol{\Theta}$. This dependence is of interest for tasks such as parameter optimization and uncertainty analysis [2,8,9] and can be assessed via 1st and 2nd order sensitivity analysis. The 1st order sensitivity of a statistical moment is its derivative with respect to the parameters $\boldsymbol{\Theta}$, e.g. $\partial\boldsymbol{\phi}(t)/\partial\boldsymbol{\Theta}$ for the macroscopic mean $\boldsymbol{\phi}(t)$. Similar to the statistical moments of state variables (1)–(3), 1st and 2nd order sensitivities of are defined as solutions for ordinary differential equations. The evolution equations for the 1st order sensitivities are obtained by differentiating the evolution equations for quantities of interest with respect to the parameters. The evolution equations for the 2nd order sensitivities are obtained by differentiating twice. For the temporal cross-covariance, the product rule needs to be applied as $\mathbb{V}(s)$ and $\mathbb{G}(s,t)$ depend on $\boldsymbol{\Theta}$. For additional information and a detailed derivation of the 1st and 2nd order sensitivity equations we refer to the LNA++ Documentation.

3 Implementation

LNA++ provides of scripts for generating and computing the LNA for user-defined CRNs provided as Systems Biology Markup Language (SBML) files [7] (Fig. 1A). LNA++ determines the stoichiometric matrix and reaction flux vector from the SBML file, and computes the elements of the ODE system for the LNA symbolically using either sympy (http://www.sympy.org) in Python or the Symbolic Math Toolbox in Matlab. In some cases it is also possible to solve for the steady-state of the means and/or variances. The resulting analytical expressions are then converted into C code using custom scripts, and combined with the LNA++ source code. LNA++ uses Blitz++ [17] for tensor manipulation and

CVODES [6] for numerical integration of the means, temporal cross-covariances, and first and second order sensitivities of both. Finally, all generated source code is compiled into a Python module or Matlab mex executable file which can be integrated directly with Python or Matlab, respectively. To determine mean, variance and temporal covariance of observables, the user can specify observation functions and noise levels.

4 Application

To illustrate a typical application of LNA++, we investigate parameter identifiability for a simple two-stage model of gene expression (see Chap. 6 in the LNA++ Documentation on https://doi.org/10.5281/zenodo.1287771, Documentation Fig. 1A). For this model, the LNA provides an exact description of the means, variances and cross-covariances of the species. The predicted protein mean and variance are in good agreement with simulations generated using the stochastic simulation algorithm (SSA) (Fig. 1B), as is the cross-covariance (Fig. 1C, Documentation Fig. 1B–C). LNA++ likewise shows excellent agreement to a finite difference approximation for both the mean (Documentation Figs. 2 and 3) and autocovariance (Documentation Figs. 4 and 5), for first and second order sensitivities, respectively. Moreover, LNA++ is able to compute sensitivities even when the finite difference approximation fails due to lack of numerical precision (see, e.g., Documentation Fig. 3, k_p and k_m).

Since LNA provides an approximation to the time-dependent probability density of the CRN, one can compute the likelihood of a particular parameter set for a set of trajectory realizations (see Documentation). To infer model parameters from data, one typically uses a numerical optimization scheme based on the likelihood function. Such schemes often depend on the sensitivity (gradient) of the likelihood function with respect to the model parameters, which can be estimated analytically or numerically, with the former typically providing better estimates than the latter due to the limited precision of numerical estimation methods. Moreover, analytically-computed sensitivities are typically faster to evaluate due to the need for fewer evaluations of the objective function (see Documentation on Zenodo). We compared parameter inference using analytical sensitivities computed using LNA++ with a numerical (finite difference-based) approach. Multi-start numerical optimization using numerical gradients resulted in many estimates with a wide range of log-likelihood values, likely due to premature stopping as a result of inaccurate estimates of the gradient. In contrast, using the analytical sensitivities lead to robust parameter estimation (Documentation Fig. 6A and B).

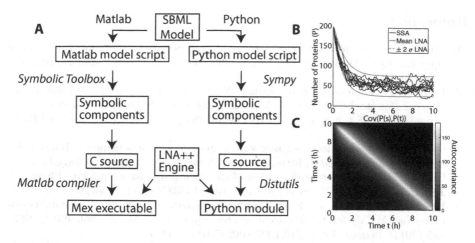

Fig. 1. LNA++ generates executable modules for user-specified chemical reaction network models, providing approximations to the mean, covariance and temporal cross-covariance along with first and second order sensitivities. (A) Models are specified using SBML and parsed in either Matlab or Python. LNA++ constructs the model symbolically and generates a Matlab executable or Python module for fast numerical simulations. Software tools employed for individual steps are written in italic next to the arrow. (B) LNA++ computes the mean and variances of species, in good agreement with stochastic simulation trajectories (SSA) [13], shown for an example system. (C) LNA++ also computes the temporal cross-covariance (shown for one species).

Using stochastic simulations, we estimated the log-likelihood surface over a range of parameters, and found that the true values used for simulations are indeed the maximum of the log-likelihood surface (Documentation Fig. 6C). Using second order sensitivities it is also possible to compute approximate confidence intervals of identified model parameters. For the considered examples, the confidence intervals are rather narrow and contain the true model parameters.

To assess scalability of the performance of LNA++ with respect to model generation and run time, we used a series of increasingly complex linear models and conclude that sensitivities computed with LNA++ are significantly faster than a finite difference approximation (Documentation Fig. 7).

5   Conclusion

We conclude that LNA++ provides a simple means for the simulation of the LNA including means and temporal cross-covariances, and their associated first and second order sensitivities. These quantities facilitate parameter inference and confidence interval estimation. LNA++ is provided as an open source package under the 3-clause BSD license, thus providing a useful tool that can be further developed by the stochastic CRN modeling community.

References

1. Engblom, S.: Computing the moments of high dimensional solutions of the master equation. Appl. Math. Comput. **180**(2), 498–515 (2006). https://doi.org/10.1016/j.amc.2005.12.032
2. Fröhlich, F., Thomas, P., Kazeroonian, A., Theis, F.J., Grima, R., Hasenauer, J.: Inference for stochastic chemical kinetics using moment equations and system size expansion. PLoS Comput. Biol. **12**(7), e1005030 (2016). https://doi.org/10.1371/journal.pcbi.1005030
3. Giagos, V.: Inference for stochastic kinetic genetic networks using the linear noise approximation, May 2011. https://rdrr.io/rforge/lnar/man/lnar-package.html
4. Gillespie, D.T.: A rigorous derivation of the chemical master equation. Physica A **188**(1), 404–425 (1992). https://doi.org/10.1016/0378-4371(92)90283-V
5. Hasenauer, J., Wolf, V., Kazeroonian, A., Theis, F.J.: Method of conditional moments (MCM) for the chemical master equation. J. Math. Biol. **69**(3), 687–735 (2014). https://doi.org/10.1007/s00285-013-0711-5
6. Hindmarsh, A.C., et al.: SUNDIALS: suite of nonlinear and differential/algebraic equation solvers. ACM Trans. Math. Softw. **31**(3), 363–396 (2005). https://doi.org/10.1145/1089014.1089020
7. Hucka, M., et al.: The systems biology markup language (SBML): a medium for representation and exchange of biochemical network models. Bioinformatics **19**(4), 524–531 (2003). https://doi.org/10.1093/bioinformatics/btg015
8. Kazeroonian, A., Fröhlich, F., Raue, A., Theis, F.J., Hasenauer, J.: CERENA: Chemical REaction network analyzer - a toolbox for the simulation and analysis of stochastic chemical kinetics. PLoS One **11**(1), e0146732 (2016). https://doi.org/10.1371/journal.pone.0146732
9. Komorowski, M., Costa, M.J., Rand, D.A., Stumpf, M.P.H.: Sensitivity, robustness, and identifiability in stochastic chemical kinetics models. Proc. Natl. Acad. Sci. U.S.A. **108**(21), 8645–8650 (2011). https://doi.org/10.1073/pnas.1015814108
10. Komorowski, M., Finkenstädt, B., Harper, C.V., Rand, D.A.: Bayesian inference of biochemical kinetic parameters using the linear noise approximation. BMC Bioinform. **10**(1), 343 (2009). https://doi.org/10.1186/1471-2105-10-343
11. Komorowski, M., Zurauskiene, J., Stumpf, M.P.H.: StochSens-MATLAB package for sensitivity analysis of stochastic chemical systems. Bioinformatics **28**(5), 731–733 (2012). https://doi.org/10.1093/bioinformatics/btr714
12. Munsky, B., Khammash, M.: The finite state projection algorithm for the solution of the chemical master equation. J. Chem. Phys. **124**(4), 044104 (2006). https://doi.org/10.1063/1.2145882
13. Sanft, K.R., Wu, S., Roh, M., Fu, J., Lim, R.K., Petzold, L.R.: StochKit2: software for discrete stochastic simulation of biochemical systems with events. Bioinformatics **27**(17), 2457–2458 (2011)
14. Stathopoulos, V., Girolami, M.A.: Markov chain Monte Carlo inference for Markov jump processes via the linear noise approximation. Philos. Trans. Ser. A **371**(1984), 20110541 (2013). https://doi.org/10.1098/rsta.2011.0541
15. Thomas, P., Matuschek, H., Grima, R.: Intrinsic noise analyzer: a software package for the exploration of stochastic biochemical kinetics using the system size expansion. PLoS One **7**(6), e38518 (2013). https://doi.org/10.1371/journal.pone.0038518
16. van Kampen, N.G.: Stochastic Processes in Physics and Chemistry, 3rd edn. North-Holland, Amsterdam (2007)
17. Veldhuizen, T.: Blitz++ User's Guide, March 2006. http://sourceforge.net/projects/blitz

Poster Abstracts

Reparametrizing the Sigmoid Model of Gene Regulation for Bayesian Inference

Martin Modrák[(✉)] [ID]

Institute of Microbiology of the Czech Academy of Sciences, Prague, Czech Republic
martin.modrak@biomed.cas.cz

Abstract. This poster describes a novel work-in-progress reparametrization of a frequently used non-linear ordinary differential equation (ODE) model for inferring gene regulations from expression data. We show that in its commonly used form, the model cannot always determine the sign of the regulatory effect as well as other parameters of the model. The proposed reparametrization makes inference over the model stable and amenable to fully Bayesian treatment with state of the art Hamiltonian Monte Carlo methods.

Complete source code and a more detailed explanation of the model is available at https://github.com/cas-bioinf/genexpi-stan.

Keywords: Gene regulation · Gene network inference Bayesian statistics

1 Introduction

Non-linear ODE models that approximate Michaelis-Menten kinetics are one of frequently used tools to infer transcriptional regulations from time series of expression data. In our attempts to reimplement a Bayesian version of a popular sigmoid ODE model [1] we noticed that very different parameter values may result in almost identical behavior, which introduces computational difficulties and questions the interpretation of previous results achieved with the model. Here we describe novel reparametrization of the model implemented in the Stan probabilistic programming language [2] that mitigates these problems.

2 The Model

The ODE model we use is based on [3]. For a single regulator y and targets $x_1, \ldots, x_n$ the model takes form

$$\frac{\mathrm{d}x_i}{\mathrm{d}t} = s_i \frac{1}{1 + e^{-\rho_i}} - d_i x_i \; ; \rho_i = w_i y + b_i \qquad (1)$$

Supported by C4Sys research infrastructure (MEYS project No: LM20150055).

M. Češka and D. Šafránek (Eds.): CMSB 2018, LNBI 11095, pp. 309–312, 2018.
https://doi.org/10.1007/978-3-319-99429-1_20

Where s_i, w_i, b_i and d_i are parameters to be fit. We call ρ_i the *regulatory input* for gene i. The model assumes that both the regulator and targets are measured at multiple time points, letting us to solve the ODE. When using microarray data we assume normal observation noise where the standard deviation has two components: one constant and one proportional to the expression. The regulator is fit using B-splines where the spline coefficients are also treated as parameters of the model which let us to both handle uncertainty in regulator measurements and obtain the regulator at arbitrarily small time resolution required for solving the ODE numerically. The last parameter of the model is the initial condition for the ODE, i.e. the value of x_i at $t = 0$. We note that the model is only weakly identifiable in its direct form (1), as very different parameter sets can result in similar solutions for x_i. Weak identifiability poses computational problems for Bayesian inference, reduces stability of maximum-likelihood estimates and limits interpretability of the model results. Critically, the sign of w_i (whether the regulator is an activator or a repressor) is not well determined for certain target profiles (Fig. 1a) and it may be impossible to determine whether $w_i \simeq 0$ (Fig. 1b). Negligible effect on x_i can also be observed under linear transformations of (w_i, b_i) when $|w_i y + b_i| \gg 0$, (Fig. 1c) and under approximately linear transformations of (s_i, d_i) (Fig. 1d) making the magnitudes of the parameters uninterpretable.

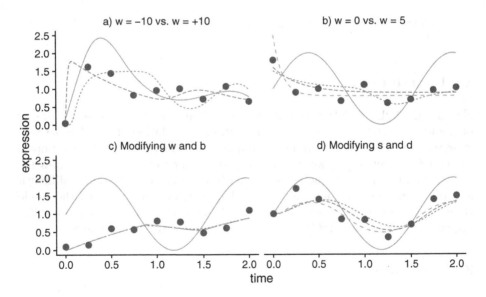

Fig. 1. Simulated examples of weakly identified parameters in the sigmoid model. Red solid line shows the expression of the regulator, dashed lines represent multiple similar solutions of the ODE corresponding to significantly different parameter values and dots represent example measured target profiles that are insufficient to distinguish between the solutions. (a) $(w, b, s, d) \in \{(10, -8, 15, 10), (-10, 3, 100, 1)\}$ (b) $(w, b, s, d) \in \{5, -1, 3, 3), (0, 0, 6, 3.4), (0, 0, 16, 10)\}$ (c) $s = 1, d = \frac{1}{2}$ and $(w, b) \in \{(5, -3), (10, -6), (50, -30)\}$ (d) $w = 1, b = -1$ and $(s, d) \in \{(10, 5), (19, 10), (180, 100)\}$ (Color figure online)

The aforementioned identifiability problems can be mitigated by (a) fixing $I_i = sgn(w_i)$ and running separate fits when both signs have to be tested and (b) replacing w_i, b_i and s_i with μ_i^ρ (mean of the regulatory input), σ_i^ρ (std. deviation of the regulatory input) and a_i - the expression of x_i in a hypothetical long-term steady state assuming the regulatory influences stay similar to those observed. To make a_i independent from the scale of the data, it is treated as relative to the maximum observed expression. Formally:

$$\mu_i^\rho = E(\rho_i) \tag{2}$$

$$\sigma_i^\rho = \text{sd}(\rho_i) \tag{3}$$

$$a_i = \frac{s_i E\left(\frac{1}{1+e^{-\rho_i}}\right)}{d_i \max \tilde{x}_i} \tag{4}$$

where E and sd correspond to the sample mean and standard deviation and $\tilde{x}_i$ is the observed expression of gene i. The values of the original parameters can be recovered from the reparametrization as:

$$w_i = I_i \frac{\sigma_i^\rho}{\text{sd}(y)} \tag{5}$$

$$b_i = \mu_i^\rho - w_i E(y) \tag{6}$$

$$s_i = \frac{a_i d_i \max \tilde{x}_i}{E\left(\frac{1}{1+e^{-\rho_i}}\right)} \tag{7}$$

Here σ_i^ρ is a more interpretable measure of the amount of influence the regulator has over i-th target, given the fixed regulatory sign I_i. It is also easier to find defensible data-independent prior distributions for the parameters. Since the logistic sigmoid is very flat outside $(-5, 5)$ and changes in ρ_i outside this interval have little impact on x_i, we can safely assume $\mu_i^\rho \sim N(0, 5)$ and $\sigma_i^\rho \sim N^+(0, 5)$ where N^+ stands for normal distribution truncated to be positive. In many cases it is also reasonable to assume that the observed expression levels of a gene are roughly the same order as expected in the long run which can be encoded in the prior distribution as $a_i \sim N^+(1, 0.5)$.

To our knowledge, this is the first time the identifiability issues are reported and handled for the sigmoid ODE model. Caution should therefore be exercised when interpreting the parameter values found in previous works using this model.

3 Using the Model

When the true status of regulatory interactions is not known, the model should be fit for each putative target separately as it implicitly assumes that all of the regulations included in the model are taking place.

When there are known regulations, the model can be "trained" by fitting those known regulations first. The posterior estimates of the expression of the regulator can then be used to decrease uncertainty when fitting putative novel targets (Fig. 2). Performance-wise, the "training" phase takes several minutes and fitting 881 putative targets on an 8-core machine takes ~1 h.

It is also important to note that model fit is in itself insufficient to determine whether a regulation is plausible. To assess plausibility, we test whether the model improves over two baselines: (a) constant synthesis ($w_i = 0$) and (b) a model where the regulator spline is not constrained by the observed regulator expression. We use the LOO-IC criterion [4] to compare the models. Plausibility within the small world of a computational model also only hints at biological reality. In our previous work with a simpler version of this model, we were able to experimentally confirm 5 out of 10 good fits as actual regulations [5].

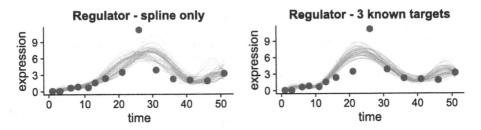

Fig. 2. Comparison of fitting the expression of a regulator only with a spline (left) and a fit using 3 known targets to reduce uncertainty (right). Curves represent samples from the posterior distribution, the points are the measured expression values. In particular, we can observe that regulator expression is consistent with a wide range of locations of the peaks while using the known targets narrows down the peak's location.

References

1. Titsias, M., et al.: Identifying targets of multiple co-regulating transcription factors from expression time-series by Bayesian model comparison. BMC Syst. Biol. **6**, 53 (2012)
2. Carpenter, B., et al.: Stan: a probabilistic programming language. J. Stat. Softw. **76**(1) (2017)
3. Vohradský, J.: Neural model of the genetic network. J. Biol. Chem. **276**(39), 36168–36173 (2001)
4. Vehtari, A., et al.: LOO: efficient leave-one-out cross-validation and WAIC for Bayesian models (2018). https://cran.r-project.org/package=loo
5. Ramaniuk, O., Černý, M., Krásný, L., Vohradský, J.: Kinetic modelling and meta-analysis of the B. subtilis SigA regulatory network during spore germination and outgrowth. Biochim. Biophys. Acta (BBA) - Gene Regul. Mech. **1860**(8), 894–904 (2017)

On the Full Control of Boolean Networks

Soumya Paul[1(✉)], Jun Pang[1,2], and Cui Su[1]

[1] Interdisciplinary Centre for Security, Reliability and Trust,
Esch-sur-Alzette, Luxembourg
[2] Faculty of Science, Technology and Communication, University of Luxembourg,
Esch-sur-Alzette, Luxembourg
{soumya.paul,jun.pang,cui.su}@uni.lu

Boolean networks (BNs), introduced by Kauffman [3], is a popular and well-established framework for modelling gene regulatory networks and their associated signalling pathways. The main advantage of this framework is that it is relatively simple and yet able to capture the important dynamical properties of the system under study, thus facilitating the modelling and analysis of large biological networks as a whole.

A Boolean network B is a pair $B = (\mathbf{x}, \mathbf{f})$, where $\mathbf{x}$ is a tuple of n variables $\mathbf{x} = (x_1, x_2, \ldots, x_n)$ and $\mathbf{f}$ is a tuple of n Boolean *update functions* $\mathbf{f} = (f_1, f_2, \ldots, f_n)$, where for every i, the function f_i, which depends on a subset of the variables in $\mathbf{x}$, governs the dynamics of x_i in time. BN is called *linear* when the functions $\mathbf{f}$ are linear. It is *non-linear* otherwise. For a BN B with n variables, its *dependency graph* is a directed graph $\mathcal{G}_B = (V, E)$ with a set V of n vertices (or nodes) for the n variables, ordered such that vertex v_i corresponds to the variable x_i. There is a directed edge from vertex v_i to v_j if and only if the function f_j depends on x_i. The *structure* of a BN B refers to the structure of its dependency graph $\mathcal{G}_B$. The variables of $\mathbf{x}$ take Boolean values. Each such tuple of values gives rise to a *state* of the BN, typically denoted as $\mathbf{s}$ or $\mathbf{t}$. For a BN with n variables, there can be a total of 2^n possible states, the elements in $\{0, 1\}^n$. The *asynchronous dynamics* of a BN B is assumed to evolve in discrete *time steps* as follows. Suppose B is in state $\mathbf{s}$ in time t. A possible *next state* to $\mathbf{s}$, i.e., a state in time $(t+1)$, is given by non-deterministically choosing exactly one i and updating the ith component of $\mathbf{s}$ by applying the function f_i and leaving the other components unchanged. This operation results in a directed graph, called the *(state) transition system* (TS) of B, denoted TS_B, whose elements are the states of B and there is a directed edge from a state $\mathbf{s}$ to a state $\mathbf{t}$ if and only if $\mathbf{t}$ is a possible next state to $\mathbf{s}$.

An *attractor* A of B is a subset of states of B that forms a bottom maximal strongly connected component (SCC) of TS_B. Attractors represent the eventual behaviour or the *steady states* of the system modelled by the BN. In biological context, attractors are hypothesised to characterise cellular phenotypes [3] and also correspond to functional cellular states such as proliferation, apoptosis differentiation etc. [1]. The identification and analysis of the attractors of a BN thus forms an integral part of the study of the corresponding biological network. *Controlling* the network means driving its dynamics from one steady state to another by modifying the parameters of the network which amounts to being able to move it between the different attractors. The *strong basin of attraction* of an attractor A of B, denoted $bas(A)$, is the subset of states of B such that

© Springer Nature Switzerland AG 2018
M. Češka and D. Šafránek (Eds.): CMSB 2018, LNBI 11095, pp. 313–317, 2018.
https://doi.org/10.1007/978-3-319-99429-1_21

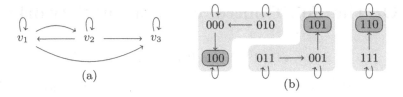

Fig. 1. Running example: the dependency graph of a B and its transition system.

there is a (possibly empty) sequence of edges from every state **s** in bas(A) to a state **t** $\in A$ and moreover there is no such sequence from **s** to any state **t**$'$ $\in A'$ for any other attractor $A' \neq A$ of B. If the current state of B is in bas(A) for some attractor A, then its dynamics is guaranteed to eventually reach A.

The full control of linear networks is a well-understood problem [2] and a number of control strategies have been developed in the literature. Recent work on network controllability has shown that full controllability and reprogramming of intercellular networks can be achieved by a minimal number of control targets [5]. However, the full control of non-linear networks is apparently more challenging predominantly due to the explosion of the potential search space with the increase in the network size. There has not been a lot of work in the study of the full control of non-linear networks. Recently, Kim et al. [4] developed a method to identify the so-called 'control kernel' which is a minimal set of nodes for fully controlling a biological network. However, their method requires the construction of the full state transition graph of the studied network and as such does not scale well for large networks.

In this work, we aim to develop a method for the *full control* of non-linear BNs with asynchronous dynamics, based both on their structural and dynamic properties. The problem is formally defined as: Given a BN B, find a *minimal subset* C of indices of the variables of B such that for any pair of attractors A_s and A_t of B, there exists a state **s** $\in A_s$ such that a subset of the variables with indices in C needs to be toggled (controlled) in **s**, in a *single step*, so that the system eventually reaches A_t. The problem can be shown to be PSPACE-hard and hence efficient algorithms for dealing with large BNs are highly unlikely. Our method is based on a decomposition-based approach for solving the corresponding minimal *target control* problem [7] which yields efficient results for many large real-life BNs having modular structures. In brief, the method analyses the structure of the BN to identify its maximal strongly connected components and uses them to decompose the vertices of the dependency graph into (possibly overlapping) subsets called *blocks* (see details in [6,7]). The blocks are sorted topologically and the full control problem is solved locally for each block in the sorted order. The local results are then combined to derive the minimal full control set for the entire network. Due to space-restriction, we describe our method in details on a running example without going into formal notations and proofs.

Consider the three-node asynchronous BN B = $(\mathbf{x}, \mathbf{f})$, where $\mathbf{x} = (x_1, x_2, x_3)$ and $\mathbf{f} = (f_1, f_2, f_3)$, such that $f_1 = \neg x_2 \vee (x_1 \wedge x_2)$, $f_2 = x_1 \wedge x_2$ and $f_3 = x_3 \wedge \neg(x_1 \wedge x_2)$. The dependency graph $\mathcal{G}_B$ and the state transition system TS_B

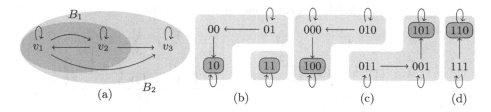

Fig. 2. The local transition systems of the blocks B_1 and B_2.

are given in Figs. 1(a) and (b), respectively. $\mathsf{TS_B}$ has three attractors $A_1 = \{(100)\}, A_2 = \{(101)\}$ and $A_3 = \{(110)\}$, shown by dark grey rectangles. Their corresponding strong basins of attractions are shown by enclosing grey regions of a lighter shade.

By definition, we know that for the BN to surely end up in an attractor A it is enough for it to be in any of the states in the strong basin of A. Thus, for example, from attractor A_2 to end up in A_3 one has to control the nodes with indices $\{2, 3\}$ or just $\{2\}$ to enter the strong basin of A_2. Table 1 notes the indices of the nodes to control for each pair of attractors of B.

To compute the minimal full control, one has to find a minimal subset C of $\{1, 2, 3\}$ such that C is a superset of at least one subset from every cell of Table 1. In this example C = $\{2, 3\}$, but the general problem is NP-hard.

Table 1. Attractor pair control indices.

	A_1	A_2	A_3
A_1		{2}, {2,3}	{3}, {1,3}, {1,2,3}
A_2	{1}, {2}, {1,2}		{2}, {2,3}
A_3	{1,3}, {2,3}, {1,2,3}	{2}, {2,3}, {1,2,3}	

We thus take advantage of our decomposition-based approach developed for the efficient computation of minimal target control for well-structured BNs [7], to compute the full control for pairs of attractors in local *blocks* and then merge them to obtain the global full control.

In the above example $\mathcal{G}_B$ has two maximal SCCs $S_1 = \{v_1, v_2\}$ and $S_2 = \{v_3\}$. Each such component S_j generates a *block* by including all the vertices from which there are incoming edges into S_j. Thus $\mathcal{G}_B$ has two blocks $B_1 = \{v_1, v_2\}$ and $B_2 = \{v_1, v_2, v_3\}$ as shown in Fig. 2(a). The vertex in B_2 depends on (has incoming edges from) vertices in B_1 whereas, B_1 has no such dependency on other blocks. Hence, they can be topologically sorted as $\{B_1, B_2\}$. In fact, the dependency relation between the blocks will always be acyclic and hence the blocks can be topologically sorted [7]. We compute the transition system TS_{B_1} of B_1 (Fig. 2(b)) and find that it has 2 attractors $A_1^1 = \{(10)\}$ and $A_1^2 = \{(11)\}$.

Table 2. Control indices for the blocks.

	A_1^1	A_1^2
A_1^1		$\{2\}$
A_1^2	$\{1\},\{2\}, \{1,2\}$	

(a)

	A_2^1	A_2^2	A_2^3
A_2^1		$\{3\}$	$\emptyset, \{3\}$
A_2^2	$\{3\}$		$\emptyset, \{3\}$
A_2^3	$\emptyset$	$\{3\}$	

(b)

The two basins of attractions $\mathsf{bas}(A_1^1)$ and $\mathsf{bas}(A_1^2)$ are used to compute two transition systems of B_2 (Figs. 2(c) and (d), respetively). The first has two attractors $A_2^1 = \{(100)\}$ and $A_2^2 = \{(101)\}$ and the second has one attractor $A_2^3 = \{(110)\}$.

It holds, as was shown in [7], that $A_1 = A_1^1 \otimes A_2^1, A_2 = A_1^1 \otimes A_2^2$ and $A_3 = A_1^2 \otimes A_2^3$ are the only attractors of the global B (where $\otimes$ is a combination operation on boolean tuples defined in [7]). Thus we can work with the transition systems of B_1 and B_2 separately to compute the minimal full control of B. For that, we construct Tables 2(a) and (b) similar to Table 1 listing the sets of indices to be controlled to move between attractors of B_1 and B_2, respectively. Here an entry of $\emptyset$ means that it is possible to move between the corresponding attractors without controlling any index. For B_2 we need only consider the indices of the vertices in $B_2 \setminus B_1$. From Table 2(a), $\mathsf{C}^1 = \{2\}$ and from Table 2(b), $\mathsf{C}^2 = \{3\}$. Combining, $\mathsf{C} = \mathsf{C}^1 \cup \mathsf{C}^2 = \{2,3\}$.

For the general case, suppose there are k blocks that are topologically sorted as $\{B_1, B_2, \ldots, B_k\}$. Then the matrix for every block B_i will involve indices of the vertices in $B_i \setminus (\bigcup_{j<i} B_j)$. The rest of the procedure is similar to the 2-block case as described here. We note that for certain BNs, the minimal global control for moving to a target attractor A_t computed by combining the minimal local control for the blocks might move the BN to a state which is not in the strong basin of attraction of A_t. We deal with this issue by augmenting the procedure to systematically rule out such problematic cases. Currently, we are implementing our approach in software and evaluating it on real-life biological networks modelled as BNs.

References

1. Huang, S.: Genomics, complexity and drug discovery: insights from Boolean network models of cellular regulation. Pharmacogenomics **2**(3), 203–222 (2001)
2. Kalman, R.E.: Mathematical description of linear dynamical systems. J. Soc. Ind. Appl. Math. **1**(2), 152–192 (1963)
3. Kauffman, S.: Homeostasis and differentiation in random genetic control networks. Nature **224**, 177–178 (1969)
4. Kim, J., Park, S.M., Cho, K.H.: Discovery of a kernel for controlling biomolecular regulatory networks. Sci. Rep. **3**, 2223 (2013)
5. Liu, Y.Y., Slotine, J.J., Barabási, A.L.: Controllability of complex networks. Nature **473**, 167–173 (2011)

6. Mizera, A., Pang, J., Qu, H., Yuan, Q.: Taming asynchrony for attractor detection in large Boolean networks. IEEE/ACM Trans. Comput. Biol. Bioinform. (2018, in press)

7. Paul, S., Su, C., Pang, J., Mizera, A.: A decomposition-based approach towards the control of Boolean networks. In: Proceedings of ACM-BCB 2018. ACM (2018, in press)

Systems Metagenomics: Applying Systems Biology Thinking to Human Microbiome Analysis

Golestan Sally Radwan[(✉)] and Hugh Shanahan

Computer Science Department, Royal Holloway University of London,
Egham TW20 0EX, UK
golestan.radwan.2016@live.rhul.ac.uk

Abstract. Metagenomics is the science of analysing the structure and function of DNA samples taken from the environment (e.g. soil or human gut) as opposed to a single organism. So far, researchers have used traditional genomics tools and pipelines applied to metagenomics analysis such as species identification, sequence alignment and assembly. In addition to being computationally expensive, these approaches lack an emphasis on the functional profile of the sample regardless of species diversity, and how it changes under different conditions. It also ignores unculturable species and genes undergoing horizontal transfer. We propose a new pipeline based on taking a "systems" approach to metagenomics analysis, in this case to analyse human gut microbiome data. Instead of identifying existing species, we examine a sample as a self-contained, open system with a distinct functional profile. The pipeline was used to analyse data from an experiment performed on the gut microbiomes of lean, obese and overweight twins. Previous analysis of this data only focused on taxonomic binning. Using our systems metagenomics approach, our analysis found two very different functional profiles for lean and obese twins, with obese ones being distinctly more diverse. There are also interesting differences in metabolic pathways which could indicate specific driving forces for obesity.

Keywords: Systems metagenomics · Population · Human microbiome
Stress response · Obesity · Function · Protein families

1 Introduction

A key goal of metagenomics is to measure the diversity of a microbial sample and hence estimate the effects of certain stresses on the organisms present. This is traditionally achieved through techniques of taxonomic binning or phylogenetic classification and requires several steps of pre-processing and assembly. In addition to being computationally expensive, these approaches lack an emphasis on the functional profile of the sample regardless of species diversity, and how it changes under different conditions. It also ignores unculturable species and genes undergoing horizontal gene transfer [3].

We propose a new pipeline for analysing short-read data which focuses on the most dominant functions of the sample, treating it as a system of genes/proteins rather than a set of individual species. Specifically, we use a k-mer approach to identify overrepresented

© Springer Nature Switzerland AG 2018
M. Češka and D. Šafránek (Eds.): CMSB 2018, LNBI 11095, pp. 318–321, 2018.
https://doi.org/10.1007/978-3-319-99429-1_22

protein family motifs in the raw short-read data [1]. Via a series of filtering and statistical methods, we exclude weak hits and false positives, then use GO-term mapping to infer the functions of the remaining motifs. By comparing two similar datasets under changing conditions we can thereby hypothesise which protein families most influence the observed function of the microbial community under study. Our method dispenses with assembly and global multiple sequence alignment, instead only performing a local alignment on chosen sequences after multiple stages of filtering and analysis, thereby minimising computational cost. The entire pipeline can run on a standard laptop with 8 GB of memory in under 3 hours per dataset.

The pipeline was used to analyse data from an experiment performed on the gut microbiomes of lean, obese and overweight twins. Previous analysis of this data focused on taxonomic binning [2]. Our analysis found two very different functional profiles for lean and obese twins, with obese ones being distinctly more diverse. The obese patient data showed almost 3 times as many prominent functional groupings based on GO-term analysis of molecular function and biological process as lean or overweight ones.

This approach can also be used to identify overrepresented novel protein families in these samples which may play a role in the gut microbiome. We have identified around 185 novel candidates which warrant further experimental research.

2 Pipeline Overview

The new pipeline consists of the following steps:

Translation: After stripping metadata from the short read files the nucleic acid sequence is directly translated into a protein sequence. There is only 1/6 chance that the sequence will be in the correct read frame and hence much data will be lost but is compensated for in the model for the data.

Frequency Vectors: Frequency vectors are computed for each k-mer (each k-mer is a protein sequence fragment). A k-mer length of 6 was chosen following several experiments including looking up protein family motifs in databases such as PRINTS [4] as well as synthetic genomes. Henceforth, these 6-mers are referred to as 'sub-motifs'. A rolling window technique was used to extract the submotifs one at a time. Any submotif with an unidentified base in it was discarded. Submotifs were then counted and written into frequency vectors along with their respective occurrences.

Null Model calculation: A statistical model was constructed to help quantify whether any given submotif would have occurred by chance with a specific frequency or whether its frequency might indicate actual over-representation of a particular sub-motif. This null model choice was built on the work outlined in [5].

Read Extraction and MSA: All submotifs that fall below the significance threshold of 1.0 are eliminated. The remaining submotifs are sorted based on the value of the log-odds ratio (i.e. most over-represented first). Then, a search is conducted on the amino acid reads to extract all short reads that contain this particular submotif. Each set of reads pertaining to one submotif are then passed onto MUSCLE [6] to construct a multiple sequence alignment (MSA).

pHMM Construction and Analysis: The resulting MSA for each submotif is then used as a basis to construct a profile Hidden Markov Model (pHMM) using the hmmbuild tool from the HMMER suite [7]. The resulting model is searched against a protein family database, in this case UniProt [8], to produce possible hits against known protein families.

Family Identification and Novel Detection: The resulting hits are sorted based on E-values and a cutoff of E ≤ 0.01 was used as a threshold for possible matches.

GO Term: GO terms for each significant UniProt hit are identified and a frequency table tabulated. For the highest frequency terms a comparison is made between the two different environments examined in this paper.

Dataset: For this paper, the dataset from [2] was used. This was a study conducted on lean, obese and overweight twins with a sample size of 46. The raw short reads in FASTQ format were downloaded from the Sequence Reads Archive through the European Bioinformatics Institute's Metagenomics portal [9].

3   Results

Of the top 40 GO terms found in both datasets, we found 31 that were common to both sets and 9 that were unique to each set. Figure 1 shows those terms plotted as a function of the difference in their frequencies (i.e. Obese − Lean frequency) and are plotted as a strictly decreasing function.

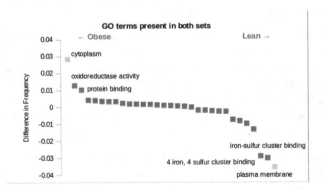

Fig. 1. Relative frequency of GO terms found in both data sets. Yellow datapoints indicate Cellular Components, orange ones indicate Molecular Functions, and green ones indicate GO terms occurring with similar frequency in both sets. (Color figure online)

4 Discussion and Future Work

Our new approach obviates the need to perform assembly on a noisy data set and no clear set of reference genomes [3]. In this study we found a greater functional diversity in the obese rather than lean data sets. In addition, the obese data set does not exhibit any abundant functional groupings with a stress response, contrary to the findings in [10]. In the obese dataset, tRNA synthesis is playing an important role which is consistent with the findings of [11]. Other terms are related to electron transport and amino acid biosynthesis. In the lean data set we see terms related to biosynthesis.

These preliminary results are promising though much more work needs to be done. While the analysis of abundant motifs is computationally efficient the matching of those motifs to known sequence data is very intensive. The present search mechanism is based on building profile HMM's on the short reads and then looking for matches in UniProt. A corresponding search based on previously known profile HMM's and then querying those against the collated motifs has to be completed. Finally, in this present analysis we have focused on identifying known functional groupings; on the other hand the above approach could also be used to identify abundant motifs that have not been observed previously. Given the wide variety of metagenomic samples this is a potentially fruitful method for identifying new protein families.

References

1. Breitwieser, F.P., Lu, J., Salzberg, S.L.: A review of methods and databases for metagenomic classification and assembly. Brief. Bioinform. **18**(5), 723–734 (2017)
2. Turnbaugh, P.J., et al.: A core gut microbiome in obese and lean twins. Nature **457**(7228), 480 (2009)
3. Simpson, J.T.: Exploring genome characteristics and sequence quality without a reference. Bioinformatics **30**(9), 1228–1235 (2014)
4. Attwood, T.K.: The PRINTS database: a resource for identification of protein families. Brief. Bioinform. **3**(3), 252–263 (2002)
5. Nei, M., Li, W.-H.: Mathematical model for studying genetic variation in terms of restriction endonucleases (molecular evolution/mitochondrial DNA/nucleotide diversity). Genetics **76** (10), 5269–5273 (1979)
6. Edgar, R.C.: MUSCLE: multiple sequence alignment with high accuracy and high throughput. Nucleic Acids Res. **32**(5), 1792–1797 (2004)
7. Finn, R.D., Clements, J., Eddy, S.R.: HMMER web server: interactive sequence similarity searching. Nucleic Acids Res. **39**(suppl_2), W29–W37 (2011)
8. Boutet, E., Lieberherr, D., Tognolli, M., Schneider, M., Bairoch, A.: UniProtKB/Swiss-Prot. In: Edwards, D. (ed.) Plant Bioinformatics, vol. 406, pp. 89–112. Humana Press, New York (2007). https://doi.org/10.1007/978-1-59745-535-0_4
9. https://www.ebi.ac.uk/metagenomics/
10. Sanz, Y., Santacruz, A., Gauffin, P.: Gut microbiota in obesity and metabolic disorders. Proc. Nutr. Soc. **69**(3), 434–441 (2010)
11. Isokpehi, R.D., et al.: Genomic evidence for bacterial determinants influencing obesity development. Int. J. Environ. Res. Public Health **14**(4), 345 (2017)

List of Accepted Posters
and Oral Presentations

Posters with flash oral presentation

- *Thomas Wright and Ian Stark.* Modelling Patterns of Gene Regulation in the Bond Calculus
- *Carolin Loos, Katharina Moeller, Fabian Fröhlich, Tim Hucho and Jan Hasenauer.* Mechanistic Hierarchical Population Model Predicts Latent Causes of Cell-to-Cell Variability
- *Jonathan Laurent, Jean Yang and Walter Fontana.* Counterfactual Resimulation for Causal Analysis of Rule-Based Models
- *David Gilbert, Monika Heiner and Leila Ghanbar.* Spatial Quorum Sensing Modelling using Coloured Hybrid Petri Nets and Simulative Model Checking
- *Matej Troják, David Šafránek, Jakub Hrabec, Jakub Šalagovič, Jan Červený, Matej Hajnal, Lukrécia Mertová, Katarína Palubová and Marek Havlík* E-cyanobacterium.org: A Web-based Platform for Systems Biology of Cyanobacteria
- *Aurélien Naldi, Céline Hernandez, Nicolas Levy, Gautier Stoll, Pedro Monteiro, Claudine Chaouiya, Thomáš Helikar, Andrei Zinovyev, Laurence Calzone, Sarah Cohen-Boulakia, Denis Thieffry and Loïc Paulevé.* The CoLoMoTo Interactive Notebook: Accessible and Reproducible Computational Analyses for Qualitative Biological Networks
- *Thomas Chatain, Stefan Haar and Loïc Paulevé* Boolean Networks: Beyond Generalized Asynchronicity
- *Pia Wilsdorf, Andreas Ruscheinski and Adelinde M. Uhrmacher.* Synergies of Simulation Model Documentation and Simulation Experiment Generation
- *Leonard Schmiester, Yannik Schälte, Fabian Fröhlich, Jan Hasenauer and Daniel Weindl.* Efficient Parameterization of Large-scale Dynamic Models using Relative Protein, Phospho-protein and Proliferation Measurements

Regular posters

- *Elife Bagci, Buket Ozahioglu, Muzaffer Erdogan, Tanju Gurel and Serbulent Yildirim.* Pore Opening in Mitochondrial Cristae through Interaction of Cytochrome c with Cardiolipin: Insights from a Mass-Action Kinetics Model
- *Anna Poskrobko and Antoni Leon Dawidowicz.* On Mathematical Model of the Bats Population's Development under the Assumption of Variable Capacity of the Habitats
- *Nikola Beneš, Luboš Brim, Martin Demko and Samuel Pastva.* Attractor Analysis in Pithya
- *Veronika Hajnová and Lenka Přibylová.* Biological and Physiological Phenomena in View of Applied Bifurcation Theory

© Springer Nature Switzerland AG 2018
M. Češka and D. Šafránek (Eds.): CMSB 2018, LNBI 11095, pp. 323–324, 2018.
https://doi.org/10.1007/978-3-319-99429-1

- *František Muzika, Lenka Schreiberová and Igor Schreiber.* Chemical Computing by Spatiotemporal Patterns in a Ring of Coupled Cells
- *Andrej Blejec, Maja Zagorščak and Kristina Gruden.* Being FAIR with pISA-tree
- *Cameron Chalk, Niels Kornerup, Wyatt Reeves and David Soloveichik.* Converging Rate-Independent Computation in Continuous Chemical Reaction Networks

Oral presentations of recently published work

- *Christoph Flamm, Jakob Lykke Andersen, Daniel Merkle and Peter Florian Stadler.* Discovering Reaction Patterns in Chemical Reaction Networks
- *Igor Schreiber, Vuk Radojkovic, František Muzika, Radovan Jurašek, Lenka Schreiberová and Jan Červený.* Reaction Network Theory as a Tool for Finding Network Motifs for Oscillatory Dynamics and Kinetic Parameter Estimation in Complex Biochemical Mechanisms
- *Jan H. van Schuppen, Jana Němcová and Kaihua Xi.* System Identification of a Continuous-time Polynomial System by a Subalgebraic Procedure

Author Index

Printed in the United States
By Bookmasters